AF713032

Advanced Modeling and Data Challenges

Mike Nguyen

Advanced Modeling and Data Challenges

Springer

Mike Nguyen
Data Sciences and Operations
University of Southern California
Los Angeles, CA, USA

ISBN 978-3-032-01718-5 ISBN 978-3-032-01719-2 (eBook)
https://doi.org/10.1007/978-3-032-01719-2

Mathematics Subject Classification: 62F03, 62P25, 62F35, 62F07, 62M20, 62P20, 62D10, 62P05

© The Editor(s) (if applicable) and The Author(s), under exclusive license to Springer Nature Switzerland AG 2025

This work is subject to copyright. All rights are solely and exclusively licensed by the Publisher, whether the whole or part of the material is concerned, specifically the rights of reprinting, reuse of illustrations, recitation, broadcasting, reproduction on microfilms or in any other physical way, and transmission or information storage and retrieval, electronic adaptation, computer software, or by similar or dissimilar methodology now known or hereafter developed.
The use of general descriptive names, registered names, trademarks, service marks, etc. in this publication does not imply, even in the absence of a specific statement, that such names are exempt from the relevant protective laws and regulations and therefore free for general use.
The publisher, the authors and the editors are safe to assume that the advice and information in this book are believed to be true and accurate at the date of publication. Neither the publisher nor the authors or the editors give a warranty, expressed or implied, with respect to the material contained herein or for any errors or omissions that may have been made. The publisher remains neutral with regard to jurisdictional claims in published maps and institutional affiliations.

This Springer imprint is published by the registered company Springer Nature Switzerland AG
The registered company address is: Gewerbestrasse 11, 6330 Cham, Switzerland

If disposing of this product, please recycle the paper.

Declarations

Competing Interests The author has no competing interests to declare that are relevant to the content of this manuscript.

Contents

Chapter 1
Introduction

This is the third book in a series designed to take the reader from statistical foundations to advanced, applied data analysis. The first volume laid the groundwork: descriptive statistics, basic inference, and the conceptual tools necessary for understanding variation, uncertainty, and evidence. The second took those principles into the realm of regression modeling, covering linear models, nonlinear relationships, and the flexible architecture of generalized and mixed effects models.

This volume is where data analysis meets the unpredictable terrain of reality.

Here, we confront the messier, more intricate problems that arise when theory meets practice. Business data is rarely clean. Variables don't always behave. Assumptions break. And yet, decisions still have to be made. This book is about developing the judgment, precision, and adaptability to work with such data-confidently, transparently, and rigorously.

Rather than introducing entirely new modeling frameworks, this volume deepens the reader's capacity to use existing tools well. We revisit data not as a passive starting point, but as an object that must often be transformed, repaired, or reimagined before analysis can begin. We explore imputation not as a mechanical fix, but as a modeling problem in itself, one that requires care and inference. Model specification is treated not as a checklist item, but as a process of discovery and discipline-an iterative negotiation between data, theory, and diagnostic feedback.

We then move into interpretation, where the stakes are highest. We ask what our models are really telling us about marginal change, about interaction, about causal direction. Concepts like moderation and mediation are examined in detail, not as abstract diagrams, but as techniques for answering urgent business questions: What drives the effect? Who does it work for? When does it fail?

This is also a book about the craft of prediction-its power, its limits, and its relationship to explanation. Throughout, we emphasize not only the models themselves, but also the mindset of responsible analysis: questioning assumptions, testing robustness, and communicating results that others can use and trust.

© The Author(s), under exclusive license to Springer Nature Switzerland AG 2025
M. Nguyen, *Advanced Modeling and Data Challenges*,
https://doi.org/10.1007/978-3-032-01719-2_1

The content is firmly grounded in business and social science applications, with examples that reflect the decisions real analysts face in marketing strategy, financial risk, customer behavior, and policy evaluation. Code is provided in R to support implementation, and each chapter is written to be both self-contained and cumulative-so that readers build a coherent workflow as they progress.

By the end of this book, the reader will be better prepared to handle the analytical ambiguity that defines real data work: not only to apply techniques, but to choose wisely among them, to defend their choices, and to adapt as conditions change.

This is not the final step in becoming a data analyst, but it may be the most important one: where technique becomes judgment, and models become decisions.

Code Replication

This book was built with R version 4.4.3 (2025-02-28 ucrt) and the following packages:

Package	Version	Source
abind	1.4-8	CRAN (R 4.4.1)
agridat	1.24	CRAN (R 4.4.3)
ape	5.8-1	CRAN (R 4.4.3)
assertthat	0.2.1	CRAN (R 4.4.3)
backports	1.5.0	CRAN (R 4.4.0)
bookdown	0.43	CRAN (R 4.4.3)
boot	1.3-31	CRAN (R 4.4.3)
broom	1.0.8	CRAN (R 4.4.3)
bslib	0.9.0	CRAN (R 4.4.2)
cachem	1.1.0	CRAN (R 4.4.2)
callr	3.7.6	CRAN (R 4.4.2)
car	3.1-3	CRAN (R 4.4.2)
carData	3.0-5	CRAN (R 4.4.2)
cellranger	1.1.0	CRAN (R 4.4.2)
cli	3.6.5	CRAN (R 4.4.3)
coda	0.19-4.1	CRAN (R 4.4.2)
colorspace	2.1-1	CRAN (R 4.4.2)
corpcor	1.6.10	CRAN (R 4.4.0)
crayon	1.5.3	CRAN (R 4.4.2)
cubature	2.1.2	CRAN (R 4.4.3)
curl	6.2.2	CRAN (R 4.4.3)
data.table	1.17.2	CRAN (R 4.4.3)
DBI	1.2.3	CRAN (R 4.4.2)
dbplyr	2.5.0	CRAN (R 4.4.2)
desc	1.4.3	CRAN (R 4.4.2)
devtools	2.4.5	CRAN (R 4.4.2)
digest	0.6.37	CRAN (R 4.4.2)

(continued)

Package	Version	Source
dplyr	1.1.4	CRAN (R 4.4.2)
ellipsis	0.3.2	CRAN (R 4.4.2)
evaluate	1.0.3	CRAN (R 4.4.2)
extrafont	NA	NA
extrafontdb	NA	NA
fansi	1.0.6	CRAN (R 4.4.2)
faraway	1.0.9	CRAN (R 4.4.3)
fastmap	1.2.0	CRAN (R 4.4.2)
forcats	1.0.0	CRAN (R 4.4.2)
foreign	0.8-90	CRAN (R 4.4.3)
fs	1.6.6	CRAN (R 4.4.3)
generics	0.1.4	CRAN (R 4.4.3)
ggplot2	3.5.2	CRAN (R 4.4.3)
glue	1.8.0	CRAN (R 4.4.2)
gtable	0.3.6	CRAN (R 4.4.2)
haven	2.5.4	CRAN (R 4.4.2)
Hmisc	5.2-3	CRAN (R 4.4.3)
hms	1.1.3	CRAN (R 4.4.2)
htmltools	0.5.8.1	CRAN (R 4.4.2)
htmlwidgets	1.6.4	CRAN (R 4.4.2)
httr	1.4.7	CRAN (R 4.4.2)
investr	1.4.2	CRAN (R 4.4.3)
jpeg	0.1-11	CRAN (R 4.4.3)
jquerylib	0.1.4	CRAN (R 4.4.2)
jsonlite	2.0.0	CRAN (R 4.4.3)
kableExtra	1.4.0	CRAN (R 4.4.3)
knitr	1.50	CRAN (R 4.4.3)
lattice	0.22-7	CRAN (R 4.4.3)
latticeExtra	0.6-30	CRAN (R 4.4.2)
lifecycle	1.0.4	CRAN (R 4.4.2)
lme4	1.1-37	CRAN (R 4.4.3)
lmerTest	3.1-3	CRAN (R 4.4.2)
lsr	0.5.2	CRAN (R 4.4.3)
ltm	1.2-0	CRAN (R 4.4.3)
lubridate	1.9.4	CRAN (R 4.4.2)
magrittr	2.0.3	CRAN (R 4.4.2)
MASS	7.3-65	CRAN (R 4.4.3)
matlib	1.0.0	CRAN (R 4.4.3)
Matrix	1.7-3	CRAN (R 4.4.3)
MCMCglmm	2.36	CRAN (R 4.4.2)
memoise	2.0.1	CRAN (R 4.4.2)

(continued)

Package	Version	Source
mgcv	1.9-3	CRAN (R 4.4.3)
minqa	1.2.8	CRAN (R 4.4.2)
modelr	0.1.11	CRAN (R 4.4.2)
munsell	0.5.1	CRAN (R 4.4.2)
nlme	3.1-168	CRAN (R 4.4.3)
nloptr	2.2.1	CRAN (R 4.4.3)
nlstools	2.1-0	CRAN (R 4.4.2)
nnet	7.3-20	CRAN (R 4.4.3)
numDeriv	2016.8-1.1	CRAN (R 4.4.0)
openxlsx	4.2.8	CRAN (R 4.4.3)
pbkrtest	0.5.4	CRAN (R 4.4.3)
pillar	1.10.2	CRAN (R 4.4.3)
pkgbuild	1.4.7	CRAN (R 4.4.3)
pkgconfig	2.0.3	CRAN (R 4.4.2)
pkgload	1.4.0	CRAN (R 4.4.2)
png	0.1-8	CRAN (R 4.4.0)
ppsr	0.0.5	CRAN (R 4.4.3)
prettyunits	1.2.0	CRAN (R 4.4.2)
processx	3.8.6	CRAN (R 4.4.3)
ps	1.9.1	CRAN (R 4.4.3)
pscl	1.5.9	CRAN (R 4.4.2)
purrr	1.0.4	CRAN (R 4.4.3)
R6	2.6.1	CRAN (R 4.4.3)
RColorBrewer	1.1-3	CRAN (R 4.4.0)
Rcpp	1.0.14	CRAN (R 4.4.2)
readr	2.1.5	CRAN (R 4.4.2)
readxl	1.4.5	CRAN (R 4.4.3)
remotes	2.5.0	CRAN (R 4.4.2)
reprex	2.1.1	CRAN (R 4.4.2)
rgl	1.3.18	CRAN (R 4.4.3)
rio	1.2.3	CRAN (R 4.4.2)
rlang	1.1.6	CRAN (R 4.4.3)
RLRsim	3.1-8	CRAN (R 4.4.3)
rmarkdown	2.29	CRAN (R 4.4.2)
rprojroot	2.0.4	CRAN (R 4.4.2)
rstudioapi	0.17.1	CRAN (R 4.4.2)
Rttf2pt1	NA	NA
rvest	1.0.4	CRAN (R 4.4.2)
sass	0.4.10	CRAN (R 4.4.3)
scales	1.4.0	CRAN (R 4.4.3)
sessioninfo	1.2.3	CRAN (R 4.4.3)

(continued)

Package	Version	Source
stringi	1.8.7	CRAN (R 4.4.3)
stringr	1.5.1	CRAN (R 4.4.2)
svglite	2.2.1	CRAN (R 4.4.3)
systemfonts	1.2.3	CRAN (R 4.4.3)
tensorA	0.36.2.1	CRAN (R 4.4.0)
testthat	3.2.3	CRAN (R 4.4.2)
tibble	3.2.1	CRAN (R 4.4.2)
tidyr	1.3.1	CRAN (R 4.4.2)
tidyselect	1.2.1	CRAN (R 4.4.2)
tidyverse	2.0.0	CRAN (R 4.4.2)
tzdb	0.5.0	CRAN (R 4.4.3)
usethis	3.1.0	CRAN (R 4.4.2)
utf8	1.2.5	CRAN (R 4.4.3)
vctrs	0.6.5	CRAN (R 4.4.2)
viridisLite	0.4.2	CRAN (R 4.4.2)
webshot	0.5.5	CRAN (R 4.4.3)
withr	3.0.2	CRAN (R 4.4.2)
xfun	0.52	CRAN (R 4.4.3)
xml2	1.3.8	CRAN (R 4.4.3)
xtable	1.8-4	CRAN (R 4.4.2)
yaml	2.3.10	CRAN (R 4.4.2)
zip	2.3.3	CRAN (R 4.4.3)

```
#> - Session info -------------------------------------------------
#>  setting  value
#>  version  R version 4.4.3 (2025-02-28 ucrt)
#>  os       Windows 11 x64 (build 26100)
#>  system   x86_64, mingw32
#>  ui       RTerm
#>  language (EN)
#>  collate  English_United States.utf8
#>  ctype    English_United States.utf8
#>  tz       America/Los_Angeles
#>  date     2025-07-14
#>  pandoc   3.1.1 @ C:/Program Files/RStudio/resources/app/
    bin/quarto/bin/tools/ (via rmarkdown)
#>  quarto   NA @ C:\\PROGRA~1\\RStudio\\resources\\app\\bin\\
    quarto\\bin\\quarto.exe
#>
#> - Packages -----------------------------------------------------
#>  package     * version date (UTC) lib source
#>  bookdown      0.43    2025-04-15 [1] CRAN (R 4.4.3)
#>  cachem        1.1.0   2024-05-16 [1] CRAN (R 4.4.2)
#>  cli           3.6.5   2025-04-23 [1] CRAN (R 4.4.3)
#>  codetools     0.2-20  2024-03-31 [1] CRAN (R 4.4.3)
#>  desc          1.4.3   2023-12-10 [1] CRAN (R 4.4.2)
```

```
#>  devtools      2.4.5   2022-10-11 [1] CRAN (R 4.4.2)
#>  digest        0.6.37  2024-08-19 [1] CRAN (R 4.4.2)
#>  ellipsis      0.3.2   2021-04-29 [1] CRAN (R 4.4.2)
#>  evaluate      1.0.3   2025-01-10 [1] CRAN (R 4.4.2)
#>  fastmap       1.2.0   2024-05-15 [1] CRAN (R 4.4.2)
#>  fs            1.6.6   2025-04-12 [1] CRAN (R 4.4.3)
#>  glue          1.8.0   2024-09-30 [1] CRAN (R 4.4.2)
#>  htmltools     0.5.8.1 2024-04-04 [1] CRAN (R 4.4.2)
#>  htmlwidgets   1.6.4   2023-12-06 [1] CRAN (R 4.4.2)
#>  httpuv        1.6.16  2025-04-16 [1] CRAN (R 4.4.3)
#>  knitr         1.50    2025-03-16 [1] CRAN (R 4.4.3)
#>  later         1.4.2   2025-04-08 [1] CRAN (R 4.4.3)
#>  lifecycle     1.0.4   2023-11-07 [1] CRAN (R 4.4.2)
#>  magrittr      2.0.3   2022-03-30 [1] CRAN (R 4.4.2)
#>  memoise       2.0.1   2021-11-26 [1] CRAN (R 4.4.2)
#>  mime          0.13    2025-03-17 [1] CRAN (R 4.4.3)
#>  miniUI        0.1.2   2025-04-17 [1] CRAN (R 4.4.3)
#>  pillar        1.10.2  2025-04-05 [1] CRAN (R 4.4.3)
#>  pkgbuild      1.4.7   2025-03-24 [1] CRAN (R 4.4.3)
#>  pkgconfig     2.0.3   2019-09-22 [1] CRAN (R 4.4.2)
#>  pkgload       1.4.0   2024-06-28 [1] CRAN (R 4.4.2)
#>  profvis       0.4.0   2024-09-20 [1] CRAN (R 4.4.2)
#>  promises      1.3.2   2024-11-28 [1] CRAN (R 4.4.2)
#>  purrr         1.0.4   2025-02-05 [1] CRAN (R 4.4.3)
#>  R6            2.6.1   2025-02-15 [1] CRAN (R 4.4.3)
#>  Rcpp          1.0.14  2025-01-12 [1] CRAN (R 4.4.2)
#>  remotes       2.5.0   2024-03-17 [1] CRAN (R 4.4.2)
#>  rlang         1.1.6   2025-04-11 [1] CRAN (R 4.4.3)
#>  rmarkdown     2.29    2024-11-04 [1] CRAN (R 4.4.2)
#>  rstudioapi    0.17.1  2024-10-22 [1] CRAN (R 4.4.2)
#>  sessioninfo   1.2.3   2025-02-05 [1] CRAN (R 4.4.3)
#>  shiny         1.10.0  2024-12-14 [1] CRAN (R 4.4.2)
#>  tibble        3.2.1   2023-03-20 [1] CRAN (R 4.4.2)
#>  tinytex       0.57    2025-04-15 [1] CRAN (R 4.4.3)
#>  urlchecker    1.0.1   2021-11-30 [1] CRAN (R 4.4.2)
#>  usethis       3.1.0   2024-11-26 [1] CRAN (R 4.4.2)
#>  vctrs         0.6.5   2023-12-01 [1] CRAN (R 4.4.2)
#>  xfun          0.52    2025-04-02 [1] CRAN (R 4.4.3)
#>  xtable        1.8-4   2019-04-21 [1] CRAN (R 4.4.2)
#>  yaml          2.3.10  2024-07-26 [1] CRAN (R 4.4.2)
#>
#>  [1] C:/Program Files/R/R-4.4.3/library
#>
#> ------------------------------------------------------------
```

Chapter 2
Data

This chapter explores the concepts of data structures essential for statistical analysis. Beginning with an overview of data types, the chapter explains how the nature of variables influences analysis strategies. Emphasis is placed on understanding the differences and implications of cross-sectional, time series, repeated cross-sectional, and panel data, with real-world examples from marketing and finance. It also offers guidance on choosing the most appropriate data structure for specific research questions, a critical step often overlooked in applied research. Finally, the chapter addresses ethical considerations in data collection and use. By the end of the chapter, readers will be equipped to critically evaluate the structure, strengths, and limitations of various types of datasets, ensuring a sound analytical foundation for subsequent modeling work.

Data can be defined broadly as any set of values, facts, or statistics that can be used for reference, analysis, and drawing inferences. In research, data drives the process of understanding phenomena, testing hypotheses, and formulating evidence-based conclusions. Choosing the right type of data (and understanding its strengths and limitations) is critical for the validity and reliability of findings.

2.1 Data Types

2.1.1 Qualitative vs. Quantitative Data

A foundational way to categorize data is by whether it is **qualitative** (non-numerical) or **quantitative** (numerical). These distinctions often guide research designs, data collection methods, and analytical techniques (Table 2.1).

© The Author(s), under exclusive license to Springer Nature Switzerland AG 2025

M. Nguyen, *Advanced Modeling and Data Challenges*,
https://doi.org/10.1007/978-3-032-01719-2_2

Table 2.1 Comparison of qualitative and quantitative research approaches

Qualitative	Quantitative
Examples: In-depth interviews, focus groups, case studies, ethnographies, open-ended questions, field notes	**Examples**: Surveys with closed-ended questions, experiments, numerical observations, structured interviews
Nature: Text-based, often descriptive, subjective interpretations	**Nature**: Numeric, more standardized, objective measures
Analysis: Thematic coding, content analysis, discourse analysis	**Analysis**: Statistical tests, regression, hypothesis testing, descriptive statistics
Outcome: Rich context, detailed understanding of phenomena	**Outcome**: Measurable facts, generalizable findings (with appropriate sampling and design)

2.1.1.1 Uses and Advantages of Qualitative Data

- **Deep Understanding**: Captures context, motivations, and perceptions in depth.
- **Flexibility**: Elicits new insights through open-ended inquiry.
- **Inductive Approaches**: Often used to build new theories or conceptual frameworks.

2.1.1.2 Uses and Advantages of Quantitative Data

- **Measurement and Comparison**: Facilitates measuring variables and comparing across groups or over time.
- **Generalizability**: With proper sampling, findings can often be generalized to broader populations.
- **Hypothesis Testing**: Permits the use of statistical methods to test specific predictions or relationships.

2.1.1.3 Limitations of Qualitative and Quantitative Data

- **Qualitative**:
  - Findings may be difficult to generalize if samples are small or non-representative.
  - Analysis can be time-consuming due to coding and interpreting text.
  - Potential for researcher bias in interpretation.
- **Quantitative**:
  - May oversimplify complex human behaviors or contextual factors by reducing them to numbers.
  - Validity depends heavily on how well constructs are operationalized.
  - Can miss underlying meanings or nuances not captured in numeric measures.

2.1.1.4 Levels of Measurement

Even within **quantitative** data, there are further distinctions based on the *level of measurement*. This classification is crucial for determining which statistical techniques are appropriate:

1. **Nominal**: Categorical data with no inherent order (e.g., gender, blood type, eye color).
2. **Ordinal**: Categorical data with a specific order or ranking but without consistent intervals between ranks (e.g., Likert scale responses: "strongly disagree," "disagree," "neutral," "agree," "strongly agree")
3. **Interval**: Numeric data with equal intervals but no true zero (e.g., temperature in Celsius or Fahrenheit).
4. **Ratio**: Numeric data with equal intervals and a meaningful zero (e.g., height, weight, income).

The level of measurement affects which statistical tests (like t-tests, ANOVA, correlations, regressions) are valid and how you can interpret differences or ratios in the data.

2.1.2 Other Ways to Classify Data

Beyond **observational structure**, there are multiple other dimensions used to classify data:

2.1.2.1 Primary vs. Secondary Data

- **Primary Data**: Collected directly by the researcher for a specific purpose (e.g., firsthand surveys, experiments, direct measurements).
- **Secondary Data**: Originally gathered by someone else for a different purpose (e.g., government census data, administrative records, previously published datasets).

2.1.2.2 Structured, Semi-structured, and Unstructured Data

- **Structured Data**: Organized in a predefined manner, typically in rows and columns (e.g., spreadsheets, relational databases).
- **Semi-Structured Data**: Contains organizational markers but not strictly tabular (e.g., JSON, XML logs, HTML).

- **Unstructured Data**: Lacks a clear, consistent format (e.g., raw text, images, videos, audio files).
  - Often analyzed using natural language processing (NLP), image recognition, or other advanced techniques.

2.1.2.3 Big Data

- Characterized by the "3 Vs": **Volume** (large amounts), **Variety** (diverse forms), and **Velocity** (high-speed generation).
- Requires specialized computational tools (e.g., Hadoop, Spark) and often cloud-based infrastructure for storage and processing.
- Can be structured or unstructured (e.g., social media feeds, sensor data, click-stream data).

2.1.2.4 Internal vs. External Data (in Organizational Contexts)

- **Internal Data**: Generated within an organization (e.g., sales records, HR data, production metrics).
- **External Data**: Sourced from outside (e.g., macroeconomic indicators, market research reports, social media analytics).

2.1.2.5 Proprietary vs. Public Datas

- **Proprietary Data**: Owned by an organization or entity, not freely available for public use.
- **Public/Open Data**: Freely accessible data provided by governments, NGOs, or other institutions (e.g., data.gov, World Bank Open Data).

2.1.3 Data by Observational Structure Over Time

Another primary way to categorize data is by *how* observations are collected over time. This classification shapes research design, analytic methods, and the types of inferences we can make. Four major types here are (Table 2.2)

1. Cross-Sectional Data
2. Time-Series Data
3. Repeated Cross-Sectional Data
4. Panel Data

Table 2.2 Comparison of data structures in empirical research

Type	Advantages	Limitations
Cross-Sectional Data	Simple, cost-effective, and good for studying distributions or correlations at a single time point	Lacks temporal information, can only infer associations, not causal links
Time-Series Data	Enables trend analysis, seasonality detection, and forecasting	Requires handling autocorrelation, stationarity issues, and structural breaks
Repeated Cross-Sectional Data	Tracks shifts in population-level parameters over time; simpler than panel data	Cannot track individual changes; comparability depends on consistent methodology
Panel Data	Allows causal inference, controls for unobserved heterogeneity, and tracks individual trajectories	Expensive, prone to attrition, and requires complex statistical methods

2.2 Cross-Sectional Data

Cross-sectional data consists of observations on **multiple entities** (e.g., individuals, firms, regions, or countries) at **a single point in time** or over a very short period, where **time is not a primary dimension** of variation.

- Each observation represents a **different entity**, rather than the same entity tracked over time.
- Unlike time series data, the order of observations does not carry temporal meaning.

Examples:

- **Labor Economics**: Wage and demographic data for 1000 workers in 2024.
- **Marketing Analytics**: Customer satisfaction ratings and purchasing behavior for 500 online shoppers surveyed in Q1 of a year.
- **Corporate Finance**: Financial statements of 1000 firms for the fiscal year 2023.

Key Characteristics:

- **Observations Are Independent (in an Ideal Setting)**: Each unit is drawn from a population with no intrinsic dependence on others.
- **No Natural Ordering**: Unlike time series data, the sequence of observations does not affect analysis.
- **Variation Occurs Across Entities, Not Over Time**: Differences in observed outcomes arise from differences between individuals, firms, or regions.

Advantages:

- **Straightforward Interpretation**: Since time effects are not present, the focus remains on relationships between variables at a single point.

- **Easier to Collect and Analyze**: Compared to time series or panel data, cross-sectional data is often simpler to collect and model.
- **Suitable for causal inference** (if exogeneity conditions hold).

Challenges:

- **Omitted Variable Bias**: Unobserved confounders may drive both the dependent and independent variables.
- **Endogeneity**: Reverse causality or measurement error can introduce bias.
- **Heteroskedasticity**: Variance of errors may differ across entities, requiring robust standard errors.

A typical cross-sectional regression model

$$y_i = \beta_0 + x_{i1}\beta_1 + x_{i2}\beta_2 + \cdots + x_{i(k-1)}\beta_{k-1} + \epsilon_i$$

where

- y_i is the outcome variable for entity i,
- x_{ij} are explanatory variables,
- ϵ_i is an error term capturing unobserved factors.

2.3 Time-Series Data

Time series data consists of observations on the *same variable(s)* recorded over multiple time periods for a single entity (or aggregated entity). These data points are typically collected at consistent intervals-hourly, daily, monthly, quarterly, or annually-allowing for the analysis of trends, patterns, and forecasting.

Examples:

- **Stock Market**: Daily closing prices of a company's stock over five years.
- **Economics**: Monthly unemployment rates in a country over a decade.
- **Macroeconomics**: Annual GDP of a country from 1960 to 2020.

Key Characteristics:

- The primary goal is to analyze trends, seasonality, cyclic patterns, and forecast future values.
- Time series data requires specialized statistical methods, such as
  - **Autoregressive integrated moving average (ARIMA)**
  - **Seasonal ARIMA (SARIMA)**
  - **Exponential smoothing**
  - **Vector autoregression (VAR)**

Advantages:

- Captures temporal patterns such as trends, seasonal fluctuations, and economic cycles.
- Essential for forecasting and policy-making, such as setting interest rates based on economic indicators.

Challenges:

- **Autocorrelation**: Observations close in time are often correlated.
- **Structural Breaks**: Sudden changes due to policy shifts or economic crises can distort analysis.
- **Seasonality**: Must be accounted for to avoid misleading conclusions.

A time series typically consists of four key components:

1. **Trend**: Long-term directional movement in the data over time.
2. **Seasonality**: Regular, periodic fluctuations (e.g., increased retail sales in December).
3. **Cyclical Patterns**: Long-term economic cycles that are irregular but recurrent.
4. **Irregular (Random) Component**: Unpredictable variations not explained by trend, seasonality, or cycles.

A general linear time series model can be expressed as

$$y_t = \beta_0 + x_{t1}\beta_1 + x_{t2}\beta_2 + \cdots + x_{t(k-1)}\beta_{k-1} + \epsilon_t$$

Some Common Model Types:

1. **Static Model**

   A simple time series regression

$$y_t = \beta_0 + x_1\beta_1 + x_2\beta_2 + x_3\beta_3 + \epsilon_t$$

2. **Finite Distributed Lag Model**

   Captures the effect of past values of an explanatory variable

$$y_t = \beta_0 + pe_t\delta_0 + pe_{t-1}\delta_1 + pe_{t-2}\delta_2 + \epsilon_t$$

- **Long-Run Propensity**: Measures the cumulative effect of explanatory variables over time

$$LRP = \delta_0 + \delta_1 + \delta_2$$

3. **Dynamic Model**

   A model incorporating lagged dependent variables

$$GDP_t = \beta_0 + \beta_1 GDP_{t-1} + \epsilon_t$$

2.3.1 Statistical Properties of Time Series Models

For time series regression, standard OLS assumptions must be carefully examined. The following conditions affect estimation:

Finite Sample Properties:

- **A1–A3**: OLS remains unbiased.
- **A1–A4**: Standard errors are consistent, and the Gauss-Markov Theorem holds (OLS is BLUE).
- **A1–A6**: Finite sample Wald tests (e.g., t-tests and F-tests) remain valid.

However, in time series settings, A3 often fails due to

- **Spurious Time Trends** (fixable by including a time trend)
- **Strict vs. Contemporaneous Exogeneity** (sometimes unavoidable)

2.3.2 Common Time Series Processes

Several key models describe different time series behaviors:

- **Autoregressive Model (AR(p))**: A process where current values depend on past values.
- **Moving Average Model (MA(q))**: A process where past error terms influence current values.
- **Autoregressive Moving Average (ARMA(p, q))**: A combination of AR and MA processes.
- **Autoregressive Conditional Heteroskedasticity (ARCH(p))**: Models time-varying volatility.
- **Generalized ARCH (GARCH(p, q))**: Extends ARCH by including past conditional variances.

2.3.3 Deterministic Time Trends

When both the dependent and independent variables exhibit trending behavior, a regression may produce **spurious** results.

Spurious Regression Example
A simple regression with trending variables

$$y_t = \alpha_0 + t\alpha_1 + v_t$$

$$x_t = \lambda_0 + t\lambda_1 + u_t$$

where

- $\alpha_1 \neq 0$ and $\lambda_1 \neq 1$,
- v_t and u_t are independent.

Despite no true relationship between x_t and y_t, estimating

$$y_t = \beta_0 + x_t\beta_1 + \epsilon_t$$

results in

- **Inconsistency**: $plim(\hat{\beta}_1) = \frac{\alpha_1}{\lambda_1}$.
- **Invalid Inference**: $|t| \to^d \infty$ for $H_0 : \beta_1 = 0$, leading to rejection of the null hypothesis as $n \to \infty$.
- **Misleading** R^2: $plim(R^2) = 1$, falsely implying perfect predictive power.

We can also rewrite the equation as

$$y_t = \beta_0 + \beta_1 x_t + \epsilon_t$$

$$\epsilon_t = \alpha_1 t + v_t$$

where $\beta_0 = \alpha_0$ and $\beta_1 = 0$. Since x_t is a deterministic function of time, ϵ_t is correlated with x_t, leading to the usual omitted variable bias.

Solutions to Spurious Trends
1. **Include a Time Trend** (t) as a Control Variable
   - Provides consistent parameter estimates and valid inference.
2. **Detrend Variables**
   - Regress both y_t and x_t on time, then use residuals in a second regression.
   - Equivalent to applying the Frisch-Waugh-Lovell Theorem.

2.3.4 *Violations of Exogeneity in Time-Series Models*

The **exogeneity assumption** (A3) plays a crucial role in ensuring unbiased and consistent estimation in time series models. However, in many cases, the assumption is **violated** due to the inherent nature of time-dependent processes.

In a standard regression framework, we assume

$$E(\epsilon_t | x_1, x_2, ..., x_T) = 0$$

which requires that the error term is uncorrelated with all past, present, and future values of the independent variables.

Common Violations of Exogeneity

1. Feedback Effect

- The error term ϵ_t **influences future values** of the independent variables.
- A classic example occurs in economic models where past shocks affect future decisions.

2. Dynamic Specification

- The dependent variable includes a **lagged version of itself** as an explanatory variable, introducing correlation between ϵ_t and past y_{t-1}.

3. Dynamic Completeness

- In finite distributed lag (FDL) models, failing to include the **correct number of lags** leads to omitted variable bias and correlation between regressors and errors.

2.3.4.1 Feedback Effect

In a simple regression model

$$y_t = \beta_0 + x_t\beta_1 + \epsilon_t$$

the standard exogeneity assumption (A3) requires

$$E(\epsilon_t | x_1, x_2, ..., x_t, x_{t+1}, ..., x_T) = 0$$

However, in the presence of feedback, past errors affect future values of x_t, leading to

$$E(\epsilon_t | x_{t+1}, ..., x_T) \neq 0$$

- This occurs when **current shocks** (e.g., economic downturns) influence **future decisions** (e.g., government spending, firm investments).
- Strict exogeneity is **violated**, as we now have dependence across time.

Implication

- Standard OLS estimators become **biased and inconsistent**.
- One common solution is using Instrumental Variables to isolate exogenous variation in x_t.

2.3.4.2 Dynamic Specification

A **dynamically specified model** includes lagged dependent variables:

$$y_t = \beta_0 + y_{t-1}\beta_1 + \epsilon_t$$

Exogeneity (A3) would require

$$E(\epsilon_t | y_1, y_2, ..., y_t, y_{t+1}, ..., y_T) = 0$$

However, since y_{t-1} depends on ϵ_{t-1} from the previous period, we obtain

$$Cov(y_{t-1}, \epsilon_t) \neq 0$$

Implication

- Strict exogeneity (A3) fails, as y_{t-1} and ϵ_t are correlated.
- OLS estimates are biased and inconsistent.
- Standard **autoregressive (AR) models** require alternative estimation techniques like **generalized method of moments** or Maximum Likelihood estimation.

2.3.4.3 Dynamic Completeness and Omitted Lags

A finite distributed lag (FDL) model

$$y_t = \beta_0 + x_t\delta_0 + x_{t-1}\delta_1 + \epsilon_t$$

assumes that the included lags **fully capture** the relationship between y_t and past values of x_t. However, if we omit relevant lags, the exogeneity assumption (A3)

$$E(\epsilon_t | x_1, x_2, ..., x_t, x_{t+1}, ..., x_T) = 0$$

fails, as unmodeled lag effects create correlation between x_{t-2} and ϵ_t.

Implication

- The regression suffers from **omitted variable bias**, making OLS estimates unreliable.

- **Solution**:
  - Include **additional lags** of x_t.
  - Use **lag selection criteria** (e.g., AIC, BIC) to determine the appropriate lag structure.

2.3.5 *Consequences of Exogeneity Violations*

If strict exogeneity (A3) fails, standard OLS assumptions no longer hold:

- OLS is biased.
- Gauss-Markov Theorem no longer applies.
- Finite Sample Properties (such as unbiasedness) are invalid.

To address these issues, we can

1. Rely on Large Sample Properties: Under certain conditions, consistency may still hold.
2. Use Weaker Forms of Exogeneity: Shift from **strict exogeneity** (A3) to **contemporaneous exogeneity** (A3a).

If strict exogeneity does not hold, we can instead assume A3a (contemporaneous exogeneity):

$$E(\mathbf{x}_t'\epsilon_t) = 0$$

This weaker assumption only requires that x_t is uncorrelated with the error in the same time period (Table 2.3).

- With contemporaneous exogeneity, ϵ_t can be correlated with past and future values of x_t.
- This allows for dynamic specifications such as

$$y_t = \beta_0 + y_{t-1}\beta_1 + \epsilon_t$$

while still maintaining **consistency** under certain assumptions.

Table 2.3 Key differences from strict exogeneity

Exogeneity type	Requirement	Allows dynamic models?
Strict exogeneity	$E(\epsilon_t \mid x_1, x_2, ..., x_T) = 0$	**No**
Contemporaneous exogeneity	$E(\mathbf{x}_t'\epsilon_t) = 0$	**Yes**

Deriving Large Sample Properties for Time Series
To establish **consistency** and **asymptotic normality**, we rely on the following assumptions:

- A1: Linearity
- A2: Full rank (no perfect multicollinearity)
- A3a: Contemporaneous exogeneity

However, the standard Weak Law of Large Numbers and Central Limit Theorem in OLS depend on A5 (random sampling), which does **not** hold in time series settings.

Since time series data exhibits dependence over time, we replace A5 (random sampling) with a weaker assumption:

- A5a: Weak dependence (stationarity)

Asymptotic Variance and Serial Correlation

- The derivation of **asymptotic variance** depends on A4 (homoskedasticity).
- However, in time series settings, we often encounter **serial correlation**:

$$\text{Cov}(\epsilon_t, \epsilon_s) \neq 0 \quad \text{for} \quad |t - s| > 0$$

- To ensure valid inference, standard errors must be corrected using methods such as Newey-West HAC estimators.

2.3.6 Highly Persistent Data

In time series analysis, a key assumption for OLS consistency is that the data-generating process exhibits A5a weak dependence (i.e., observations are not too strongly correlated over time). However, when y_t and x_t are highly persistent, standard OLS assumptions break down.

If a time series is **not weakly dependent**, it means

- y_t and y_{t-h} remain strongly correlated even for large lags ($h \to \infty$).
- A5a (weak dependence assumption) fails, leading to
  - **OLS inconsistency**.
  - **No valid limiting distribution** (asymptotic normality does not hold).

Example
A classic example of a highly persistent process is a random walk:

$$y_t = y_{t-1} + u_t$$

or with drift:

$$y_t = \alpha + y_{t-1} + u_t$$

where u_t is a white noise error term.

- y_t does not revert to a mean-it has an infinite variance as $t \to \infty$.
- **Shocks accumulate**, making standard regression analysis unreliable.

2.3.6.1 Solution: First Differencing

A common way to transform non-stationary series into stationary ones is through first differencing:

$$\Delta y_t = y_t - y_{t-1} = u_t$$

- If u_t is a weakly dependent process (i.e., $I(0)$, stationary), then y_t is said to be **difference-stationary** or **integrated of order 1, $I(1)$**.
- If both y_t and x_t follow a random walk ($I(1)$), we estimate

$$\Delta y_t = (\Delta \mathbf{x}_t \beta) + (\epsilon_t - \epsilon_{t-1})$$
$$\Delta y_t = \Delta \mathbf{x}_t \beta + \Delta u_t$$

  This ensures **OLS estimation remains valid.**

2.3.7 Unit Root Testing

To formally determine whether a time series contains a **unit root** (i.e., is non-stationary), we test

$$y_t = \alpha + \rho y_{t-1} + u_t$$

Hypothesis Testing

- $H_0 : \rho = 1$ (unit root, non-stationary)
  - OLS is **not consistent or asymptotically normal.**
- $H_a : \rho < 1$ (stationary process)
  - OLS is **consistent and asymptotically normal.**

Key Issues

- The usual **t-test is not valid** because OLS under H_0 does not have a standard distribution.
- Instead, specialized tests such as Dickey-Fuller and Augmented Dickey-Fuller tests are required.

2.3.7.1 Dickey-Fuller Test for Unit Roots

The **Dickey-Fuller test** transforms the original equation by subtracting y_{t-1} from both sides:

$$\Delta y_t = \alpha + \theta y_{t-1} + v_t$$

where

$$\theta = \rho - 1$$

- **Null Hypothesis** $(H_0 : \theta = 0)$ $\rightarrow$ Implies $\rho = 1$ (unit root, non-stationary).
- **Alternative** $(H_a : \theta < 0)$ $\rightarrow$ Implies $\rho < 1$ (stationary).

Since y_t follows a non-standard asymptotic distribution under H_0, Dickey and Fuller derived specialized critical values.

Decision Rule

- If the test statistic is more negative than the critical value, reject H_0 $\rightarrow$ y_t is stationary.
- Otherwise, fail to reject H_0 $\rightarrow$ y_t has a unit root (non-stationary).

The standard DF test may fail due to two key limitations:

1. **Simplistic Dynamic Relationship**
    - The DF test assumes only one lag in the autoregressive structure.
    - However, in reality, higher-order lags of Δy_t may be needed.

Solution

Use the Augmented Dickey-Fuller test, which includes extra lags:

$$\Delta y_t = \alpha + \theta y_{t-1} + \gamma_1 \Delta y_{t-1} + \cdots + \gamma_p \Delta y_{t-p} + v_t$$

- Under H_0, Δy_t follows an AR(1) process.
- Under H_a, y_t follows an AR(2) or higher process.

Including lags of Δy_t ensures a better-specified model.

2. **Ignoring Deterministic Time Trends**

If a series exhibits a deterministic trend, failing to include it biases the unit root test.

Example
If y_t grows over time, a test without a trend component will falsely detect a unit root.

Solution
Include a deterministic time trend (t) in the regression:

$$\Delta y_t = \alpha + \theta y_{t-1} + \delta t + v_t$$

- Allows for quadratic relationships with time.
- Changes the critical values, requiring an adjusted statistical test.

2.3.7.2 Augmented Dickey-Fuller Test

The **ADF test** generalizes the DF test by allowing for

1. **Lags of** Δy_t (to correct for serial correlation)
2. **Time trends** (to handle deterministic trends)

Regression Equation

$$\Delta y_t = \alpha + \theta y_{t-1} + \delta t + \gamma_1 \Delta y_{t-1} + \cdots + \gamma_p \Delta y_{t-p} + v_t$$

where $\theta = 1 - \rho$.

Hypotheses
- $H_0 : \theta = 0$ (unit root: non-stationary)
- $H_a : \theta < 0$ (stationary)

2.3.8 Newey-West Standard Errors

Newey-West standard errors, also known as **heteroskedasticity and autocorrelation consistent (HAC) estimators**, provide valid inference when errors exhibit both heteroskedasticity (i.e., A4 homoskedasticity assumption is violated) and serial correlation. These standard errors adjust for dependence in the error structure, ensuring that hypothesis tests remain valid.

Key Features
- **Accounts for Autocorrelation**: Handles time dependence in error terms.

- **Accounts for Heteroskedasticity**: Allows for non-constant variance across observations.
- **Ensures Positive Semi-definiteness**: Downweights longer-lagged covariances to maintain mathematical validity.

The estimator is computed as

$$\hat{B} = T^{-1}\sum_{t=1}^{T} e_t^2 \mathbf{x_t'x_t} + \sum_{h=1}^{g}\left(1-\frac{h}{g+1}\right)T^{-1}\sum_{t=h+1}^{T} e_t e_{t-h}(\mathbf{x_t'x_{t-h}} + \mathbf{x_{t-h}'x_t})$$

where

- T is the sample size,
- g is the chosen **lag truncation parameter** (bandwidth),
- e_t are the residuals from the OLS regression,
- $\mathbf{x}_t$ are the explanatory variables.

Choosing the Lag Length (g)
Selecting an appropriate lag truncation parameter (g) is crucial for balancing **efficiency** and **bias**. Common guidelines include

- **Yearly Data**: $g = 1$ or 2 usually suffices.
- **Quarterly Data**: $g = 4$ or 8 accounts for seasonal dependencies.
- **Monthly Data**: $g = 12$ or 14 captures typical cyclical effects.

Alternatively, data-driven methods can be used:

- **Newey-West Rule**: $g = \lfloor 4(T/100)^{2/9} \rfloor$
- **Alternative Heuristic**: $g = \lfloor T^{1/4} \rfloor$

```r
# Load necessary libraries
library(sandwich)
library(lmtest)
#> Loading required package: zoo
#>
#> Attaching package: 'zoo'
#> The following objects are masked from 'package:base':
#>
#>     as.Date, as.Date.numeric

# Simulate data
set.seed(42)
T <- 100  # Sample size
time <- 1:T
x <- rnorm(T)
epsilon <- arima.sim(n = T, list(ar = 0.5))  # Autocorrelated errors
y <- 2 + 3 * x + epsilon  # True model

# Estimate OLS model
model <- lm(y ~ x)
```

```
# Compute Newey-West standard errors
lag_length <- floor(4 * (T / 100) ^ (2 / 9)) # Newey-West rule
nw_se <- NeweyWest(model, lag = lag_length, prewhite = FALSE)

# Display robust standard errors
coeftest(model, vcov = nw_se)
#>
#> t test of coefficients:
#>
#>             Estimate Std. Error t value  Pr(>|t|)
#> (Intercept)  1.71372    0.13189  12.993 < 2.2e-16 ***
#> x            3.15831    0.13402  23.567 < 2.2e-16 ***
#> ---
#> Signif. codes:
#> 0 '***' 0.001 '**' 0.01 '*' 0.05 '.' 0.1 ' ' 1
```

2.3.8.1 Testing for Serial Correlation

Serial correlation (also known as **autocorrelation**) occurs when error terms are correlated across time:

$$E(\epsilon_t \epsilon_{t-h}) \neq 0 \quad \text{for some } h \neq 0$$

Steps for Detecting Serial Correlation

1. **Estimate an OLS Regression**:
   - Run the regression of y_t on $\mathbf{x}_t$ and obtain residuals e_t.
2. **Test for Autocorrelation in Residuals**:
   - Regress e_t on $\mathbf{x}_t$ and its lagged residual e_{t-1}:

   $$e_t = \gamma_0 + \mathbf{x}_t' \gamma + \rho e_{t-1} + v_t$$

   - Test whether ρ is significantly different from zero.
3. **Decision Rule**:
   - If ρ is statistically significant at the 5% level, reject the null hypothesis of no serial correlation.

Higher-Order Serial Correlation
To test for **higher-order autocorrelation**, extend the previous regression:

$$e_t = \gamma_0 + \mathbf{x}_t' \gamma + \rho_1 e_{t-1} + \rho_2 e_{t-2} + \cdots + \rho_p e_{t-p} + v_t$$

- **Jointly test** $\rho_1 = \rho_2 = \cdots = \rho_p = 0$ using an **F-test**.
- If the null is rejected, autocorrelation of order p is present.

Step 1: Estimate an OLS Regression and Obtain Residuals

```
# Load necessary libraries
library(lmtest)
library(sandwich)

# Generate some example data
set.seed(123)
n <- 100
x <- rnorm(n)
y <- 1 + 0.5 * x + rnorm(n)  # True model: y = 1 + 0.5*x + e

# Estimate the OLS regression
model <- lm(y ~ x)

# Obtain residuals
residuals <- resid(model)
```

Step 2: Test for Autocorrelation in Residuals

```
# Create lagged residuals
lagged_residuals <- c(NA, residuals[-length(residuals)])

# Regress residuals on x and lagged residuals
autocorr_test_model <- lm(residuals ~ x + lagged_residuals)

# Summary of the regression
summary(autocorr_test_model)
#>
#> Call:
#> lm(formula = residuals ~ x + lagged_residuals)
#>
#> Residuals:
#>      Min       1Q   Median       3Q      Max
#> -1.94809 -0.72539 -0.08105  0.58503  3.12941
#>
#> Coefficients:
#>                   Estimate Std. Error t value Pr(>|t|)
#> (Intercept)       0.008175   0.098112   0.083    0.934
#> x                -0.002841   0.107167  -0.027    0.979
#> lagged_residuals -0.127605   0.101746  -1.254    0.213
#>
#> Residual standard error: 0.9707 on 96 degrees of freedom
#>   (1 observation deleted due to missingness)
#> Multiple R-squared:  0.01614,	Adjusted R-squared:  -0.004354
#> F-statistic: 0.7876 on 2 and 96 DF,  p-value: 0.4579

# Test if the coefficient of lagged_residuals is significant
rho <- coef(autocorr_test_model)["lagged_residuals"]
rho_p_value <-
```

```
    summary(autocorr_test_model)$coefficients["lagged_residuals",
    ↪  "Pr(>|t|)"]

# Decision Rule
if (rho_p_value < 0.05) {
    cat("Reject the null hypothesis: There is evidence of serial
    ↪  correlation.\n")
} else {
    cat("Fail to reject the null hypothesis: No evidence of serial
    ↪  correlation.\n")
}
#> Fail to reject the null hypothesis: No evidence of serial correlation.
```

Step 3: Testing for Higher-Order Serial Correlation

```
# Number of lags to test
p <- 2  # Example: testing for 2nd order autocorrelation

# Create a matrix of lagged residuals
lagged_residuals_matrix <- sapply(1:p, function(i) c(rep(NA, i),
↪  residuals[1:(n - i)]))

# Regress residuals on x and lagged residuals
higher_order_autocorr_test_model <- lm(residuals ~ x +
↪  lagged_residuals_matrix)

# Summary of the regression
summary(higher_order_autocorr_test_model)
#>
#> Call:
#> lm(formula = residuals ~ x + lagged_residuals_matrix)
#>
#> Residuals:
#>     Min      1Q  Median      3Q     Max
#> -1.9401 -0.7290 -0.1036  0.6359  3.0253
#>
#> Coefficients:
#>                           Estimate Std. Error t value
#> (Intercept)               0.006263   0.099104   0.063
#> x                         0.010442   0.108370   0.096
#> lagged_residuals_matrix1 -0.140426   0.103419  -1.358
#> lagged_residuals_matrix2 -0.107385   0.103922  -1.033
#>                          Pr(>|t|)
#> (Intercept)                 0.950
#> x                           0.923
#> lagged_residuals_matrix1    0.178
#> lagged_residuals_matrix2    0.304
#>
#> Residual standard error: 0.975 on 94 degrees of freedom
#>   (2 observations deleted due to missingness)
#> Multiple R-squared:  0.02667,	Adjusted R-squared:  -0.004391
#> F-statistic: 0.8587 on 3 and 94 DF,  p-value: 0.4655
```

```
# Joint F-test for the significance of lagged residuals
f_test <- car::linearHypothesis(higher_order_autocorr_test_model,
                                paste0("lagged_residuals_matrix", 1:p, " =
                                ↪ 0"))

# Print the F-test results
print(f_test)
#>
#> Linear hypothesis test:
#> lagged_residuals_matrix1 = 0
#> lagged_residuals_matrix2 = 0
#>
#> Model 1: restricted model
#> Model 2: residuals ~ x + lagged_residuals_matrix
#>
#>   Res.Df    RSS Df Sum of Sq      F Pr(>F)
#> 1     96 91.816
#> 2     94 89.368  2    2.4479 1.2874 0.2808

# Decision Rule
if (f_test$`Pr(>F)`[2] < 0.05) {
  cat("Reject the null hypothesis: There is evidence of higher-order
  ↪  serial correlation.\n")
} else {
  cat("Fail to reject the null hypothesis: No evidence of higher-order
  ↪  serial correlation.\n")
}
#> Fail to reject the null hypothesis: No evidence of higher-order serial
↪  correlation.
```

Corrections for Serial Correlation

If serial correlation is detected, the following adjustments should be made (Table 2.4):

Table 2.4 Strategies for addressing serial correlation in regression models

Problem	Solution
Mild serial correlation	Use Newey-West standard errors
Severe serial correlation	Use Generalized Least Squares or Prais-Winsten transformation
Autoregressive structure in errors	Model as an **ARMA** process
Higher-order serial correlation	Include lags of dependent variable or use HAC estimators with higher lag orders

2.4 Repeated Cross-Sectional Data

Repeated cross-sectional data consists of **multiple independent cross-sections** collected at different points in time. Unlike panel data, where the same individuals are tracked over time, repeated cross-sections **draw a fresh sample in each wave**.

This approach allows researchers to analyze **aggregate trends over time**, but it does **not track individual-level changes**.

Examples

- **General Social Survey (GSS) (U.S.)**—Conducted every two years with a new sample of respondents.
- **Political Opinion Polls**—Monthly voter surveys to track shifts in public sentiment.
- **National Health Surveys**—Annual studies with fresh samples to monitor **population-wide** health trends.
- **Educational Surveys**—Sampling different groups of students each year to assess learning outcomes.

2.4.1 Key Characteristics

1. **Fresh Sample in Each Wave**
   - Each survey represents an **independent** cross-section.
   - No respondent is tracked across waves.
2. **Population-Level Trends Over Time**
   - Researchers can study **how the distribution of characteristics (e.g., income, attitudes, behaviors) changes over time**.
   - However, individual trajectories **cannot** be observed.
3. **Sample Design Consistency**
   - To ensure comparability across waves, researchers must maintain **consistent**
     - Sampling methods
     - Questionnaire design
     - Definitions of key variables

2.4.2 Statistical Modeling for Repeated Cross-Sections

Since repeated cross-sections do not track the same individuals, specific regression methods are used to **analyze changes over time**.

1. **Pooled Cross-Sectional Regression (Time Fixed Effects)**

Combines multiple survey waves into a single dataset while controlling for time effects

$$y_i = \mathbf{x}_i\beta + \delta_1 y_1 + ... + \delta_T y_T + \epsilon_i$$

where

- y_i is the outcome for individual i,
- $\mathbf{x}_i$ are explanatory variables,
- y_t are time period dummies,
- δ_t captures the **average change** in outcomes across time periods.

Key Features

- Allows for different intercepts across time periods, capturing shifts in baseline outcomes.
- Tracks overall population trends without assuming a constant effect of $\mathbf{x}_i$ over time.

2. **Allowing for Structural Change in Pooled Cross-Sections (Time-Dependent Effects)**

To test whether relationships between variables change over time (**structural breaks**), interactions between time dummies and explanatory variables can be introduced:

$$y_i = \mathbf{x}_i\beta + \mathbf{x}_i y_1\gamma_1 + ... + \mathbf{x}_i y_T\gamma_T + \delta_1 y_1 + ... + \delta_T y_T + \epsilon_i$$

- **Interacting** x_i with time period dummies allows for
  - **Different slopes** for each time period.
  - **Time-dependent effects** of explanatory variables.

Practical Application

- If $\mathbf{x}_i$ represents education level and y_t represents survey year, an interaction term can test whether the effect of education on income has changed over time.
- Structural break tests help determine whether such time-varying effects are statistically significant.
- Useful for **policy analysis**, where a policy might impact certain subgroups differently across time.

3. **Difference-in-Means Over Time**

A simple approach to comparing **aggregate trends**

$$\bar{y}_t - \bar{y}_{t-1}$$

- Measures whether the average outcome has changed over time

- Common in policy evaluations (e.g., assessing the effect of minimum wage increases on average income)

4. **Synthetic Cohort Analysis**

Since repeated cross-sections do not track individuals, a **synthetic cohort** can be created by grouping observations based on shared characteristics:

- Example: If education levels are collected over multiple waves, we can track **average income changes** within education groups to approximate trends.

2.4.3 Advantages of Repeated Cross-Sectional Data

Refer to Table 2.5 for the advantages of repeated cross-sectional data.

2.4.4 Disadvantages of Repeated Cross-Sectional Data

Refer to Table 2.6 for the disadvantages of repeated cross-sectional data.

Table 2.5 Advantages of repeated cross-sectional data

Advantage	Explanation
Tracks population trends	Useful for studying shifts in demographics, attitudes, and economic conditions over time
Lower cost than panel data	Tracking individuals across multiple waves (as in panel studies) is expensive and prone to attrition
No attrition bias	Unlike panel surveys, where respondents drop out over time, each wave draws a new representative sample
Easier implementation	Organizations can design a single survey protocol and repeat it at set intervals without managing panel retention

Table 2.6 Disadvantages of repeated cross-sectional data

Disadvantage	Explanation
No individual-level transitions	Cannot track **how specific individuals change** over time (e.g., income mobility, changes in attitudes)
Limited causal inference	Since we observe different people in each wave, we cannot directly infer individual **cause-and-effect** relationships
Comparability issues	Small differences in survey design (e.g., question wording or sampling frame) can make it difficult to compare across waves

To ensure valid comparisons across time

- **Consistent Sampling**: Each wave should use the same **sampling frame and methodology**.
- **Standardized Questions**: Small variations in **question wording** can introduce inconsistencies.
- **Weighting Adjustments**: If sampling strategies change, apply **survey weights** to maintain representativeness.
- **Accounting for Structural Changes**: Economic, demographic, or social changes may impact comparability.

2.5 Panel Data

Panel data (also called **longitudinal data**) consists of **observations of the same entities over multiple time periods**. Unlike repeated cross-sections, where new samples are drawn in each wave, panel data **tracks the same individuals, households, firms, or regions over time**, enabling richer statistical analysis.

Panel data combines **cross-sectional variation (differences across entities)** and **time series variation (changes within entities over time)**.

Examples

- **Panel Study of Income Dynamics**-Follows households annually, collecting data on income, employment, and expenditures.
- **Medical Longitudinal Studies**-Tracks the same patients over months or years to study disease progression.
- **Firm-Level Financial Data**-Follows a set of companies over multiple years through financial statements.
- **Student Achievement Studies**-Follows the same students across different grade levels to assess academic progress.

Structure

- N entities (individuals, firms, etc.) observed over T time periods.
- The dataset can be
  - **Balanced Panel**: All entities are observed in every time period.
  - **Unbalanced Panel**: Some entities have missing observations for certain periods.

Types of Panels

- **Short Panel**: Many individuals (N) but few time periods (T).
- **Long Panel**: Many time periods (T) but few individuals (N).
- **Both Large**: Large N and T (e.g., firm-level data over decades).

Table 2.7 Advantages of panel data

Advantage	Explanation
Captures individual trajectories	Allows for studying how individuals or firms evolve over time
Controls for unobserved heterogeneity	Fixed effects models remove time-invariant individual characteristics
Stronger causal inference	Difference-in-differences and FE models improve causal interpretation
More efficient estimates	Exploits both cross-sectional and time series variation

Table 2.8 Disadvantages of panel data

Disadvantage	Explanation
Higher cost and complexity	Tracking individuals over time is resource-intensive
Attrition bias	If certain individuals drop out systematically, results may be biased
Measurement errors	Errors accumulate over time, leading to potential biases

2.5.1 Advantages of Panel Data

Refer to Table 2.7 for the advantages of panel data.

2.5.2 Disadvantages of Panel Data

Refer to Table 2.8 for the disadvantages of panel data.

2.5.3 Sources of Variation in Panel Data

Since we observe **both individuals and time periods**, we distinguish three types of variation (Table 2.9):

- **Overall Variation**: Differences across both time and individuals.
- **Between Variation**: Differences **between** individuals (cross-sectional variation).
- **Within Variation**: Differences **within** individuals (time variation).

Table 2.9 Decomposition of variance in panel data

Estimate	Formula
Individual mean	$\bar{x}_i = \frac{1}{T}\sum_t x_{it}$
Overall mean	$\bar{x} = \frac{1}{NT}\sum_i \sum_t x_{it}$
Overall variance	$s_O^2 = \frac{1}{NT-1}\sum_i \sum_t (x_{it} - \bar{x})^2$
Between variance	$s_B^2 = \frac{1}{N-1}\sum_i (\bar{x}_i - \bar{x})^2$
Within variance	$s_W^2 = \frac{1}{NT-1}\sum_i \sum_t (x_{it} - \bar{x}_i)^2$

Note: $s_O^2 \approx s_B^2 + s_W^2$

2.5.4 *Pooled OLS Estimator*

The pooled ordinary least squares estimator is the simplest way to estimate relationships in panel data. It treats panel data as a **large cross-sectional dataset**, ignoring individual-specific effects and time dependence.

The pooled OLS model is specified as

$$y_{it} = \mathbf{x}_{it}\beta + \epsilon_{it}$$

where

- y_{it} is the dependent variable for individual i at time t,
- $\mathbf{x}_{it}$ is a vector of explanatory variables,
- β is the vector of coefficients to be estimated,
- $\epsilon_{it} = c_i + u_{it}$ is the composite error term.
  - c_i is the unobserved individual heterogeneity.
  - u_{it} is the idiosyncratic shock.

By treating all observations as **independent**, pooled OLS **assumes no systematic differences across individuals** beyond what is captured by $\mathbf{x}_{it}$.

For **pooled OLS** to be **consistent and unbiased**, the following conditions must hold:

1. Linearity in Parameters (A1)
   - The relationship between y_{it} and $\mathbf{x}_{it}$ is correctly specified as linear.
2. Full Rank Condition (A2)
   - The regressors are not perfectly collinear across individuals and time.
3. Strict Exogeneity (A3)
   - No correlation between regressors and error terms

$$E(\epsilon_{it}|\mathbf{x}_{it}) = 0$$

- This ensures that OLS remains unbiased.

4. Homoskedasticity (A4)
   - Constant variance of errors

$$\text{Var}(\epsilon_{it}|\mathbf{x}_{it}) = \sigma^2$$

   - If violated (heteroskedasticity exists), standard errors must be adjusted using a clustered robust approach, but OLS is still consistent.

5. No Autocorrelation Across Time (A5)
   - The error term should not be correlated over time for a given individual:

$$E(\epsilon_{it}, \epsilon_{is}) = 0, \quad \forall t \neq s$$

   - If this assumption fails, clustered standard errors are needed.

6. Random Sampling (A6)
   - Observations are independent across individuals:

$$(y_{i1}, ..., y_{iT}, x_{i1}, ..., x_{iT}) \perp (y_{j1}, ..., y_{jT}, x_{j1}, ..., x_{jT}) \quad \forall i \neq j$$

   - This assumption is often reasonable but not always valid (e.g., in firm-level or country-level panel data).

   For **pooled OLS to be consistent**, we require A3a:

$$E\left(\mathbf{x}_{it}'(c_i + u_{it})\right) = 0$$

which holds if and only if

1. **Exogeneity of** u_{it} (time-varying error):

$$E\left(\mathbf{x}_{it}'u_{it}\right) = 0$$

   - Ensures that regressors are **not correlated** with the random error component.

2. **Random Effects Assumption** (time-invariant error):

$$E(\mathbf{x}_{it}'c_i) = 0$$

   - Ensures that **unobserved heterogeneity** (c_i) is uncorrelated with regressors. If this assumption fails, pooled OLS suffers from **omitted variable bias**.

Implication

- If c_i is correlated with $\mathbf{x}_{it}$, pooled OLS is **biased and inconsistent**.
- If c_i is uncorrelated with $\mathbf{x}_{it}$, pooled OLS is **consistent but inefficient**.

Variance Decomposition in Panel Data

Since panel data contains both **between-entity** and **within-entity** variation, the total variance can be decomposed into

$$s_O^2 = s_B^2 + s_W^2$$

where

- s_O^2 = **Overall variance** (variation over time and across individuals),
- s_B^2 = **Between variance** (variation between individuals),
- s_W^2 = **Within variance** (variation within individuals over time).

Key Insight

- Pooled OLS does not separate within-individual and between-individual variation.
- Fixed effects models eliminate c_i and use only within-individual variation.
- Random effects models use both within and between variation.

Robust Inference in Pooled OLS

If the standard assumptions fail, adjustments are necessary:

1. **Heteroskedasticity:** If A4 (homoskedasticity) is violated, standard errors must be adjusted using
   (a) White's Robust Standard Errors (for cross-sectional heteroskedasticity).
   (b) Cluster-Robust Standard Errors (for panel-specific heteroskedasticity).
2. **Serial Correlation (Autocorrelation):** If errors are correlated across time
   (a) Use Newey-West Standard Errors for time dependence.
   (b) Use Clustered Standard Errors at the individual level.
3. **Multicollinearity:** If regressors are highly correlated
   (a) Remove redundant variables.
   (b) Use Variance Inflation Factor diagnostics.

Comparing Pooled OLS with Alternative Panel Models

See Table 2.10.

Table 2.10 Comparing pooled OLS with alternative panel models

Model	Assumption about c_i	Uses within variation?	Uses between variation?	Best when
Pooled OLS	Assumes c_i is uncorrelated with x_{it}	Yes	Yes	No individual heterogeneity
Fixed effects	Removes c_i via demeaning	Yes	No	c_i is correlated with x_{it}
Random effects	Assumes c_i is uncorrelated with x_{it}	Yes	Yes	c_i is uncorrelated with x_{it}

When to Use Pooled OLS?

- If individual heterogeneity is negligible.
- If panel is short (T is small) and cross-section is large (N is big).
- If random effects assumption holds ($E(\mathbf{x}'_{it}c_i) = 0$).

If these conditions fail, Fixed Effects or Random Effects models should be used instead.

2.5.5 *Individual-Specific Effects Model*

In panel data, unobserved heterogeneity can arise when **individual-specific factors** (c_i) influence the dependent variable. These effects can be

- **Correlated with Regressors** ($E(\mathbf{x}'_{it}c_i) \neq 0$): Use the Fixed Effects estimator.
- **Uncorrelated with Regressors** ($E(\mathbf{x}'_{it}c_i) = 0$): Use the Random Effects estimator.

  The general model is

$$y_{it} = \mathbf{x}_{it}\beta + c_i + u_{it}$$

where

- c_i is the **individual-specific effect** (time-invariant),
- u_{it} is the **idiosyncratic error** (time-variant).

Model	Assumption on c_i	Uses within variation?	Uses between variation?	Best when
Fixed effects	c_i is correlated with x_{it}	Yes	No	Unobserved heterogeneity bias present
Random effects	c_i is uncorrelated with x_{it}	Yes	Yes	No correlation with regressors

2.5.6 *Random Effects Estimator*

The random effects (RE) estimator is a Feasible Generalized Least Squares method used in panel data analysis. It assumes that individual-specific effects (c_i) are uncorrelated with the explanatory variables ($\mathbf{x}_{it}$), allowing for estimation using both within-group (time variation) and between-group (cross-sectional variation).

The standard **random effects model** is

$$y_{it} = \mathbf{x}_{it}\beta + c_i + u_{it}$$

where

- y_{it} is the dependent variable for entity i at time t,
- $\mathbf{x}_{it}$ is a vector of explanatory variables,
- β represents the coefficients of interest,
- c_i is the **unobserved individual-specific effect** (time-invariant),
- u_{it} is the **idiosyncratic error** (time-varying).

In contrast to the Fixed Effects model, which eliminates c_i by demeaning the data, the random effects model treats c_i as a random variable and incorporates it into the error structure.

2.5.6.1 Key Assumptions for Random Effects

For the random effects estimator to be **consistent**, the following assumptions must hold:

1. **Exogeneity of the Time-Varying Error** (u_{it}) (A3a)

The idiosyncratic error term (u_{it}) must be **uncorrelated with regressors**:

$$E(\mathbf{x}_{it}'u_{it}) = 0$$

This assumption ensures that within-period variation in the regressors does not systematically affect the error term.

2. **Exogeneity of Individual-Specific Effects** (c_i) (A3a)

A crucial assumption of the RE model is that the individual effect (c_i) is uncorrelated with the explanatory variables:

$$E(\mathbf{x}_{it}'c_i) = 0$$

This means that the individual-specific unobserved characteristics do not systematically affect the choice of explanatory variables.

- **If this assumption fails**, the RE model produces **biased and inconsistent estimates** due to omitted variable bias.
- **If this assumption holds**, RE is **more efficient than FE** because it retains **both within-group and between-group variation**.

3. **No Serial Correlation in** u_{it}

The error term (u_{it}) must be **uncorrelated across time**:

$$E(u_{it}u_{is}) = 0, \quad \forall t \neq s$$

If this assumption fails

- Standard errors will be incorrect.
- Generalized Least Squares adjustments or cluster-robust standard errors are required.

2.5.6.2 Efficiency of Random Effects

The RE estimator is a GLS estimator, meaning it is BLUE (best linear unbiased estimator) under homoskedasticity (Table 2.11).

- When the variance of errors differs across individuals, RE can still be used but must be adjusted with robust standard errors.
- If errors are correlated over time, standard Newey-West or cluster-robust standard errors should be applied.

To efficiently estimate β, we transform the RE model using Generalized Least Squares.

Define the **quasi-demeaned transformation**:

$$\tilde{y}_{it} = y_{it} - \theta \bar{y}_i$$

$$\tilde{\mathbf{x}}_{it} = \mathbf{x}_{it} - \theta \bar{\mathbf{x}}_i$$

where

$$\theta = 1 - \sqrt{\frac{\sigma_u^2}{T\sigma_c^2 + \sigma_u^2}}$$

- If $\theta = 1$, RE becomes the FE estimator.
- If $\theta = 0$, RE becomes Pooled OLS.

Table 2.11 Efficiency of random effects estimator under assumption A4

Scenario	Efficiency of RE
A4 (homoskedasticity) holds	RE is the most efficient estimator
A4 fails (heteroskedasticity or serial correlation)	RE remains more efficient than pooled OLS but is no longer optimal

The final RE regression equation is

$$\tilde{y}_{it} = \tilde{\mathbf{x}}_{it}\beta + \tilde{u}_{it}$$

which is estimated using GLS.

2.5.7 Fixed Effects Estimator

Also known as the **within estimator**, the **FE model** controls for individual-specific effects by removing them through transformation.

Key Assumption
If the RE assumption fails ($E(\mathbf{x}'_{it}c_i) \neq 0$), then

- **Pooled OLS and RE become biased and inconsistent** (due to omitted variable bias).
- **FE is still consistent** because it eliminates c_i.

However, FE only corrects bias from time-invariant factors and does not handle time-variant omitted variables.

Challenges with FE
- **Bias with Lagged Dependent Variables**: FE is biased in dynamic models (Nickell, 1981; Narayanan & Nair, 2013).
- **Exacerbates Measurement Error**: FE can worsen **errors-in-variables bias**.

2.5.7.1 Demean (Within) Transformation

To remove c_i, we take the individual mean of the regression equation:

$$y_{it} = \mathbf{x}_{it}\beta + c_i + u_{it}$$

Averaging over time (T)

$$\bar{y}_i = \bar{\mathbf{x}}_i\beta + c_i + \bar{u}_i$$

Subtracting the second equation from the first (i.e., **within transformation**)

$$(y_{it} - \bar{y}_i) = (\mathbf{x}_{it} - \bar{\mathbf{x}}_i)\beta + (u_{it} - \bar{u}_i)$$

This transformation

- Eliminates c_i, solving omitted variable bias.
- Only uses within-individual variation.

The transformed regression is estimated via OLS:

$$y_{it} - \bar{y}_i = (\mathbf{x}_{it} - \bar{\mathbf{x}}_i)\beta + d_1\delta_1 + \cdots + d_{T-2}\delta_{T-2} + (u_{it} - \bar{u}_i)$$

where

- d_t is a time dummy variable, which equals 1 if the observation in the time periods t, and 0 otherwise. This variable is for period $t = 1, \ldots, T - 1$ (one period omitted to avoid perfect multicollinearity).
- δ_t is the coefficient on the time dummy, capturing aggregate shocks that affect all individuals in period t.

Key Conditions for Consistency

- Strict exogeneity (A3):

$$E\left[(\mathbf{x}_{it} - \bar{\mathbf{x}}_i)'(u_{it} - \bar{u}_i)\right] = 0$$

- Time-invariant variables are dropped (e.g., gender, ethnicity). If you're interested in the effect of these time-invariant variables, consider using either OLS or the between estimator.
- Cluster-robust standard errors should be used.

2.5.7.2 Dummy Variable Approach

The Dummy Variable Approach is an alternative way to estimate Fixed Effects in panel data. Instead of transforming the data by demeaning (Within Transformation), this method explicitly includes individual dummy variables to control for entity-specific heterogeneity.

The general FE model is

$$y_{it} = \mathbf{x}_{it}\beta + c_i + u_{it}$$

where

- c_i is the unobserved, **time-invariant individual effect** (e.g., ability, cultural preferences, managerial style).
- u_{it} is the **idiosyncratic error term** (fluctuates over time and across individuals).

To estimate this model using the Dummy Variable Approach, we include a separate dummy variable for each individual:

$$y_{it} = \mathbf{x}_{it}\beta + d_1\delta_1 + \ldots + d_{T-2}\delta_{T-2} + c_1\gamma_1 + \ldots + c_{n-1}\gamma_{n-1} + u_{it}$$

where

- c_i is now modeled explicitly as a dummy variable ($c_i\gamma_i$) for each individual.

- d_t are time dummies, capturing time-specific shocks.
- δ_t are coefficients on time dummies, controlling for common time effects.

Interpretation of the Dummy Variables

- The dummy variable c_i takes a value of 1 for individual i and 0 otherwise:

$$c_i = \begin{cases} 1 & \text{if observation is for individual } i \\ 0 & \text{otherwise} \end{cases}$$

- These N dummy variables absorb all individual-specific variation, ensuring that only within-individual (over-time) variation remains.

Advantages of the Dummy Variable Approach

1. Easy to Interpret: Explicitly includes entity-specific effects, making it easier to understand how individual heterogeneity is modeled.
2. Equivalent to the Within (Demean) Transformation: Mathematically, this approach produces the same coefficient estimates as the within transformation.
3. Allows for Inclusion of Time Dummies: The model can easily incorporate time dummies (d_t) to control for period-specific shocks.

Limitations of the Dummy Variable Approach

1. **Computational Complexity with Large** N
   - Adding N dummy variables significantly increases the number of parameters estimated.
   - If N is very large (e.g., 10,000 individuals), this approach can be computationally expensive.
2. **Standard Errors Are Incorrectly Estimated**
   - The standard errors for c_i dummy variables are often incorrectly calculated, as they absorb all within-individual variation.
   - This is why the within transformation (demeaning approach) is generally preferred.
3. **Consumes Degrees of Freedom**
   - Introducing N additional parameters reduces degrees of freedom, which can lead to overfitting.

2.5.7.3 First-Difference Approach

An alternative way to eliminate individual-specific effects (c_i) is to **take first differences across time**, rather than subtracting the individual mean.

The FE model

$$y_{it} = \mathbf{x}_{it}\beta + c_i + u_{it}$$

Since c_i is constant over time, taking the first difference

$$y_{it} - y_{i(t-1)} = (\mathbf{x}_{it} - \mathbf{x}_{i(t-1)})\beta + (u_{it} - u_{i(t-1)})$$

This transformation **removes** c_i completely, leaving a model that can be estimated using Pooled OLS.

Advantages of the First-Difference Approach

1. **Eliminates Individual Effects** (c_i)
   - Since c_i is time-invariant, differencing removes it from the equation.
2. **Works Well with Few Time Periods** (T is Small)
   - If T is small, first-differencing is often preferred over the Within Transformation, as it does not require averaging over many periods.
3. **Less Computationally Intensive**
   - Unlike the Dummy Variable Approach, which requires estimating N additional parameters, the First-Difference Approach reduces the dimensionality of the problem.

Limitations of the First-Difference Approach

1. **Cannot Handle Missing Observations Well**
   - If data is missing in period $t - 1$ for an individual, then the corresponding first-difference observation is lost.
   - This can significantly reduce sample size in unbalanced panels.
2. **Reduces Number of Observations by One**
   - Since first differences require $y_{i(t-1)}$ to exist, the model loses one time period ($T - 1$ observations per individual instead of T).
3. **Can Introduce Serial Correlation**
   - Since we are differencing $u_{it} - u_{i(t-1)}$, the error term now exhibits autocorrelation.
   - This means standard OLS assumptions (independent errors) no longer hold, requiring the use of robust standard errors.

2.5.7.4 Comparison of FE Approaches

Refer to Table 2.12 for comparison of FE approaches.

Key Insights

1. The Dummy Variable Approach explicitly models c_i but is computationally expensive for large N.

Table 2.12 Comparison of FE approaches

Approach	How it works	Pros	Cons
Dummy Variable	Includes individual dummies (c_i) in regression	Intuitive, easy to interpret	Computationally expensive for large N, standard errors may be incorrect
Within Transformation (Demeaning)	Subtracts individual mean from each variable	Computationally efficient, correct standard errors	Cannot estimate time-invariant variables
First-Difference Approach	Takes time differences to remove c_i	Simple, works well for small T	Reduces sample size, introduces autocorrelation

2. The Within (Demean) Transformation is the most commonly used FE method because it is computationally efficient and produces correct standard errors.
3. The First-Difference Approach is useful when T is small, but it reduces sample size and introduces autocorrelation.
4. If data has many missing values, the First-Difference Approach is not recommended due to its sensitivity to gaps in observations.
5. Time dummies (d_t) can be included in any FE model to control for time shocks that affect all individuals.
6. FE only exploits within variation, meaning only status changes contribute to β estimates.
7. With limited status changes, standard errors explode (small number of switchers leads to high variance).
8. Treatment effect is non-directional but can be parameterized.
9. Switchers vs. non-switchers
   - If switchers differ fundamentally, the FE estimator may still be biased.
   - Descriptive statistics on switchers/non-switchers help verify robustness.

2.5.7.5 Variance of Errors in FE

FE reduces variation by removing c_i, which affects **error variance**:

$$\hat{\sigma}_\epsilon^2 = \frac{\text{SSR}_{\text{OLS}}}{NT - K}$$

$$\hat{\sigma}_u^2 = \frac{\text{SSR}_{\text{FE}}}{NT - (N + K)} = \frac{\text{SSR}_{\text{FE}}}{N(T - 1) - K}$$

Implication

- The variance of the error may increase or decrease because
  - SSR can increase (since FE eliminates between variation).
  - Degrees of freedom decrease (as more parameters are estimated).

2.5.7.6 Fixed Effects Examples

2.5.7.6.1 Intergenerational Mobility—Blau (1999)

Research Questions

- Does **transferring resources** to low-income families improve **upward mobility** for children?
- What are the **mechanisms** of intergenerational mobility?

Mechanisms for Intergenerational Mobility

There are multiple pathways through which **parental income influences child outcomes**:

1. Genetics (Ability Endowment)
   - If mobility is purely genetic, policy cannot affect outcomes.
2. Environmental Indirect Effects
   - Family background, peer influences, and school quality.
3. Environmental Direct Effects
   - Parental investments in education, health, and social capital.
4. Financial Transfers
   - Direct monetary support, inheritance, and wealth accumulation.

   One way to measure the impact of income on human capital accumulation is

$$\frac{\%\Delta \text{Human Capital}}{\%\Delta \text{Parental Income}}$$

where human capital includes education, skills, and job market outcomes.

Income is measured in different ways to capture its **long-term effects**:

1. **Total household income**
2. **Wage income**
3. **Non-wage income**
4. **Annual vs. permanent income** (important distinction for long-term analysis)

Key control variables must be **exogenous** to avoid bias. Bad Controls are those that are **jointly determined with the dependent variable** (e.g., mother's labor force participation).

Exogenous Controls
- Mother's race
- Birth location
- Parental education
- Household structure at age 14

The estimated model is

$$Y_{ijt} = X_{jt}\beta_i + I_{jt}\alpha_i + \epsilon_{ijt}$$

where

- i = Test (e.g., academic test score).
- j = Individual (child).
- t = Time.
- X_{jt} = Observable child characteristics.
- I_{jt} = Parental income.
- ϵ_{ijt} = Error term.

Grandmother's Fixed Effects Model
Since a child (j) is **nested** within a mother (m), and a mother is **nested** within a grandmother (g), we estimate

$$Y_{ijgmt} = X_{it}\beta_i + I_{jt}\alpha_i + \gamma_g + u_{ijgmt}$$

where

- g = Grandmother, m = Mother, j = Child, and t = Time.
- γ_g captures both grandmother and mother fixed effects.
- The nested structure controls for genetic and fixed family environment effects.
- **Cluster standard errors** at the family level to **account for correlation in errors** across generations.

Pros of Grandmother FE Model
- Controls for genetics + fixed family background.
- Allows estimation of income effects independent of family background.

Cons
- Might not fully control for unobserved heterogeneity.
- Measurement errors in income can exaggerate attenuation bias.

2.5.7.6.2 Fixed Effects in Teacher Quality Studies—Babcock (2010)

The study investigates

- How teacher quality influences student performance.
- Whether students adjust their course selection behavior based on past grading experiences.
- How to properly estimate teacher fixed effects while addressing selection bias and measurement error.

The initial model estimates student performance (T_{ijct}) based on class expectations and student characteristics:

$$T_{ijct} = \alpha_0 + S_{jct}\alpha_1 + X_{ijct}\alpha_2 + u_{ijct}$$

where

- T_{ijct} = Student test score.
- S_{jct} = Class-level grading expectation (e.g., expected GPA in the course).
- X_{ijct} = Individual student characteristics.
- i = Student, j = Instructor, c = Course, and t = Time.
- u_{ijct} = Idiosyncratic error term.

A key issue in this model is that **grading expectations** may not be randomly assigned. If students select into courses based on grading expectations, **simultaneity bias** can arise.

To control for **instructor and course heterogeneity**, the model introduces **teacher-course fixed effects** (μ_{jc}):

$$T_{ijct} = \beta_0 + S_{jct}\beta_1 + X_{ijct}\beta_2 + \mu_{jc} + \epsilon_{ijct}$$

where

- μ_{jc} is a unique fixed effect for each instructor-course combination.
- This controls for instructor-specific grading policies and course difficulty.
- It differs from a simple instructor effect (θ_j) and course effect (δ_c) because it captures interaction effects.

Implications of Instructor-Course Fixed Effects

1. **Reduces Bias from Course Shopping**
    - Students may select courses based on grading expectations.
    - Including μ_{jc} controls for the fact that some instructors systematically assign easier grades.
2. **Shifts in Student Expectations**
    - Even if course content remains constant, students adjust their expectations based on past grading experiences.

- This influences their future course selection behavior.

Identification Strategy
A key challenge in estimating teacher effects is **endogeneity** from

1. **Simultaneity Bias**

    1. Grading expectations (S_{jct}) and student performance may be jointly determined.
    2. If grading expectations are based on past student performance, **OLS will be biased**.

2. **Unobserved Teacher Characteristics**

    - Some teachers may have an innate ability to motivate students, leading to higher student performance independent of observable teacher traits.

To address these concerns, the model first controls for **observable teacher characteristics**:

$$Y_{ijt} = X_{it}\beta_1 + \text{Teacher Experience}_{jt}\beta_2 + \text{Teacher Education}_{jt}\beta_3 \\ + \text{Teacher Score}_{it}\beta_4 + \cdots + \epsilon_{ijt}$$

However, if **teacher characteristics are correlated with unobserved ability**, we replace them with **teacher fixed effects**:

$$Y_{ijt} = X_{it}\alpha + \Gamma_{it}\theta_j + u_{ijt}$$

where

- θ_j = **teacher fixed effect**, capturing all time-invariant teacher characteristics.
- Γ_{it} represents within-teacher variation.

To further analyze teacher impact, we express student test scores as

$$Y_{ijt} = X_{it}\gamma + \epsilon_{ijt}$$

where

- γ represents the between and within variation.
- e_{ijt} is the prediction error.

Decomposing the error term

$$e_{ijt} = T_{it}\delta_j + \tilde{e}_{ijt}$$

where

- δ_j = Group-level teacher effect.
- $\tilde{e}_{ijt}$ = Residual error.

To control for **prior student performance**, we introduce **lagged test scores**:

$$Y_{ijkt} = Y_{ijkt-1} + X_{it}\beta + T_{it}\tau_j + (W_i + P_k + \epsilon_{ijkt})$$

where

- Y_{ijkt-1} = Lagged student test score.
- τ_j = Teacher fixed effect.
- W_i = Student fixed effect.
- P_k = School fixed effect.
- $u_{ijkt} = W_i + P_k + \epsilon_{ijkt}$.

A major issue is **selection bias**:

- If students **sort into better teachers**, the teacher effect (τ) may be **overestimated**.
- Bias in τ for teacher j is

$$\frac{1}{N_j}\sum_{i=1}^{N_j}(W_i + P_k + \epsilon_{ijkt})$$

- where N_j is the number of students in class with teacher j.
- **Smaller class sizes** → **higher bias** in teacher effect estimates because $\frac{1}{N_j}\sum_{i=1}^{N_j}\epsilon_{ijkt} \neq 0$ will inflate the teacher fixed effect. If we use the random teacher effects instead, τ will still contain bias, and we do not know the direction of the bias.

If teachers **switch schools**, we can separately estimate:

- Teacher fixed effects (τ_j)
- School fixed effects (P_k)

The **mobility web** refers to the network of teacher transitions across schools, which helps in identifying both teacher and school fixed effects.

- Thin mobility web: Few teachers switch schools, making it harder to separate teacher effects from school effects.
- Thick mobility web: Many teachers switch schools, improving identification of teacher quality independent of school characteristics.

The **panel data model** capturing student performance over time is

$$Y_{ijkt} = Y_{ijk(t-1)}\alpha + X_{it}\beta + T_{it}\tau + P_k + \epsilon_{ijkt}$$

where

- Y_{ijkt} = Student performance at time t.
- $Y_{ijk(t-1)}$ = Lagged student test score.
- X_{it} = Student characteristics.
- T_{it} = Teacher effect (τ).
- P_k = School fixed effect.
- ϵ_{ijkt} = Idiosyncratic error term.

If we apply fixed effects (demeaning transformation)

$$Y_{ijkt} - \bar{Y}_{ijk} = (X_{it} - \bar{X}_i)\beta + (T_{it} - \bar{T}_i)\tau + (P_k - \bar{P}) + (\epsilon_{ijkt} - \bar{\epsilon}_{ijk})$$

- This transformation removes teacher fixed effects (τ).
- If we want to explicitly estimate τ, we must include teacher fixed effects before demeaning.

The paper argues that controlling for school fixed effects (P_k) ensures no selection bias, meaning students are **randomly assigned** within schools.

A key claim in the paper is that teacher quality (τ) does not depend on the number of students per teacher (N_j).

To test this, we examine the variance of estimated teacher effects:

$$\text{var}(\tau)$$

If

$$\text{var}(\tau) = 0$$

this implies **teacher quality does not impact student performance**.

To empirically test this, the study analyzes

$$\frac{1}{N_j} \sum_{i=1}^{N_j} \epsilon_{ijkt}$$

which represents **teacher-level average residuals**.

Key Finding

- The variance of teacher effects remains stable across different class sizes (N_j).
- This suggests that random assignment of students across teachers is not biasing τ.

Since teacher effects (τ_j) are **estimated with error** (Spin-off of Measurement Error), we decompose them as

$$\hat{\tau}_j = \tau_j + \lambda_j$$

where

- τ_j = True teacher effect.
- λ_j = Measurement error (e.g., sampling error, estimation noise).

Assuming τ_j and λ_j are uncorrelated

$$\text{cov}(\tau_j, \lambda_j) = 0$$

this means that **the randomness in student assignments does not systematically bias teacher quality estimates**.

The total observed variance in estimated teacher effects is

$$\text{var}(\hat{\tau}) = \text{var}(\tau) + \text{var}(\lambda)$$

Rearranging

$$\text{var}(\tau) = \text{var}(\hat{\tau}) - \text{var}(\lambda)$$

Since we observe $\text{var}(\hat{\tau})$, we need to estimate $\text{var}(\lambda)$.

Measurement error variance ($\text{var}(\lambda)$) can be approximated using the average squared standard error of teacher effects:

$$\text{var}(\lambda) = \frac{1}{J}\sum_{j=1}^{J}\hat{\sigma}_j^2$$

where $\hat{\sigma}_j^2$ is the squared standard error of teacher j (which depends on sample size N_j).

The **signal-to-noise ratio** (or **reliability**) of teacher effect estimates is

$$\frac{\text{var}(\tau)}{\text{var}(\hat{\tau})} = \text{Reliability}$$

where

- **Higher reliability** indicates that most of the variation comes from true teacher effects (τ) rather than noise.
- **Lower reliability** suggests that a large portion of variation is due to **measurement error**.

The proportion of **error variance** in estimated teacher effects is

$$1 - \frac{\text{var}(\tau)}{\text{var}(\hat{\tau})} = \text{Noise}$$

Even if true teacher quality depends on class size (N_j), our method for estimating λ remains unaffected.

To check whether **teacher effects are biased by sampling error**, we regress estimated teacher effects ($\hat{\tau}_j$) on teacher characteristics (X_j):

$$\hat{\tau}_j = \beta_0 + X_j\beta_1 + \epsilon_j$$

If teacher characteristics **do not predict sampling error**, then

$$R^2 \approx 0$$

This would confirm that **teacher characteristics are uncorrelated with measurement error**, validating the identification strategy.

2.5.8 Tests for Assumptions in Panel Data Analysis

We typically don't test heteroskedasticity explicitly because robust covariance matrix estimation is used. However, other key assumptions should be tested before choosing the appropriate panel model.

```
library("plm")
data("EmplUK", package="plm")
data("Produc", package="plm")
data("Grunfeld", package="plm")
data("Wages", package="plm")
```

2.5.8.1 Poolability Test

Tests whether coefficients are the same across individuals (also known as an F-test of stability or Chow test).

- H_0: All individuals have the same coefficients (i.e., equal coefficients for all individuals).
- H_a: Different individuals have different coefficients.

Notes:

- A fixed effects model assumes different intercepts per individual.
- A random effects model assumes a common intercept.

```
library(plm)
plm::pooltest(inv ~ value + capital,
              data = Grunfeld,
              model = "within")
```

```
#>
#>  F statistic
#>
#> data:  inv ~ value + capital
#> F = 5.7805, df1 = 18, df2 = 170, p-value = 1.219e-10
#> alternative hypothesis: unstability
```

If the null is rejected, we should not use a pooled OLS model.

2.5.8.2 Testing for Individual and Time Effects

Checks for the presence of individual or time effects, or both.

- **Types of tests:**
  - honda: Default test for individual effects (Honda, 1985)
  - bp: Breusch-Pagan test for unbalanced panels (Breusch & Pagan, 1980)
  - kw: King-Wu test for unbalanced panels with two-way effects (King & Wu, 1997)
  - ghm: Gourieroux, Holly, and Monfort test for two-way effects (Gourieroux et al., 1982)

```
pFtest(inv ~ value + capital,
       data = Grunfeld,
       effect = "twoways")
#>
#>  F test for twoways effects
#>
#> data:  inv ~ value + capital
#> F = 17.403, df1 = 28, df2 = 169, p-value < 2.2e-16
#> alternative hypothesis: significant effects
pFtest(inv ~ value + capital,
       data = Grunfeld,
       effect = "individual")
#>
#>  F test for individual effects
#>
#> data:  inv ~ value + capital
#> F = 49.177, df1 = 9, df2 = 188, p-value < 2.2e-16
#> alternative hypothesis: significant effects
pFtest(inv ~ value + capital,
       data = Grunfeld,
       effect = "time")
#>
#>  F test for time effects
#>
#> data:  inv ~ value + capital
#> F = 0.23451, df1 = 19, df2 = 178, p-value = 0.9997
#> alternative hypothesis: significant effects
```

If the null hypothesis is rejected, a fixed effects model is more appropriate.

2.5.8.3 Cross-Sectional Dependence (Contemporaneous Correlation)

Tests whether residuals across entities are correlated.

```
pcdtest(inv ~ value + capital,
        data = Grunfeld, model = "within")
#>
#>  Pesaran CD test for cross-sectional dependence in
#>  panels
#>
#> data:  inv ~ value + capital
#> z = 4.6612, p-value = 3.144e-06
#> alternative hypothesis: cross-sectional dependence
```

2.5.8.3.1 Global Cross-Sectional Dependence

2.5.8.3.2 Local Cross-Sectional Dependence

Uses a spatial weight matrix **w**.

```
pcdtest(inv ~ value + capital,
        data = Grunfeld,
        model = "within",
        w = weight_matrix)
```

If the null is rejected, cross-sectional correlation exists and should be addressed.

2.5.8.4 Serial Correlation in Panel Data

- **Null Hypothesis**: There is no serial correlation.
- Serial correlation is typically observed in macro panels with long time series (large N and T). It is less relevant in micro panels with short time series (small T and large N).
- **Sources of Serial Correlation:**
  - **Unobserved Individual Effects:** Time-invariant error components.
  - **Idiosyncratic Error Terms:** Often modeled as an autoregressive process (e.g., AR(1)).
  - Typically, "serial correlation" refers to the second type (idiosyncratic errors).

Types of Serial Correlation Tests:

- **Marginal Tests:** Test for one type of dependence at a time but may be biased toward rejection.
- **Joint Tests:** Detect both sources of dependence but do not distinguish the source of the problem.
- **Conditional Tests:** Assume one dependence structure is correctly specified and test for additional departures.

2.5.8.4.1 Unobserved Effects Test

- A semi-parametric test for unobserved effects, with the test statistic $W \sim N$ regardless of the error distribution.
- Null hypothesis (H_0): No unobserved effects ($\sigma_\mu^2 = 0$), which supports using pooled OLS.
- Under H_0: The covariance matrix of residuals is diagonal (no off-diagonal correlations).
- Robustness: The test is robust to both unobserved individual effects and serial correlation.

```
library(plm)
data("Produc", package = "plm")

# Wooldridge test for unobserved individual effects
pwtest(log(gsp) ~ log(pcap) + log(pc) + log(emp) + unemp,
       data = Produc)
#>
#>  Wooldridge's test for unobserved individual effects
#>
#> data:  formula
#> z = 3.9383, p-value = 8.207e-05
#> alternative hypothesis: unobserved effect
```

Interpretation
If we reject H_0, pooled OLS is inappropriate due to the presence of unobserved effects.

2.5.8.4.2 Locally Robust Tests for Serial Correlation and Random Effects

- Joint LM test for random effects and serial correlation
- A Lagrange Multiplier test to jointly detect
  - Random effects (panel-level variance components).
  - Serial correlation (time series dependence).

- Null Hypothesis: Normality and homoskedasticity of idiosyncratic errors (Baltagi & Li, 1991, 1995)
  - This is equivalent to assuming there is no presence of serial correlation and random effects.

```
# Baltagi and Li's joint test for serial correlation and random effects
pbsytest(log(gsp) ~ log(pcap) + log(pc) + log(emp) + unemp,
         data = Produc,
         test = "j")
#>
#>  Baltagi and Li AR-RE joint test
#>
#> data:  formula
#> chisq = 4187.6, df = 2, p-value < 2.2e-16
#> alternative hypothesis: AR(1) errors or random effects
```

Interpretation: If we reject H_0, either serial correlation, random effects, or both are present. But we don't know the source of dependence.

To distinguish the source of dependence, we use either (both tests assume normality and homoskedasticity) (Bera et al., 2001):

BSY Test for Serial Correlation

```
pbsytest(log(gsp) ~ log(pcap) + log(pc) + log(emp) + unemp,
         data = Produc)
#>
#>  Bera, Sosa-Escudero and Yoon locally robust test
#>
#> data:  formula
#> chisq = 52.636, df = 1, p-value = 4.015e-13
#> alternative hypothesis: AR(1) errors sub random effects
```

BSY Test for Random Effects

```
pbsytest(log(gsp) ~ log(pcap) + log(pc) + log(emp) + unemp,
         data = Produc,
         test = "re")
#>
#>  Bera, Sosa-Escudero and Yoon locally robust test
#>  (one-sided)
#>
#> data:  formula
#> z = 57.914, p-value < 2.2e-16
#> alternative hypothesis: random effects sub AR(1) errors
```

If serial correlation is "known" to be absent (based on the BSY test), the **LM test for random effects** is superior.

```
plmtest(inv ~ value + capital,
        data = Grunfeld,
        type = "honda")
#>
#>  Lagrange Multiplier Test - (Honda)
#>
#> data:  inv ~ value + capital
#> normal = 28.252, p-value < 2.2e-16
#> alternative hypothesis: significant effects
```

If random effects are absent (based on the BSY test), we use **Breusch-Godfrey's serial correlation test** (Breusch, 1978; Godfrey, 1978).

```
lmtest::bgtest()
```

If Random Effects are Present: Use Baltagi and Li's Test
Baltagi and Li's test detects serial correlation in AR(1) and MA(1) processes under random effects.

- Null Hypothesis (H_0): Uncorrelated errors.
- Note:
  - The test has power only against positive serial correlation (one-sided).
  - It is applicable only to balanced panels.

```
pbltest(log(gsp) ~ log(pcap) + log(pc) + log(emp) + unemp,
        data = Produc,
        alternative = "onesided")
#>
#>  Baltagi and Li one-sided LM test
#>
#> data:  log(gsp) ~ log(pcap) + log(pc) + log(emp) + unemp
#> z = 21.69, p-value < 2.2e-16
#> alternative hypothesis: AR(1)/MA(1) errors in RE panel model
```

2.5.8.4.3 General Serial Correlation Tests

- Applicable to **random effects, pooled OLS, and fixed effects** models
- Can test for **higher-order serial correlation**

```
# Baltagi-Griffin test for higher-order serial correlation
plm::pbgtest(plm::plm(inv ~ value + capital,
                      data = Grunfeld,
                      model = "within"),
             order = 2)
#>
```

```
#>  Breusch-Godfrey/Wooldridge test for serial
#>  correlation in panel models
#>
#> data:  inv ~ value + capital
#> chisq = 42.587, df = 2, p-value = 5.655e-10
#> alternative hypothesis: serial correlation in idiosyncratic errors
```

For short panels (Small T, Large N), use Wooldridge's test:

```
pwartest(log(emp) ~ log(wage) + log(capital),
         data = EmplUK)
#>
#>  Wooldridge's test for serial correlation in FE panels
#>
#> data:  plm.model
#> F = 312.3, df1 = 1, df2 = 889, p-value < 2.2e-16
#> alternative hypothesis: serial correlation
```

2.5.8.5 Unit Roots and Stationarity in Panel Data

2.5.8.5.1 Dickey-Fuller Test for Stochastic Trends

- Purpose: Tests for the presence of a unit root (non-stationarity) in a time series.
- Null Hypothesis (H_0): The series is non-stationary (i.e., it has a unit root).
- Alternative Hypothesis (H_A): The series is stationary (no unit root).
- Decision Rule:
  - If the test statistic is less than the critical value (or $p < 0.05$), reject H_0, indicating stationarity.
  - If the test statistic is greater than the critical value (or $p \geq 0.05$), fail to reject H_0, suggesting the presence of a unit root.

```
library(tseries)
#> Registered S3 method overwritten by 'quantmod':
#>   method            from
#>   as.zoo.data.frame zoo

# Example: Test for unit root in GDP data
adf.test(Produc$gsp, alternative = "stationary")
#> Warning in adf.test(Produc$gsp, alternative =
#> "stationary"): p-value smaller than printed p-value
#>
#>  Augmented Dickey-Fuller Test
#>
#> data:  Produc$gsp
#> Dickey-Fuller = -6.5425, Lag order = 9, p-value =
```

```
#> 0.01
#> alternative hypothesis: stationary
```

If we reject H_0, the series is **stationary** and does not exhibit a stochastic trend.

2.5.8.5.2 Levin-Lin-Chu Unit Root Test

- Purpose: Tests for the presence of a unit root in a panel dataset.
- Null Hypothesis (H_0): The series has a unit root (non-stationary).
- Alternative Hypothesis (H_A): The series is stationary.
- Assumptions: Requires large N (cross-sections) and moderate T (time periods).
- Decision Rule: If the test statistic is less than the critical value or $p < 0.05$, reject H_0 (evidence of stationarity).

```
library(tseries)
library(plm)

# Levin-Lin-Chu (LLC) Unit Root Test
purtest(Grunfeld, test = "levinlin")
#>
#>  Levin-Lin-Chu Unit-Root Test (ex. var.: None)
#>
#> data:  Grunfeld
#> z = 0.39906, p-value = 0.6551
#> alternative hypothesis: stationarity
```

If we reject H_0, the series is **stationary**.

2.5.8.6 Heteroskedasticity in Panel Data

2.5.8.6.1 Breusch-Pagan Test

- Purpose: Detects heteroskedasticity in regression residuals.
- Null Hypothesis (H_0): The data is homoskedastic (constant variance).
- Alternative Hypothesis (H_A): The data exhibits heteroskedasticity (non-constant variance).
- Decision Rule:
  - If the p-value is small (e.g., $p < 0.05$), reject H_0, suggesting heteroskedasticity.
  - If the p-value is large ($p \geq 0.05$), fail to reject H_0, implying homoskedasticity.

```
library(lmtest)

# Fit a panel model (pooled OLS)
```

```r
model <- lm(log(gsp) ~ log(pcap) + log(pc) + log(emp) + unemp,
            data = Produc)

# Breusch-Pagan Test for Heteroskedasticity
bptest(model)
#>
#>  studentized Breusch-Pagan test
#>
#> data:  model
#> BP = 80.033, df = 4, p-value < 2.2e-16
```

If heteroskedasticity is detected, we need to adjust for it using **robust standard errors**.

2.5.8.6.2 Robust Covariance Matrix Estimation (Sandwich Estimator)

If heteroskedasticity is present, robust covariance matrix estimation is recommended. Different estimators apply depending on whether serial correlation is also an issue (Table 2.13).

```r
library(plm)

# Fit a fixed effects model
fe_model <- plm(log(gsp) ~ log(pcap) + log(pc) + log(emp) + unemp,
                data = Produc,
                model = "within")

# Compute robust standard errors using Arellano's method
coeftest(fe_model, vcov = vcovHC(fe_model, method = "arellano"))
#>
#> t test of coefficients:
#>
#>             Estimate Std. Error t value  Pr(>|t|)
#> log(pcap) -0.0261497  0.0603262 -0.4335   0.66480
#> log(pc)    0.2920069  0.0617425  4.7294 2.681e-06 ***
#> log(emp)   0.7681595  0.0816652  9.4062 < 2.2e-16 ***
#> unemp     -0.0052977  0.0024958 -2.1226   0.03411 *
#> ---
#> Signif. codes:
#> 0 '***' 0.001 '**' 0.01 '*' 0.05 '.' 0.1 ' ' 1
```

Using a robust covariance matrix corrects for heteroskedasticity and/or serial correlation, ensuring valid inference.

Table 2.13 Choosing the correct robust covariance matrix estimator

Estimator	Corrects for heteroskedasticity?	Corrects for serial correlation?	Recommended for
`"white1"`	Yes	No	Random effects
`"white2"`	Yes (common variance within groups)	No	Random effects
`"arellano"`	Yes	Yes	Fixed effects

2.5.9 Model Selection in Panel Data

Panel data models must be chosen based on the structure of the data and underlying assumptions. This section provides guidance on selecting between Pooled OLS, Random Effects, and Fixed Effects models.

2.5.9.1 Pooled OLS vs. Random Effects

The choice between POLS and RE depends on whether there are unobserved individual effects.

Breusch-Pagan Lagrange Multiplier Test

- **Purpose:** Tests whether a random effects model is preferable to a pooled OLS model.
- **Null Hypothesis** (H_0): Variance across entities is zero (i.e., no panel effect $\rightarrow$ POLS is preferred).
- **Alternative Hypothesis** (H_A): There is significant panel-level variation $\rightarrow$ **RE is preferable to POLS.**
- **Decision Rule:** If $p < 0.05$, reject H_0, indicating that RE is preferred.

```
library(plm)

# Breusch-Pagan LM Test
plmtest(plm(inv ~ value + capital, data = Grunfeld,
            model = "pooling"), type = "bp")
#>
#>  Lagrange Multiplier Test - (Breusch-Pagan)
#>
#> data:  inv ~ value + capital
#> chisq = 798.16, df = 1, p-value < 2.2e-16
#> alternative hypothesis: significant effects
```

If the test is significant, RE is more appropriate than POLS.

Table 2.14 Implications of the exogeneity assumption for Fixed and Random Effects estimators

Hypothesis	If true
$H_0 : \text{Cov}(c_i, \mathbf{x_{it}}) = 0$	$\hat{\beta}_{\text{RE}}$ *is* ***consistent and efficient****, while* $\hat{\beta}_{\text{FE}}$ is **consistent**
$H_0 : \text{Cov}(c_i, \mathbf{x_{it}}) \neq 0$	$\hat{\beta}_{\text{RE}}$ *is* ***inconsistent****, while* $\hat{\beta}_{\text{FE}}$ remains **consistent**

2.5.9.2 Fixed Effects vs. Random Effects

The choice between **FE and RE** depends on whether the individual-specific effects are correlated with the regressors (Table 2.14).

Key Assumptions and Properties

Hausman Test

- **Purpose:** Determines whether FE or RE is appropriate.

  For the Hausman test to work, you need to assume that

- Strict exogeneity holds
- A4 to hold for u_{it}

  Then

- Hausman test statistic: $H = (\hat{\beta}_{\text{RE}} - \hat{\beta}_{\text{FE}})'(V(\hat{\beta}_{\text{RE}}) - V(\hat{\beta}_{\text{FE}}))(\hat{\beta}_{\text{RE}} - \hat{\beta}_{\text{FE}}) \sim \chi^2_{n(X)}$ where $n(X)$ is the number of parameters for the time-varying regressors.
- **Null Hypothesis (H_0):** RE estimator is consistent and efficient.
- **Alternative Hypothesis (H_A):** RE estimator is inconsistent, meaning FE should be used.
- **Decision Rule:**
  - If $p < 0.05$: Reject H_0, meaning FE is preferred.
  - If $p \geq 0.05$: Fail to reject H_0, meaning RE can be used.

```
library(plm)

# Fit FE and RE models
fe_model <- plm(inv ~ value + capital, data = Grunfeld, model = "within")
re_model <- plm(inv ~ value + capital, data = Grunfeld, model = "random")

# Hausman test
phtest(fe_model, re_model)
#>
#>  Hausman Test
#>
#> data:  inv ~ value + capital
#> chisq = 2.3304, df = 2, p-value = 0.3119
#> alternative hypothesis: one model is inconsistent
```

If the null hypothesis is rejected, use FE. If not, RE is appropriate.

2.5.9.3 Summary of Model Assumptions and Consistency

All three estimators (POLS, RE, FE) require:

- A1 Linearity
- A2 Full rank
- A5 Data generation (random sampling) for individuals

However, additional assumptions determine whether the estimator is **consistent and efficient**.

POLS

- **Consistent if**
  - A3a Exogeneity holds: $E(\mathbf{x}_{it}' u_{it}) = 0$
  - RE assumption holds: $E(\mathbf{x}_{it}' c_i) = 0$
- If A4 Homoskedasticity does not hold: Use cluster-robust standard errors, but POLS is not efficient.

RE

- **Consistent if**
  - A3a Exogeneity holds: $E(\mathbf{x}_{it}' u_{it}) = 0$
  - RE assumption holds: $E(\mathbf{x}_{it}' c_i) = 0$
- If A4 Homoskedasticity holds: RE is most efficient.
- If A4 Homoskedasticity does not hold: Use cluster-robust standard errors. RE remains more efficient than POLS but is not the most efficient.

FE

- **Consistent if**
  - A3a Exogeneity holds: $E((\mathbf{x}_{it} - \bar{\mathbf{x}}_{it})'(u_{it} - \bar{u}_{it})) = 0$
- **Limitations:**
  - Cannot estimate the effects of time-constant variables.
  - A4 Homoskedasticity generally does not hold, so cluster-robust SEs are required (Table 2.15).

Table 2.15 Estimator selection guide

Estimator/true model	POLS	RE	FE
POLS	Consistent	Consistent	Inconsistent
FE	Consistent	Consistent	Consistent
RE	Consistent	Consistent	Inconsistent

2.5.10 Alternative Estimators

Other estimators are available depending on model violations and additional considerations:

- **Basic Estimators**: Standard POLS, RE, FE.
- **Instrumental Variable Estimator**: Used for endogeneity.
- **Variable Coefficient Estimator**: Allows varying coefficients.
- **Generalized Method of Moments Estimator**: For dynamic panel models.
- **General Feasible GLS Estimator**: Accounts for heteroskedasticity and serial correlation.
- **Means Groups Estimator**: Averages individual-specific estimates.
- **Common Correlated Effects Mean Group Estimator**: Accounts for cross-sectional dependence.
- **Limited Dependent Variable Estimators**: Used for binary or censored data.

2.5.11 Application

2.5.11.1 `plm` Package

The `plm` package in R is designed for panel data analysis, allowing users to estimate various models, including **pooled OLS**, **fixed effects**, **random effects**, and other specifications commonly used in econometrics.

For a detailed guide, refer to

- The official vignette on `plm` functions
- The model component reference

```
# Load the package
library("plm")

data("Produc", package = "plm")

# Display first few rows
head(Produc)
#>     state year region     pcap     hwy   water    util
#> 1 ALABAMA 1970      6 15032.67 7325.80 1655.68 6051.20
#> 2 ALABAMA 1971      6 15501.94 7525.94 1721.02 6254.98
#> 3 ALABAMA 1972      6 15972.41 7765.42 1764.75 6442.23
#> 4 ALABAMA 1973      6 16406.26 7907.66 1742.41 6756.19
#> 5 ALABAMA 1974      6 16762.67 8025.52 1734.85 7002.29
#> 6 ALABAMA 1975      6 17316.26 8158.23 1752.27 7405.76
#>         pc   gsp    emp unemp
#> 1 35793.80 28418 1010.5   4.7
#> 2 37299.91 29375 1021.9   5.2
#> 3 38670.30 31303 1072.3   4.7
#> 4 40084.01 33430 1135.5   3.9
```

```
#> 5 42057.31 33749 1169.8    5.5
#> 6 43971.71 33604 1155.4    7.7
```

To specify panel data, we define the **individual (cross-sectional) and time identifiers**:

```
# Convert data to panel format
pdata <- pdata.frame(Produc, index = c("state", "year"))

# Check structure
summary(pdata)
#>          state          year          region
#>  ALABAMA    : 17   1970    : 48   5      :136
#>  ARIZONA    : 17   1971    : 48   8      :136
#>  ARKANSAS   : 17   1972    : 48   4      :119
#>  CALIFORNIA : 17   1973    : 48   1      :102
#>  COLORADO   : 17   1974    : 48   3      : 85
#>  CONNECTICUT: 17   1975    : 48   6      : 68
#>  (Other)    :714   (Other):528   (Other):170
#>       pcap             hwy              water
#>  Min.   : 2627   Min.   : 1827   Min.   :  228.5
#>  1st Qu.:  7097  1st Qu.: 3858   1st Qu.:  764.5
#>  Median : 17572  Median : 7556   Median : 2266.5
#>  Mean   : 25037  Mean   :10218   Mean   : 3618.8
#>  3rd Qu.: 27692  3rd Qu.:11267   3rd Qu.: 4318.7
#>  Max.   :140217  Max.   :47699   Max.   :24592.3
#>
#>       util               pc               gsp
#>  Min.   :  538.5   Min.   :  4053   Min.   :  4354
#>  1st Qu.: 2488.3   1st Qu.: 21651   1st Qu.: 16503
#>  Median : 7008.8   Median : 40671   Median : 39987
#>  Mean   :11199.5   Mean   : 58188   Mean   : 61014
#>  3rd Qu.:11598.5   3rd Qu.: 64796   3rd Qu.: 68126
#>  Max.   :80728.1   Max.   :375342   Max.   :464550
#>
#>       emp              unemp
#>  Min.   :  108.3   Min.   : 2.800
#>  1st Qu.:  475.0   1st Qu.: 5.000
#>  Median : 1164.8   Median : 6.200
#>  Mean   : 1747.1   Mean   : 6.602
#>  3rd Qu.: 2114.1   3rd Qu.: 7.900
#>  Max.   :11258.0   Max.   :18.000
#>
```

The `plm` package allows for the estimation of several different panel data models.

1. **Pooled OLS Estimator**

A simple **pooled OLS** model assumes a common intercept and ignores individual-specific effects.

```
pooling <- plm(log(gsp) ~ log(pcap) + log(emp) + unemp,
               data = pdata,
               model = "pooling")
summary(pooling)
#> Pooling Model
#>
#> Call:
#> plm(formula = log(gsp) ~ log(pcap) + log(emp) + unemp, data = pdata,
#>     model = "pooling")
#>
#> Balanced Panel: n = 48, T = 17, N = 816
#>
#> Residuals:
#>       Min.    1st Qu.     Median    3rd Qu.       Max.
#> -0.302260 -0.085204 -0.018166  0.051783  0.500144
#>
#> Coefficients:
#>                Estimate Std. Error t-value Pr(>|t|)
#> (Intercept)  2.2124123  0.0790988 27.9703  < 2e-16 ***
#> log(pcap)    0.4121307  0.0216314 19.0525  < 2e-16 ***
#> log(emp)     0.6205834  0.0199495 31.1078  < 2e-16 ***
#> unemp       -0.0035444  0.0020539 -1.7257  0.08478 .
#> ---
#> Signif. codes:
#> 0 '***' 0.001 '**' 0.01 '*' 0.05 '.' 0.1 ' ' 1
#>
#> Total Sum of Squares:    849.81
#> Residual Sum of Squares: 13.326
#> R-Squared:      0.98432
#> Adj. R-Squared: 0.98426
#> F-statistic: 16990.2 on 3 and 812 DF, p-value: < 2.22e-16
```

2. **Between Estimator**

This estimator takes the average over time for each entity, reducing within-group variation.

```
between <- plm(log(gsp) ~ log(pcap) + log(emp) + unemp,
               data = pdata, model = "between")
summary(between)
#> Oneway (individual) effect Between Model
#>
#> Call:
#> plm(formula = log(gsp) ~ log(pcap) + log(emp) + unemp, data = pdata,
#>     model = "between")
#>
#> Balanced Panel: n = 48, T = 17, N = 816
#> Observations used in estimation: 48
#>
#> Residuals:
#>       Min.    1st Qu.     Median    3rd Qu.       Max.
#> -0.172055 -0.086456 -0.013203  0.038100  0.394336
```

```
#>
#> Coefficients:
#>               Estimate Std. Error t-value  Pr(>|t|)
#> (Intercept)  2.0784403  0.3277756  6.3410 1.063e-07 ***
#> log(pcap)    0.4585009  0.0892620  5.1366 6.134e-06 ***
#> log(emp)     0.5751005  0.0828921  6.9379 1.410e-08 ***
#> unemp       -0.0031585  0.0145683 -0.2168    0.8294
#> ---
#> Signif. codes:
#> 0 '***' 0.001 '**' 0.01 '*' 0.05 '.' 0.1 ' ' 1
#>
#> Total Sum of Squares:    48.875
#> Residual Sum of Squares: 0.65861
#> R-Squared:      0.98652
#> Adj. R-Squared: 0.98561
#> F-statistic: 1073.73 on 3 and 44 DF, p-value: < 2.22e-16
```

3. **First-Differences Estimator**

The **first-differences** model eliminates time-invariant effects by differencing adjacent periods.

```
firstdiff <- plm(log(gsp) ~ log(pcap) + log(emp) + unemp,
                 data = pdata, model = "fd")
summary(firstdiff)
#> Oneway (individual) effect First-Difference Model
#>
#> Call:
#> plm(formula = log(gsp) ~ log(pcap) + log(emp) + unemp, data = pdata,
#>     model = "fd")
#>
#> Balanced Panel: n = 48, T = 17, N = 816
#> Observations used in estimation: 768
#>
#> Residuals:
#>       Min.    1st Qu.     Median    3rd Qu.       Max.
#> -0.0846921 -0.0108511  0.0016861  0.0124968  0.1018911
#>
#> Coefficients:
#>               Estimate Std. Error t-value  Pr(>|t|)
#> (Intercept)  0.0101353  0.0013206  7.6749 5.058e-14 ***
#> log(pcap)   -0.0167634  0.0453958 -0.3693     0.712
#> log(emp)     0.8212694  0.0362737 22.6409 < 2.2e-16 ***
#> unemp       -0.0061615  0.0007516 -8.1978 1.032e-15 ***
#> ---
#> Signif. codes:
#> 0 '***' 0.001 '**' 0.01 '*' 0.05 '.' 0.1 ' ' 1
#>
#> Total Sum of Squares:    1.0802
#> Residual Sum of Squares: 0.33394
#> R-Squared:      0.69086
```

```
#> Adj. R-Squared: 0.68965
#> F-statistic: 569.123 on 3 and 764 DF, p-value: < 2.22e-16
```

4. **Fixed Effects (Within) Estimator**

   Controls for time-invariant heterogeneity by demeaning data within individuals.

```
fixed <- plm(log(gsp) ~ log(pcap) + log(emp) + unemp,
             data = pdata, model = "within")
summary(fixed)
#> Oneway (individual) effect Within Model
#>
#> Call:
#> plm(formula = log(gsp) ~ log(pcap) + log(emp) + unemp, data = pdata,
#>     model = "within")
#>
#> Balanced Panel: n = 48, T = 17, N = 816
#>
#> Residuals:
#>       Min.    1st Qu.     Median    3rd Qu.       Max.
#> -0.1253873 -0.0248746 -0.0054276  0.0184698  0.2026394
#>
#> Coefficients:
#>              Estimate  Std. Error t-value Pr(>|t|)
#> log(pcap)  0.03488447  0.03092191  1.1281   0.2596
#> log(emp)   1.03017988  0.02161353 47.6636   <2e-16 ***
#> unemp     -0.00021084  0.00096121 -0.2194   0.8264
#> ---
#> Signif. codes:
#> 0 '***' 0.001 '**' 0.01 '*' 0.05 '.' 0.1 ' ' 1
#>
#> Total Sum of Squares:    18.941
#> Residual Sum of Squares: 1.3077
#> R-Squared:      0.93096
#> Adj. R-Squared: 0.92645
#> F-statistic: 3438.48 on 3 and 765 DF, p-value: < 2.22e-16
```

5. **Random Effects Estimator**

   Accounts for unobserved heterogeneity by modeling it as a random component.

```
random <- plm(log(gsp) ~ log(pcap) + log(emp) + unemp,
              data = pdata, model = "random")
summary(random)
#> Oneway (individual) effect Random Effect Model
#>    (Swamy-Arora's transformation)
#>
#> Call:
#> plm(formula = log(gsp) ~ log(pcap) + log(emp) + unemp, data = pdata,
#>     model = "random")
#>
```

```
#> Balanced Panel: n = 48, T = 17, N = 816
#>
#> Effects:
#>                    var  std.dev share
#> idiosyncratic 0.001709 0.041345 0.103
#> individual    0.014868 0.121934 0.897
#> theta: 0.918
#>
#> Residuals:
#>       Min.    1st Qu.     Median    3rd Qu.       Max.
#> -0.1246674 -0.0268273 -0.0049657  0.0214145  0.2389889
#>
#> Coefficients:
#>               Estimate Std. Error z-value Pr(>|z|)
#> (Intercept) 3.10569727 0.14715985 21.1042   <2e-16 ***
#> log(pcap)   0.03708054 0.02747015  1.3498   0.1771
#> log(emp)    1.00937552 0.02103951 47.9752   <2e-16 ***
#> unemp       0.00004806 0.00092301  0.0521   0.9585
#> ---
#> Signif. codes:
#> 0 '***' 0.001 '**' 0.01 '*' 0.05 '.' 0.1 ' ' 1
#>
#> Total Sum of Squares:    24.523
#> Residual Sum of Squares: 1.4425
#> R-Squared:      0.94118
#> Adj. R-Squared: 0.94096
#> Chisq: 12992.5 on 3 DF, p-value: < 2.22e-16
```

Model Selection and Diagnostic Tests

1. **Lagrange Multiplier Test for Random Effects**

   The **Breusch-Pagan LM test** compares random effects with pooled OLS.

- **Null Hypothesis:** OLS is preferred.
- **Alternative Hypothesis:** Random effects model is appropriate.

```
plmtest(pooling, effect = "individual", type = "bp")
#>
#>  Lagrange Multiplier Test - (Breusch-Pagan)
#>
#> data:  log(gsp) ~ log(pcap) + log(emp) + unemp
#> chisq = 4567.1, df = 1, p-value < 2.2e-16
#> alternative hypothesis: significant effects
```

   Other test types: `"honda"`, `"kw"`, `"ghm"`. Other effects: `"time"`, `"twoways"`.

2. **Cross-Sectional Dependence Tests**

   Breusch-Pagan LM test for cross-sectional dependence

```r
pcdtest(fixed, test = "lm")
#>
#>  Breusch-Pagan LM test for cross-sectional dependence
#>  in panels
#>
#> data:  log(gsp) ~ log(pcap) + log(emp) + unemp
#> chisq = 6490.4, df = 1128, p-value < 2.2e-16
#> alternative hypothesis: cross-sectional dependence
```

Pesaran's CD statistic

```r
pcdtest(fixed, test = "cd")
#>
#>  Pesaran CD test for cross-sectional dependence in
#>  panels
#>
#> data:  log(gsp) ~ log(pcap) + log(emp) + unemp
#> z = 37.13, p-value < 2.2e-16
#> alternative hypothesis: cross-sectional dependence
```

3. **Serial Correlation Test (Panel Version of the Breusch-Godfrey Test)**

   Used to check for autocorrelation in panel data.

```r
pbgtest(fixed)
#>
#>  Breusch-Godfrey/Wooldridge test for serial
#>  correlation in panel models
#>
#> data:  log(gsp) ~ log(pcap) + log(emp) + unemp
#> chisq = 476.92, df = 17, p-value < 2.2e-16
#> alternative hypothesis: serial correlation in idiosyncratic errors
```

4. **Stationarity Test (Augmented Dickey-Fuller Test)**

   Checks whether a time series variable is stationary.

```r
library(tseries)
adf.test(pdata$gsp, k = 2)
#> Warning in adf.test(pdata$gsp, k = 2): p-value smaller than
#> printed p-value
#>
#>  Augmented Dickey-Fuller Test
#>
#> data:  pdata$gsp
#> Dickey-Fuller = -5.9028, Lag order = 2, p-value =
#> 0.01
#> alternative hypothesis: stationary
```

5. **F-Test for Fixed Effects vs. Pooled OLS**

- **Null Hypothesis:** Pooled OLS is appropriate.
- **Alternative Hypothesis:** Fixed effects model is preferred.

```
pFtest(fixed, pooling)
#>
#>  F test for individual effects
#>
#> data:  log(gsp) ~ log(pcap) + log(emp) + unemp
#> F = 149.58, df1 = 47, df2 = 765, p-value < 2.2e-16
#> alternative hypothesis: significant effects
```

6. **Hausman Test for Fixed vs. Random Effects**

- **Null Hypothesis:** Random effects are appropriate.
- **Alternative Hypothesis:** Fixed effects are preferred (RE assumptions are violated).

```
phtest(random, fixed)
#>
#>  Hausman Test
#>
#> data:  log(gsp) ~ log(pcap) + log(emp) + unemp
#> chisq = 84.924, df = 3, p-value < 2.2e-16
#> alternative hypothesis: one model is inconsistent
```

Heteroskedasticity and Robust Standard Errors

1. **Breusch-Pagan Test for Heteroskedasticity**

   Tests whether heteroskedasticity is present in the panel dataset (Table 2.16).

```
library(lmtest)
bptest(log(gsp) ~ log(pcap) + log(emp) + unemp, data = pdata)
#>
#>  studentized Breusch-Pagan test
#>
#> data:  log(gsp) ~ log(pcap) + log(emp) + unemp
#> BP = 98.223, df = 3, p-value < 2.2e-16
```

2. **Correcting for Heteroskedasticity**

   If heteroskedasticity is detected, use robust standard errors:

For Random Effects Model

```
# Original coefficients
coeftest(random)
#>
#> t test of coefficients:
#>
```

```
#>                Estimate Std. Error t value Pr(>|t|)
#> (Intercept) 3.10569727 0.14715985 21.1042   <2e-16 ***
#> log(pcap)   0.03708054 0.02747015  1.3498   0.1774
#> log(emp)    1.00937552 0.02103951 47.9752   <2e-16 ***
#> unemp       0.00004806 0.00092301  0.0521   0.9585
#> ---
#> Signif. codes:
#> 0 '***' 0.001 '**' 0.01 '*' 0.05 '.' 0.1 ' ' 1

# Heteroskedasticity-consistent standard errors
coeftest(random, vcovHC)
#>
#> t test of coefficients:
#>
#>                Estimate Std. Error t value Pr(>|t|)
#> (Intercept) 3.10569727 0.23261788 13.3511   <2e-16 ***
#> log(pcap)   0.03708054 0.06125725  0.6053   0.5451
#> log(emp)    1.00937552 0.06395880 15.7817   <2e-16 ***
#> unemp       0.00004806 0.00215219  0.0223   0.9822
#> ---
#> Signif. codes:
#> 0 '***' 0.001 '**' 0.01 '*' 0.05 '.' 0.1 ' ' 1

# Different HC types
t(sapply(c("HC0", "HC1", "HC2", "HC3", "HC4"), function(x)
    sqrt(diag(vcovHC(random, type = x)))
))
#>     (Intercept)  log(pcap)   log(emp)       unemp
#> HC0   0.2326179 0.06125725 0.06395880 0.002152189
#> HC1   0.2331901 0.06140795 0.06411614 0.002157484
#> HC2   0.2334857 0.06161618 0.06439057 0.002160392
#> HC3   0.2343595 0.06197939 0.06482756 0.002168646
#> HC4   0.2342815 0.06235576 0.06537813 0.002168867
```

- **HC0**: Default heteroskedasticity-consistent (White's estimator).
- **HC1, HC2, HC3**: Recommended for small samples.
- **HC4**: Useful for small samples with influential observations.

For Fixed Effects Model

```
# Original coefficients
coeftest(fixed)
#>
#> t test of coefficients:
#>
#>              Estimate  Std. Error t value Pr(>|t|)
#> log(pcap)  0.03488447  0.03092191  1.1281   0.2596
#> log(emp)   1.03017988  0.02161353 47.6636   <2e-16 ***
#> unemp     -0.00021084  0.00096121 -0.2194   0.8264
#> ---
#> Signif. codes:
#> 0 '***' 0.001 '**' 0.01 '*' 0.05 '.' 0.1 ' ' 1
```

```
# Heteroskedasticity-consistent standard errors
coeftest(fixed, vcovHC)
#>
#> t test of coefficients:
#>
#>               Estimate  Std. Error t value Pr(>|t|)
#> log(pcap)  0.03488447  0.06661083  0.5237   0.6006
#> log(emp)   1.03017988  0.06413365 16.0630   <2e-16 ***
#> unemp     -0.00021084  0.00217453 -0.0970   0.9228
#> ---
#> Signif. codes:
#> 0 '***' 0.001 '**' 0.01 '*' 0.05 '.' 0.1 ' ' 1

# Arellano method for robust errors
coeftest(fixed, vcovHC(fixed, method = "arellano"))
#>
#> t test of coefficients:
#>
#>               Estimate  Std. Error t value Pr(>|t|)
#> log(pcap)  0.03488447  0.06661083  0.5237   0.6006
#> log(emp)   1.03017988  0.06413365 16.0630   <2e-16 ***
#> unemp     -0.00021084  0.00217453 -0.0970   0.9228
#> ---
#> Signif. codes:
#> 0 '***' 0.001 '**' 0.01 '*' 0.05 '.' 0.1 ' ' 1

# Different HC types
t(sapply(c("HC0", "HC1", "HC2", "HC3", "HC4"), function(x)
    sqrt(diag(vcovHC(fixed, type = x)))
))
#>      log(pcap)   log(emp)       unemp
#> HC0 0.06661083 0.06413365 0.002174525
#> HC1 0.06673362 0.06425187 0.002178534
#> HC2 0.06689078 0.06441024 0.002182114
#> HC3 0.06717278 0.06468886 0.002189747
#> HC4 0.06742431 0.06496436 0.002193150
```

Variance Component Structure

Beyond the standard random effects model, the `plm` package provides additional methods for estimating **variance component models** and instrumental variable techniques for dealing with endogeneity in panel data.

Table 2.16 Summary of model selection

Test	Null hypothesis (H)	Decision rule
LM test	OLS is appropriate	Reject H → use RE
Hausman test	Random effects preferred	reject H → Use FE
pFtest	OLS is appropriate	Reject H → use FE
Breusch-Pagan	No heteroskedasticity	Reject H → use robust SE

Different estimators for the **variance component structure** exist in the literature, and `plm` allows users to specify them through the `random.method` argument.

Random Effects Estimators

- `"swar"` (*default*): Swamy and Arora estimator (Swamy & Arora, 1972).
- `"walhus"`: Wallace and Hussain estimator (Wallace & Hussain, 1969).
- `"amemiya"`: Amemiya estimator (Amemiya, 1971).
- `"nerlove"`: Nerlove estimator (Nerlove, 1971) (**Note:** Not available for two-way random effects).

Effects in Panel Models

- **Individual effects** (*default*).
- **Time effects** (`effect = "time"`).
- **Two-way effects** (`effect = "twoways"`).

```
amemiya <- plm(
    log(gsp) ~ log(pcap) + log(emp) + unemp,
    data = pdata,
    model = "random",
    random.method = "amemiya",
    effect = "twoways"
)

summary(amemiya)
#> Twoways effects Random Effect Model
#>    (Amemiya's transformation)
#>
#> Call:
#> plm(formula = log(gsp) ~ log(pcap) + log(emp) + unemp, data = pdata,
#>     effect = "twoways", model = "random", random.method = "amemiya")
#>
#> Balanced Panel: n = 48, T = 17, N = 816
#>
#> Effects:
#>                    var  std.dev share
#> idiosyncratic 0.001228 0.035039 0.028
#> individual    0.041201 0.202981 0.941
#> time          0.001359 0.036859 0.031
#> theta: 0.9582 (id) 0.8641 (time) 0.8622 (total)
#>
#> Residuals:
#>        Min.     1st Qu.      Median     3rd Qu.        Max.
#> -0.13796209 -0.01951506 -0.00053384  0.01807398  0.20452581
#>
#> Coefficients:
#>               Estimate Std. Error z-value  Pr(>|z|)
#> (Intercept)  3.9581876  0.1767036 22.4001 < 2.2e-16 ***
#> log(pcap)    0.0378443  0.0253963  1.4902  0.136184
#> log(emp)     0.8891887  0.0227677 39.0548 < 2.2e-16 ***
#> unemp       -0.0031568  0.0011240 -2.8086  0.004976 **
#> ---
```

```
#> Signif. codes:
#> 0 '***' 0.001 '**' 0.01 '*' 0.05 '.' 0.1 ' ' 1
#>
#> Total Sum of Squares:    5.3265
#> Residual Sum of Squares: 0.98398
#> R-Squared:      0.81527
#> Adj. R-Squared: 0.81458
#> Chisq: 3583.53 on 3 DF, p-value: < 2.22e-16
```

The ercomp() function retrieves estimates of the **variance components** in a random effects model. Below, we extract the variance decomposition using Amemiya's method:

```
ercomp(log(gsp) ~ log(pcap) + log(emp) + unemp,
       data = pdata,
       method = "amemiya",
       effect = "twoways")
#>                    var  std.dev share
#> idiosyncratic 0.001228 0.035039 0.028
#> individual    0.041201 0.202981 0.941
#> time          0.001359 0.036859 0.031
#> theta: 0.9582 (id) 0.8641 (time) 0.8622 (total)
```

This output includes

- Variance of the **individual effect**.
- Variance of the **time effect** (if applicable).
- Variance of the **idiosyncratic error**.

Checking Panel Data Balance

Panel datasets may be balanced (each individual has observations for all time periods) or unbalanced (some individuals are missing observations). The punbalancedness() function measures the degree of balance in the data, with values closer to 1 indicating a balanced panel (Ahrens & Pincus, 1981).

```
punbalancedness(random)
#> gamma    nu
#>     1     1
```

Instrumental Variables in Panel Data

Instrumental variables (IV) are used to address **endogeneity**, which arises when regressors are correlated with the error term. plm provides various IV estimation methods through the inst.method argument.

Instrumental Variable Estimators

- "bvk": Balestra-Varadharajan-Krishnakumar estimator (*default*) (Balestra & Varadharajan-Krishnakumar, 1987).
- "baltagi": Baltagi estimator (Baltagi, 1981).

- "am": Amemiya-MaCurdy estimator (Amemiya & MaCurdy, 1986).
- "bms": Breusch-Mizon-Schmidt estimator (Breusch et al., 1989).

Other Estimators in Panel Data Models

Beyond standard fixed effects and random effects models, the `plm` package provides additional estimation techniques tailored for heterogeneous coefficients, dynamic panel models, and feasible generalized least squares (FGLS) methods.

Variable Coefficients Model (pvcm)

The variable coefficients model (VCM) allows coefficients to vary across cross-sectional units, accounting for unobserved heterogeneity more flexibly.

Two Estimation Approaches

- **Fixed Effects (`within`)**: Assumes coefficients are constant over time but vary across individuals.
- **Random Effects (`random`)**: Assumes coefficients are drawn from a random distribution.

```
fixed_pvcm <-
    pvcm(log(gsp) ~ log(pcap) + log(emp) + unemp,
         data = pdata,
         model = "within")
random_pvcm <-
    pvcm(log(gsp) ~ log(pcap) + log(emp) + unemp,
         data = pdata,
         model = "random")

summary(fixed_pvcm)
#> Oneway (individual) effect No-pooling model
#>
#> Call:
#> pvcm(formula = log(gsp) ~ log(pcap) + log(emp) + unemp, data = pdata,
#>     model = "within")
#>
#> Balanced Panel: n = 48, T = 17, N = 816
#>
#> Residuals:
#>         Min.      1st Qu.       Median      3rd Qu.
#> -0.075247625 -0.013247956  0.000666934  0.013852996
#>         Max.
#>  0.118966807
#>
#> Coefficients:
#>   (Intercept)         log(pcap)           log(emp)
#>  Min.   :-3.8868   Min.   :-1.11962   Min.   :0.3790
#>  1st Qu.: 0.9917   1st Qu.:-0.38475   1st Qu.:0.8197
#>  Median : 2.9848   Median :-0.03147   Median :1.1506
#>  Mean   : 2.8079   Mean   :-0.06028   Mean   :1.1656
#>  3rd Qu.: 4.3553   3rd Qu.: 0.25573   3rd Qu.:1.3779
#>  Max.   :12.8800   Max.   : 1.16922   Max.   :2.4276
#>      unemp
```

```
#>  Min.   :-1.597e-02
#>  1st Qu.:-5.319e-03
#>  Median : 5.335e-05
#>  Mean   : 9.024e-04
#>  3rd Qu.: 8.374e-03
#>  Max.   : 2.507e-02
#>
#> Total Sum of Squares: 15729
#> Residual Sum of Squares: 0.40484
#> Multiple R-Squared: 0.99997
summary(random_pvcm)
#> Oneway (individual) effect Random coefficients model
#>
#> Call:
#> pvcm(formula = log(gsp) ~ log(pcap) + log(emp) + unemp, data = pdata,
#>     model = "random")
#>
#> Balanced Panel: n = 48, T = 17, N = 816
#>
#> Residuals:
#>     Min.  1st Qu.   Median     Mean  3rd Qu.     Max.
#> -0.23364 -0.03401  0.05558  0.09811  0.19349  1.14326
#>
#> Estimated mean of the coefficients:
#>                Estimate  Std. Error z-value  Pr(>|z|)
#> (Intercept)  2.79030044  0.53104167  5.2544 1.485e-07 ***
#> log(pcap)   -0.04195768  0.08621579 -0.4867    0.6265
#> log(emp)     1.14988911  0.07225221 15.9149 < 2.2e-16 ***
#> unemp        0.00031135  0.00163864  0.1900    0.8493
#> ---
#> Signif. codes:
#> 0 '***' 0.001 '**' 0.01 '*' 0.05 '.' 0.1 ' ' 1
#>
#> Estimated variance of the coefficients:
#>             (Intercept) log(pcap)   log(emp)       unemp
#> (Intercept)  11.2648882 -1.335932  0.2035824  0.00827707
#> log(pcap)    -1.3359322  0.287021 -0.1872915 -0.00345298
#> log(emp)      0.2035824 -0.187291  0.2134845  0.00336374
#> unemp         0.0082771 -0.003453  0.0033637  0.00009425
#>
#> Total Sum of Squares: 15729
#> Residual Sum of Squares: 40.789
#> Multiple R-Squared: 0.99741
#> Chisq: 739.334 on 3 DF, p-value: < 2.22e-16
#> Test for parameter homogeneity: Chisq = 21768.8 on 188 DF, p-value: <
↪   2.22e-16
```

Generalized Method of Moments Estimator (`pgmm`)

The **generalized method of moments** estimator is commonly used for **dynamic panel models**, especially when

- There is concern over **endogeneity** in lagged dependent variables.
- Instrumental variables are used for estimation.

```r
library(plm)

# estimates a dynamic labor demand function using one-step GMM,
# applying lagged variables as instruments
z2 <- pgmm(
    log(emp) ~ lag(log(emp), 1) + lag(log(wage), 0:1) +
        lag(log(capital), 0:1) |
        lag(log(emp), 2:99) +
        lag(log(wage), 2:99) + lag(log(capital), 2:99),
    data = EmplUK,
    effect = "twoways",
    model = "onestep",
    transformation = "ld"
)

summary(z2, robust = TRUE)
#> Twoways effects One-step model System GMM
#>
#> Call:
#> pgmm(formula = log(emp) ~ lag(log(emp), 1) + lag(log(wage), 0:1) +
#>     lag(log(capital), 0:1) | lag(log(emp), 2:99) + lag(log(wage),
#>     2:99) + lag(log(capital), 2:99), data = EmplUK, effect =
↪  "twoways",
#>     model = "onestep", transformation = "ld")
#>
#> Unbalanced Panel: n = 140, T = 7-9, N = 1031
#>
#> Number of Observations Used: 1642
#> Residuals:
#>       Min.    1st Qu.     Median       Mean    3rd Qu.
#> -0.7530341 -0.0369030  0.0000000  0.0002882  0.0466069
#>       Max.
#>  0.6001503
#>
#> Coefficients:
#>                          Estimate Std. Error z-value
#> lag(log(emp), 1)         0.935605   0.026295 35.5810
#> lag(log(wage), 0:1)0    -0.630976   0.118054 -5.3448
#> lag(log(wage), 0:1)1     0.482620   0.136887  3.5257
#> lag(log(capital), 0:1)0  0.483930   0.053867  8.9838
#> lag(log(capital), 0:1)1 -0.424393   0.058479 -7.2572
#>                          Pr(>|z|)
#> lag(log(emp), 1)        < 2.2e-16 ***
#> lag(log(wage), 0:1)0    9.050e-08 ***
#> lag(log(wage), 0:1)1    0.0004224 ***
#> lag(log(capital), 0:1)0 < 2.2e-16 ***
#> lag(log(capital), 0:1)1 3.952e-13 ***
#> ---
#> Signif. codes:
#> 0 '***' 0.001 '**' 0.01 '*' 0.05 '.' 0.1 ' ' 1
#>
#> Sargan test: chisq(100) = 118.763 (p-value = 0.097096)
#> Autocorrelation test (1): normal = -4.808434 (p-value = 1.5212e-06)
```

```
#> Autocorrelation test (2): normal = -0.2800133 (p-value = 0.77947)
#> Wald test for coefficients: chisq(5) = 11174.82 (p-value = < 2.22e-16)
#> Wald test for time dummies: chisq(7) = 14.71138 (p-value = 0.039882)
```

Explanation of Arguments

- `log(emp) ~ lag(log(emp), 1) + lag(log(wage), 0:1) + lag(log(capital), 0:1)`
  → Specifies the **dynamic model**, where `log(emp)` depends on its first lag and contemporaneous plus lagged values of `log(wage)` and `log(capital)`.
- `| lag(log(emp), 2:99) + lag(log(wage), 2:99) + lag(log(capital), 2:99)`
  → Instruments for endogenous regressors, using **further lags**.
- `effect = "twoways"`
  → Includes both **individual and time effects**.
- `model = "onestep"`
  → Uses **one-step GMM** (alternative: `"twostep"` for efficiency gain).
- `transformation = "ld"`
  → Uses **lagged differences** as transformation.

Generalized Feasible Generalized Least Squares Models (`pggls`) The FGLS estimator (`pggls`) is robust against

- **Intragroup heteroskedasticity**.
- **Serial correlation** (within groups).

However, it assumes **no cross-sectional correlation** and is most suitable when NNN (cross-sectional units) is much larger than TTT (time periods), i.e., **long panels**.

Random Effects FGLS Model

```
zz <- pggls(
    log(emp) ~ log(wage) + log(capital),
    data = EmplUK,
    model = "pooling"
)

summary(zz)
#> Oneway (individual) effect General FGLS model
#>
#> Call:
#> pggls(formula = log(emp) ~ log(wage) + log(capital), data = EmplUK,
#>     model = "pooling")
#>
#> Unbalanced Panel: n = 140, T = 7-9, N = 1031
#>
#> Residuals:
#>      Min.   1st Qu.    Median      Mean   3rd Qu.      Max.
#> -1.80696 -0.36552  0.06181  0.03230  0.44279  1.58719
```

```
#>
#> Coefficients:
#>                 Estimate Std. Error z-value  Pr(>|z|)
#> (Intercept)     2.023480   0.158468 12.7690 < 2.2e-16 ***
#> log(wage)      -0.232329   0.048001 -4.8401 1.298e-06 ***
#> log(capital)    0.610484   0.017434 35.0174 < 2.2e-16 ***
#> ---
#> Signif. codes:
#> 0 '***' 0.001 '**' 0.01 '*' 0.05 '.' 0.1 ' ' 1
#> Total Sum of Squares: 1853.6
#> Residual Sum of Squares: 402.55
#> Multiple R-squared: 0.78283
```

Fixed Effects FGLS Model

```
zz <- pggls(
    log(emp) ~ log(wage) + log(capital),
    data = EmplUK,
    model = "within"
)

summary(zz)
#> Oneway (individual) effect Within FGLS model
#>
#> Call:
#> pggls(formula = log(emp) ~ log(wage) + log(capital), data = EmplUK,
#>     model = "within")
#>
#> Unbalanced Panel: n = 140, T = 7-9, N = 1031
#>
#> Residuals:
#>          Min.       1st Qu.        Median       3rd Qu.
#> -0.508362414 -0.074254395 -0.002442181  0.076139063
#>          Max.
#>  0.601442300
#>
#> Coefficients:
#>                 Estimate Std. Error z-value  Pr(>|z|)
#> log(wage)      -0.617617   0.030794 -20.056 < 2.2e-16 ***
#> log(capital)    0.561049   0.017185  32.648 < 2.2e-16 ***
#> ---
#> Signif. codes:
#> 0 '***' 0.001 '**' 0.01 '*' 0.05 '.' 0.1 ' ' 1
#> Total Sum of Squares: 1853.6
#> Residual Sum of Squares: 17.368
#> Multiple R-squared: 0.99063
```

Key Considerations

- Efficient under the assumption of homoskedasticity.
- Inefficient if there is group-wise heteroskedasticity.
- Ideal for large-N, small-T panels.
- For a Summary of alternative panel data estimators, refer to Table 2.17.

Table 2.17 Summary of alternative panel data estimators

Estimator	Method	Application
Variable coefficients (pvcm)	Fixed (within), random (random)	Allows coefficients to vary across individuals
GMM (pgmm)	One-step, two-step	Used in dynamic models with endogeneity
Feasible GLS (pggls)	Fixed (within), random (pooling)	Handles heteroskedasticity and serial correlation but assumes no cross-sectional correlation

Table 2.18 Available estimation functions in fixest

Function	Model type
feols	Fixed effects OLS (linear regression)
feglm	Generalized linear models (GLMs)
femlm	Maximum likelihood estimation (MLE)
feNmlm	Non-linear models (non-linear in RHS parameters)
fepois	Poisson fixed-effects regression
fenegbin	Negative binomial fixed-effects regression

Note: These functions work only for fixest objects

2.5.11.2 fixest Package

The fixest package provides **efficient** and **flexible** methods for estimating **fixed effects** and generalized linear models in panel data. It is optimized for handling large datasets with high-dimensional fixed effects and allows for multiple model estimation, robust standard errors, and split-sample estimation.

For further details, refer to the official fixest vignette (Table 2.18).

```
library(fixest)
data(airquality)

# Setting a variable dictionary for output labeling
setFixest_dict(
    c(
        Ozone   = "Ozone (ppb)",
        Solar.R = "Solar Radiation (Langleys)",
        Wind    = "Wind Speed (mph)",
        Temp    = "Temperature"
    )
)

# Fixed effects OLS with stepwise estimation and clustering
est <- feols(
    Ozone ~ Solar.R + sw0(Wind + Temp) | csw(Month, Day),
    data = airquality,
    cluster = ~ Day
)
```

```
#> NOTE: 42 observations removed because of NA values (LHS: 37, RHS: 7).
#>       |-> this msg only concerns the variables common to all
↪  estimations

# Display results
etable(est)
#>                                          est.1
#> Dependent Var.:                    Ozone (ppb)
#>
#> Solar Radiation (Langleys) 0.1148*** (0.0234)
#> Wind Speed (mph)
#> Temperature
#> Fixed-Effects:             ------------------
#> Month                                     Yes
#> Day                                        No
#> __________________________ __________________
#> S.E.: Clustered                       by: Day
#> Observations                              111
#> R2                                    0.31974
#> Within R2                             0.12245
#>
#>                                          est.2
#> Dependent Var.:                    Ozone (ppb)
#>
#> Solar Radiation (Langleys)   0.0522* (0.0202)
#> Wind Speed (mph)           -3.109*** (0.7986)
#> Temperature                 1.875*** (0.3671)
#> Fixed-Effects:             ------------------
#> Month                                     Yes
#> Day                                        No
#> __________________________ __________________
#> S.E.: Clustered                       by: Day
#> Observations                              111
#> R2                                    0.63686
#> Within R2                             0.53154
#>
#>                                         est.3
#> Dependent Var.:                   Ozone (ppb)
#>
#> Solar Radiation (Langleys) 0.1078** (0.0329)
#> Wind Speed (mph)
#> Temperature
#> Fixed-Effects:             -----------------
#> Month                                    Yes
#> Day                                      Yes
#> __________________________ _________________
#> S.E.: Clustered                      by: Day
#> Observations                             111
#> R2                                   0.58018
#> Within R2                            0.12074
#>
#>                                          est.4
#> Dependent Var.:                    Ozone (ppb)
```

```
#>
#> Solar Radiation (Langleys)   0.0509* (0.0236)
#> Wind Speed (mph)            -3.289*** (0.7777)
#> Temperature                  2.052*** (0.2415)
#> Fixed-Effects:              ------------------
#> Month                                     Yes
#> Day                                       Yes
#> ___________________________ __________________
#> S.E.: Clustered                       by: Day
#> Observations                              111
#> R2                                    0.81604
#> Within R2                             0.61471
#> ---
#> Signif. codes: 0 '***' 0.001 '**' 0.01 '*' 0.05 '.' 0.1 ' ' 1

# Output in LaTeX format
# etable(est, tex = TRUE)

# Extract fixed-effects coefficients
fixedEffects <- fixef(est[[1]])
summary(fixedEffects)
#> Fixed_effects coefficients
#> Number of fixed-effects for variable Month is 5.
#>   Mean = 19.6 Variance = 272
#>
#> COEFFICIENTS:
#>   Month:     5     6     7     8     9
#>          3.219 8.288 34.26 40.12 12.13

# View fixed effects for one dimension
fixedEffects$Month
#>         5         6         7         8         9
#>  3.218876  8.287899 34.260812 40.122257 12.130971
```

```
# Plot fixed effects
plot(fixedEffects)
```

This example demonstrates

- Fixed effects estimation (`| csw(Month, Day)`).
- Stepwise selection (`sw0(Wind + Temp)`).
- Clustering of standard errors (`cluster = ~ Day`).
- Extracting and plotting fixed effects (Fig. 2.1).

Multiple Model Estimation

1. **Estimating Multiple Dependent Variables (LHS)**

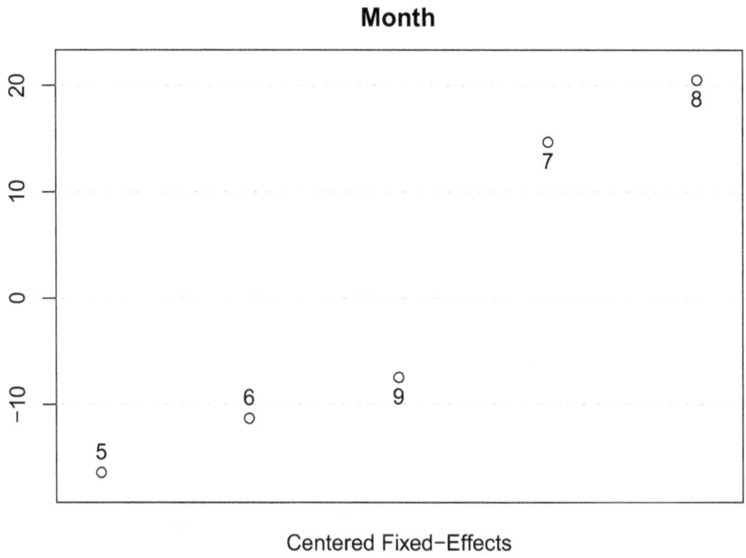

Fig. 2.1 Centered FE plot

Use `feols()` to estimate models with multiple dependent variables simultaneously:

```
etable(feols(c(Sepal.Length, Sepal.Width) ~
               Petal.Length + Petal.Width,
             data = iris))
#>                   feols(c(Sepal.L..1 feols(c(Sepal.Le..2
#> Dependent Var.:         Sepal.Length         Sepal.Width
#>
#> Constant           4.191*** (0.0970)   3.587*** (0.0937)
#> Petal.Length      0.5418*** (0.0693) -0.2571*** (0.0669)
#> Petal.Width        -0.3196* (0.1605)    0.3640* (0.1550)
#> _______________ __________________ ___________________
#> S.E. type                       IID                 IID
#> Observations                    150                 150
#> R2                          0.76626             0.21310
#> Adj. R2                     0.76308             0.20240
#> ---
#> Signif. codes: 0 '***' 0.001 '**' 0.01 '*' 0.05 '.' 0.1 ' ' 1
```

Alternatively, define a list of dependent variables and loop over them:

```
depvars <- c("Sepal.Length", "Sepal.Width")

res <- lapply(depvars, function(var) {
    feols(xpd(..lhs ~ Petal.Length + Petal.Width, ..lhs = var), data =
    ↪  iris)
})
```

```
etable(res)
#>                          model 1            model 2
#> Dependent Var.:      Sepal.Length        Sepal.Width
#>
#> Constant          4.191*** (0.0970)   3.587*** (0.0937)
#> Petal.Length    0.5418*** (0.0693) -0.2571*** (0.0669)
#> Petal.Width      -0.3196* (0.1605)    0.3640* (0.1550)
#> _______________ __________________ ___________________
#> S.E. type                      IID                 IID
#> Observations                   150                 150
#> R2                         0.76626             0.21310
#> Adj. R2                    0.76308             0.20240
#> ---
#> Signif. codes: 0 '***' 0.001 '**' 0.01 '*' 0.05 '.' 0.1 ' ' 1
```

2. **Estimating Multiple Specifications (RHS)**

   Use **stepwise functions** to estimate different model specifications efficiently (Table 2.19).

   Options to write the functions

- `sw` (stepwise): sequentially analyze each element
  - `y ~ sw(x1, x2)` will be estimated as `y ~ x1` and `y ~ x2`
- `sw0` (stepwise 0): similar to `sw` but also estimate a model without the elements in the set first
  - `y ~ sw(x1, x2)` will be estimated as `y ~ 1` and `y ~ x1` and `y ~ x2`
- `csw` (cumulative stepwise): sequentially add each element of the set to the formula
  - `y ~ csw(x1, x2)` will be estimated as `y ~ x1` and `y ~ x1 + x2`
- `csw0` (cumulative stepwise 0): similar to `csw` but also estimate a model without the elements in the set first
  - `y ~ csw(x1, x2)` will be estimated as `y~ 1` `y ~ x1` and `y ~ x1 + x2`
- `mvsw` (multiverse stepwise): all possible combinations of the elements in the set (it will get large very quickly).
  - `mvsw(x1, x2, x3)` will be `sw0(x1, x2, x3, x1 + x2, x1 + x3, x2 + x3, x1 + x2 + x3)`

```
# Example: Cumulative Stepwise Estimation
etable(feols(Ozone ~ csw(Solar.R, Wind, Temp), data = airquality))
#> NOTE: 42 observations removed because of NA values (LHS: 37, RHS: 7).
#>        |-> this msg only concerns the variables common to all
↪  estimations
```

```
#>                              feols(Ozone ~ c..1
#> Dependent Var.:                      Ozone (ppb)
#>
#> Constant                        18.60** (6.748)
#> Solar Radiation (Langleys) 0.1272*** (0.0328)
#> Wind Speed (mph)
#> Temperature
#> __________________________ __________________
#> S.E. type                                  IID
#> Observations                               111
#> R2                                     0.12134
#> Adj. R2                                0.11328
#>
#>                              feols(Ozone ~ c..2
#> Dependent Var.:                      Ozone (ppb)
#>
#> Constant                       77.25*** (9.068)
#> Solar Radiation (Langleys) 0.1004*** (0.0263)
#> Wind Speed (mph)            -5.402*** (0.6732)
#> Temperature
#> __________________________ __________________
#> S.E. type                                  IID
#> Observations                               111
#> R2                                     0.44949
#> Adj. R2                                0.43930
#>
#>                              feols(Ozone ~ c..3
#> Dependent Var.:                      Ozone (ppb)
#>
#> Constant                        -64.34** (23.05)
#> Solar Radiation (Langleys)  0.0598* (0.0232)
#> Wind Speed (mph)            -3.334*** (0.6544)
#> Temperature                  1.652*** (0.2535)
#> __________________________ __________________
#> S.E. type                                  IID
#> Observations                               111
#> R2                                     0.60589
#> Adj. R2                                0.59484
#> ---
#> Signif. codes: 0 '***' 0.001 '**' 0.01 '*' 0.05 '.' 0.1 ' ' 1
```

Table 2.19 Overview of stepwise model selection functions and their behaviors

Stepwise function	Description
`sw(x1, x2)`	Sequentially estimates models with each element separately
`sw0(x1, x2)`	Same as `sw()`, but also estimates a baseline model without the elements
`csw(x1, x2)`	Sequentially adds each element to the formula
`csw0(x1, x2)`	Same as `csw()`, but also includes a baseline model
`mvsw(x1, x2, x3)`	Estimates all possible combinations of the variables

Split-Sample Estimation

Estimate separate regressions for different subgroups in the dataset using `fsplit`.

```
etable(feols(Ozone ~ Solar.R + Wind,
             fsplit = ~ Month,
             data = airquality))
#> NOTE: 42 observations removed because of NA values (LHS: 37, RHS: 7).
#>                                feols(Ozone ~ S..1
#> Sample (Month)                       Full sample
#> Dependent Var.:                      Ozone (ppb)
#>
#> Constant                        77.25*** (9.068)
#> Solar Radiation (Langleys) 0.1004*** (0.0263)
#> Wind Speed (mph)               -5.402*** (0.6732)
#> ___________________________ __________________
#> S.E. type                                    IID
#> Observations                                 111
#> R2                                       0.44949
#> Adj. R2                                  0.43930
#>
#>                            feols(Ozone ..2 feols(Ozone ~..3
#> Sample (Month)                           5                6
#> Dependent Var.:                Ozone (ppb)      Ozone (ppb)
#>
#> Constant                    50.55* (18.30)    8.997 (16.83)
#> Solar Radiation (Langleys) 0.0294 (0.0379) 0.1518. (0.0676)
#> Wind Speed (mph)           -2.762* (1.300)  -0.6177 (1.674)
#> ___________________________ _______________ ________________
#> S.E. type                              IID              IID
#> Observations                            24                9
#> R2                                 0.22543          0.52593
#> Adj. R2                            0.15166          0.36790
#>
#>                            feols(Ozone ~ ..4
#> Sample (Month)                             7
#> Dependent Var.:                  Ozone (ppb)
#>
#> Constant                    88.39*** (20.81)
#> Solar Radiation (Langleys)  0.1135. (0.0582)
#> Wind Speed (mph)           -6.319*** (1.559)
#> ___________________________ _________________
#> S.E. type                                IID
#> Observations                              26
#> R2                                   0.52423
#> Adj. R2                              0.48286
#>
#>                            feols(Ozone ~ ..5
#> Sample (Month)                             8
#> Dependent Var.:                  Ozone (ppb)
#>
#> Constant                    95.76*** (19.83)
#> Solar Radiation (Langleys) 0.2146** (0.0654)
#> Wind Speed (mph)           -8.228*** (1.528)
```

```
#> _______________________________ ____________________
#> S.E. type                                        IID
#> Observations                                      23
#> R2                                           0.70640
#> Adj. R2                                      0.67704
#>
#>                                  feols(Ozone ~ ..6
#> Sample (Month)                                     9
#> Dependent Var.:                          Ozone (ppb)
#>
#> Constant                           67.10*** (14.35)
#> Solar Radiation (Langleys)          0.0373 (0.0463)
#> Wind Speed (mph)                   -4.161*** (1.071)
#> _______________________________ ____________________
#> S.E. type                                        IID
#> Observations                                      29
#> R2                                           0.38792
#> Adj. R2                                      0.34084
#> ---
#> Signif. codes: 0 '***' 0.001 '**' 0.01 '*' 0.05 '.' 0.1 ' ' 1
```

This estimates models separately for each **Month** in the dataset.

Robust Standard Errors in `fixest` `fixest` supports a variety of **robust standard error estimators**, including

- `iid`: Errors are homoskedastic, independent, and identically distributed.
- `hetero`: Errors are heteroskedastic using White correction.
- `cluster`: Errors are correlated within the cluster groups.
- `newey_west`: (Newey & West, 1986) use for time series or panel data. Errors are heteroskedastic and serially correlated.
  - `vcov = newey_west ~ id + period` where `id` is the subject id and `period` is time period of the panel.
  - To specify lag period to consider `vcov = newey_west(2) ~ id + period` where we're considering 2 lag periods.
- `driscoll_kraay` (Driscoll & Kraay, 1998) use for panel data. Errors are cross-sectionally and serially correlated.
  - `vcov = discoll_kraay ~ period`
- `conley`: (Conley, 1999) for cross-section data. Errors are spatially correlated.
  - `vcov = conley ~ latitude + longitude`
  - To specify the distance cutoff, `vcov = vcov_conley(lat = "lat", lon = "long", cutoff = 100, distance = "spherical")`, which will use the `conley()` helper function.
- `hc`: from the `sandwich` package.
  - `vcov = function(x) sandwich::vcovHC(x, type = "HC1"))`

To let R know which SE estimation you want to use, insert `vcov = vcov_type ~ variables`.

Example: Newey-West Standard Errors

```
etable(feols(
    Ozone ~ Solar.R + Wind,
    data = airquality,
    vcov = newey_west ~ Month + Day
))
#> NOTE: 42 observations removed because of NA values (LHS: 37, RHS: 7).
#>                              feols(Ozone ~ So..
#> Dependent Var.:                     Ozone (ppb)
#>
#> Constant                      77.25*** (10.03)
#> Solar Radiation (Langleys) 0.1004*** (0.0258)
#> Wind Speed (mph)            -5.402*** (0.8353)
#> ___________________________ __________________
#> S.E. type                     Newey-West (L=2)
#> Observations                               111
#> R2                                     0.44949
#> Adj. R2                                0.43930
#> ---
#> Signif. codes: 0 '***' 0.001 '**' 0.01 '*' 0.05 '.' 0.1 ' ' 1
```

Specify the number of lag periods to consider:

```
etable(feols(
    Ozone ~ Solar.R + Wind,
    data = airquality,
    vcov = newey_west(2) ~ Month + Day
))
#> NOTE: 42 observations removed because of NA values (LHS: 37, RHS: 7).
#>                              feols(Ozone ~ So..
#> Dependent Var.:                     Ozone (ppb)
#>
#> Constant                      77.25*** (10.03)
#> Solar Radiation (Langleys) 0.1004*** (0.0258)
#> Wind Speed (mph)            -5.402*** (0.8353)
#> ___________________________ __________________
#> S.E. type                     Newey-West (L=2)
#> Observations                               111
#> R2                                     0.44949
#> Adj. R2                                0.43930
#> ---
#> Signif. codes: 0 '***' 0.001 '**' 0.01 '*' 0.05 '.' 0.1 ' ' 1
```

Conley Spatial Correlation: `vcov = conley ~ latitude + longitude`

To specify a distance cutoff: `vcov = vcov_conley(lat = "lat", lon = "long", cutoff = 100, distance = "spherical")`

Using Standard Errors from the `sandwich` Package

```
etable(feols(
    Ozone ~ Solar.R + Wind,
    data = airquality,
    vcov = function(x)
        sandwich::vcovHC(x, type = "HC1")
))
#> NOTE: 42 observations removed because of NA values (LHS: 37, RHS: 7).
#>                              feols(Ozone ~ So..
#> Dependent Var.:                     Ozone (ppb)
#>
#> Constant                      77.25*** (9.590)
#> Solar Radiation (Langleys) 0.1004*** (0.0231)
#> Wind Speed (mph)            -5.402*** (0.8134)
#> __________________________ __________________
#> S.E. type                  vcovHC(type="HC1")
#> Observations                              111
#> R2                                    0.44949
#> Adj. R2                               0.43930
#> ---
#> Signif. codes: 0 '***' 0.001 '**' 0.01 '*' 0.05 '.' 0.1 ' ' 1
```

Small Sample Correction
Apply small sample adjustments using `ssc()`:

```
etable(feols(
    Ozone ~ Solar.R + Wind,
    data = airquality,
    ssc = ssc(adj = TRUE, cluster.adj = TRUE)
))
#> NOTE: 42 observations removed because of NA values (LHS: 37, RHS: 7).
#>                              feols(Ozone ~ So..
#> Dependent Var.:                     Ozone (ppb)
#>
#> Constant                      77.25*** (9.068)
#> Solar Radiation (Langleys) 0.1004*** (0.0263)
#> Wind Speed (mph)            -5.402*** (0.6732)
#> __________________________ __________________
#> S.E. type                                 IID
#> Observations                              111
#> R2                                    0.44949
#> Adj. R2                               0.43930
#> ---
#> Signif. codes: 0 '***' 0.001 '**' 0.01 '*' 0.05 '.' 0.1 ' ' 1
```

This corrects for **bias** when working with small samples.

2.6 Choosing the Right Type of Data

Selecting the appropriate data type depends on

1. **Research Questions**: Do you need to understand changes over time at the individual level (panel) or just a snapshot comparison at one point (cross-sectional)?
2. **Resources**: Longitudinal or panel studies can be resource-intensive.
3. **Time Constraints**: If you need fast results, cross-sectional or repeated cross-sectional might be more practical.
4. **Analytical Goals**: Time series forecasting, causal inference, or descriptive comparison each has different data requirements.
5. **Availability**: Sometimes only secondary or repeated cross-sectional data is available, which constrains the design.

2.7 Data Quality and Ethical Considerations

Regardless of data type, **data quality** is crucial. Poor data-be it incomplete, biased, or improperly measured-can lead to incorrect conclusions. Researchers should

- **Ensure Validity and Reliability**: Use well-designed instruments and consistent measurement techniques.
- **Address Missing Data**: Apply appropriate imputation methods if feasible.
- **Manage Attrition** (in Panel Data): Consider weighting or sensitivity analyses to deal with dropouts.
- **Check Representativeness**: Especially in cross-sectional and repeated cross-sectional surveys, ensure sampling frames match the target population.
- **Protect Confidentiality and Privacy**: Particularly in panel studies with repeated contact, store data securely and follow ethical guidelines.
- **Obtain Proper Consent**: Inform participants about study details, usage of data, and rights to withdraw.

Chapter 3
Variable Transformation

This chapter introduces the theoretical rationale and practical application of variable transformation techniques to improve model fit, satisfy assumptions, and enhance interpretability. It begins with transformations for continuous variables, including standardization, log transformations, power transformations (e.g., Box-Cox), and winsorization. These methods are presented in the context of their effects on skewness, variance stabilization, and coefficient interpretation. The chapter also addresses transformation strategies for categorical variables, including dummy coding, effects coding, and treatment contrasts. Readers will learn when transformations are appropriate, when they might obscure interpretation, and how they interact with model specification. The emphasis throughout is on practical implementation and the analytical reasoning behind transformation decisions.

3.1 Continuous Variables

Transforming continuous variables can be useful for various reasons, including

- **Changing the scale of variables** to make them more interpretable or comparable
- **Reducing skewness** to approximate a normal distribution, which can improve statistical inference
- **Stabilizing variance** in cases of heteroskedasticity
- **Enhancing interpretability** in business applications (e.g., logarithmic transformations for financial data)

© The Author(s), under exclusive license to Springer Nature Switzerland AG 2025
M. Nguyen, *Advanced Modeling and Data Challenges*,
https://doi.org/10.1007/978-3-032-01719-2_3

3.1.1 Standardization (Z-Score Normalization)

A common transformation to center and scale data

$$x_i' = \frac{x_i - \bar{x}}{s}$$

where

- x_i is the original value,
- $\bar{x}$ is the sample mean,
- s is the sample standard deviation.

When to Use
- When variables have different units of measurement and need to be on a common scale.
- When a few large numbers dominate the dataset.

3.1.2 Min-Max Scaling (Normalization)

Rescales data to a fixed range, typically [0, 1]

$$x_i' = \frac{x_i - x_{\min}}{x_{\max} - x_{\min}}$$

When to Use
- When working with fixed-interval data (e.g., percentages, proportions).
- When preserving relative relationships between values is important.
- Caution: This method is sensitive to outliers, as extreme values determine the range.

3.1.3 Square Root and Cube Root Transformations

Useful for handling **positive skewness** and **heteroskedasticity**

- **Square Root:** Reduces moderate skewness and variance.
- **Cube Root:** Works on more extreme skewness and allows negative values.

Common Use Cases
- **Frequency count data** (e.g., website visits, sales transactions).
- **Data with many small values or zeros** (e.g., income distributions in microfinance).

Table 3.1 Common log transformations

Formula	When to use
$x_i' = \log(x_i)$	When all values are positive
$x_i' = \log(x_i + 1)$	When data contains zeros
$x_i' = \log(x_i + c)$	Choosing c depends on context
$x_i' = \frac{x_i}{\lvert x_i \rvert} \log \lvert x_i \rvert$	When data contains negative values
$x_i'^{\lambda} = \log\left(x_i + \sqrt{x_i^2 + \lambda}\right)$	Generalized log transformation

3.1.4 Logarithmic Transformation

Logarithmic transformations are particularly useful for handling highly skewed data. They compress large values while expanding small values, which helps with heteroskedasticity and normality assumptions (Table 3.1).

Selecting the constant c is critical:

- If c is too large, it can obscure the true nature of the data.
- If c is too small, the transformation might not effectively reduce skewness.

From a statistical modeling perspective

- **For inference-based models**, the choice of c can significantly impact the fit. See Ekwaru and Veugelers (2018).
- **In causal inference** (e.g., DID, IV), improper log transformations (e.g., logging zero values) can introduce bias (Chen & Roth, 2024).

When is Log Transformation Problematic?

- When **zero values have a meaningful interpretation** (e.g., income of unemployed individuals).
- When **data are censored** (e.g., income data truncated at reporting thresholds).
- When **measurement error exists** (e.g., rounding errors from survey responses).

If zeros are **small but meaningful** (e.g., revenue from startups), then using $\log(x + c)$ may be acceptable.

```
library(tidyverse)
#> -- Attaching core tidyverse packages ---- tidyverse 2.0.0 --
#> v dplyr     1.1.4     v readr     2.1.5
#> v forcats   1.0.0     v stringr   1.5.1
#> v ggplot2   3.5.2     v tibble    3.2.1
#> v lubridate 1.9.4     v tidyr     1.3.1
#> v purrr     1.0.4
#> -- Conflicts -------------------- tidyverse_conflicts() --
#> x dplyr::filter() masks stats::filter()
#> x dplyr::lag()    masks stats::lag()
#> i Use the conflicted package (<http://conflicted.r-lib.org/>) to force
↪   all conflicts to become errors
```

```r
# Load dataset
cars = datasets::cars

# Original values
head(cars$speed)
#> [1] 4 4 7 7 8 9

# Log transformation (basic)
log(cars$speed) %>% head()
#> [1] 1.386294 1.386294 1.945910 1.945910 2.079442 2.197225

# Log transformation for zero-inflated data
log1p(cars$speed) %>% head()
#> [1] 1.609438 1.609438 2.079442 2.079442 2.197225 2.302585
```

3.1.5 *Exponential Transformation*

The exponential transformation is useful when data exhibit negative skewness or when an underlying logarithmic trend is suspected, such as in survival analysis and decay models.

When to Use

- Negatively skewed distributions.
- Processes that follow an exponential trend (e.g., population growth, depreciation of assets).

3.1.6 *Power Transformation*

Power transformations help adjust skewness, particularly for negatively skewed data (Fig. 3.1).

When to Use

- When variables have a negatively skewed distribution.
- When the relationship between variables is non-linear.

Common power transformations include

- Square transformation: x^2 (moderate adjustment).
- Cubic transformation: x^3 (stronger adjustment).
- Fourth-root transformation: $x^{1/4}$ (more subtle than square root).

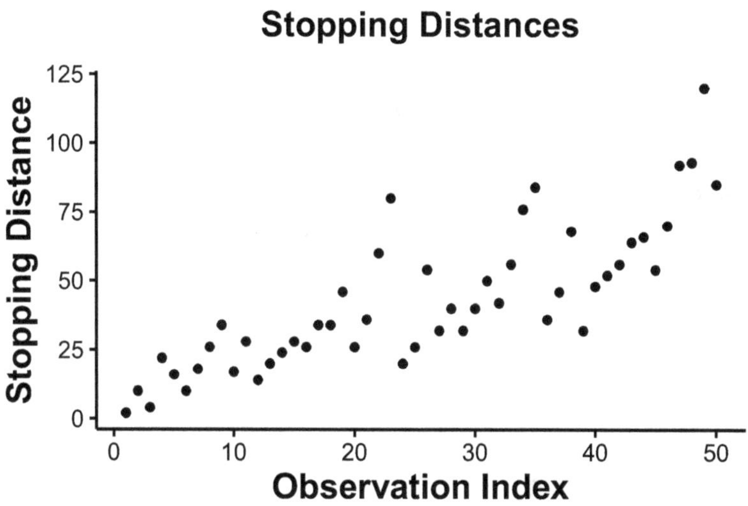

Fig. 3.1 Stopping distances

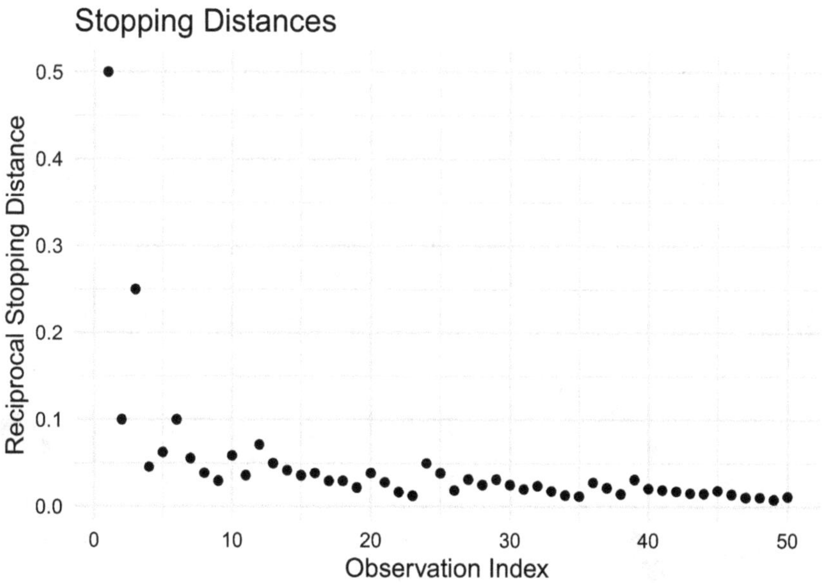

Fig. 3.2 Reciprocal stopping distances

3.1.7 Inverse (Reciprocal) Transformation

The inverse transformation is useful for handling platykurtic (flat) distributions or positively skewed data (Fig. 3.2).

Formula

$$x_i' = \frac{1}{x_i}$$

When to Use

- Reducing extreme values in positively skewed distributions.
- Ratio data (e.g., speed = distance/time).
- When the variable has a natural lower bound (e.g., time to completion).

```
# data(cars)
cars = datasets::cars

# Original distribution
head(cars$dist)
#> [1]  2 10  4 22 16 10
```

```
library(ggplot2)
ggplot(cars, aes(x = seq_along(dist), y = dist)) +
  geom_point(size = 2) +
  labs(
    title = "Stopping Distances",
    x     = "Observation Index",
    y     = "Stopping Distance"
  ) +
    causalverse::ama_theme()
#> Warning: replacing previous import 'data.table::wday' by
#> 'lubridate::wday' when loading 'causalverse'
#> Warning: replacing previous import 'data.table::second' by
#> 'lubridate::second' when loading 'causalverse'
#> Warning: replacing previous import 'data.table::isoweek' by
#> 'lubridate::isoweek' when loading 'causalverse'
#> Warning: replacing previous import 'data.table::yday' by
#> 'lubridate::yday' when loading 'causalverse'
#> Warning: replacing previous import 'data.table::hour' by
#> 'lubridate::hour' when loading 'causalverse'
#> Warning: replacing previous import 'data.table::year' by
#> 'lubridate::year' when loading 'causalverse'
#> Warning: replacing previous import 'data.table::month' by
#> 'lubridate::month' when loading 'causalverse'
#> Warning: replacing previous import 'data.table::week' by
#> 'lubridate::week' when loading 'causalverse'
#> Warning: replacing previous import 'data.table::minute' by
#> 'lubridate::minute' when loading 'causalverse'
#> Warning: replacing previous import 'data.table::mday' by
#> 'lubridate::mday' when loading 'causalverse'
#> Warning: replacing previous import 'data.table::quarter' by
#> 'lubridate::quarter' when loading 'causalverse'
#> Warning: replacing previous import 'formattable::digits' by
#> 'xtable::digits' when loading 'causalverse'
```

```
# Reciprocal transformation
library(ggplot2)
ggplot(cars, aes(x = seq_along(dist), y = 1/dist)) +
  geom_point(size = 2) +
  labs(
    title = "Stopping Distances",
    x     = "Observation Index",
    y     = "Reciprocal Stopping Distance"
  ) +
  theme_minimal(base_size = 14)
```

3.1.8 *Hyperbolic Arcsine Transformation*

The arcsinh (inverse hyperbolic sine) transformation is useful for handling proportion variables (0–1) and skewed distributions. It behaves similarly to the logarithmic transformation but has the advantage of handling zero and negative values (Figs. 3.3 and 3.4).

Formula

$$\text{arcsinh}(Y) = \log\left(\sqrt{1+Y^2}+Y\right)$$

When to Use

- Proportion variables (e.g., market share, probability estimates).

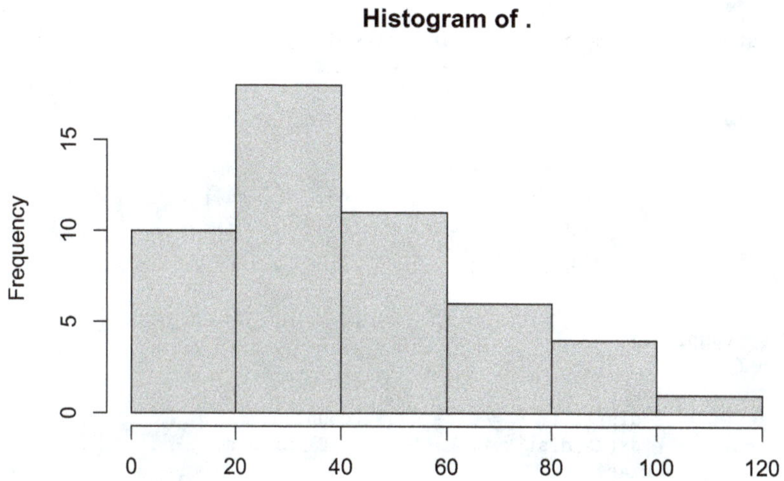

Fig. 3.3 Histogram plot

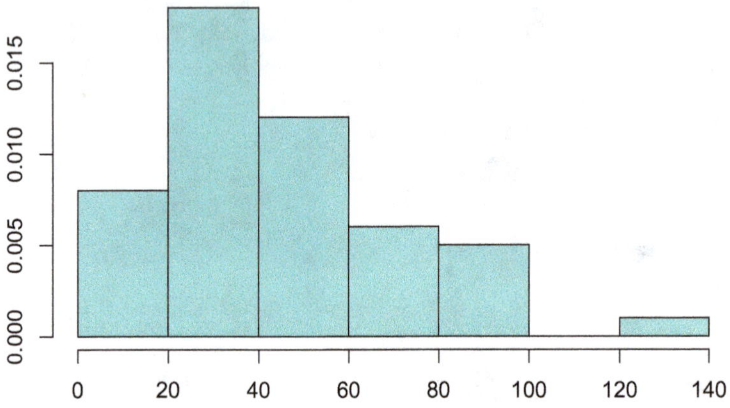

Fig. 3.4 Alternative histogram plot

- Data with extreme skewness where log transformation is problematic.
- Variables containing zeros or negative values (unlike log, arcsinh handles zeros naturally).
- Alternative to log transformation for handling zeros.

```
# Visualize original distribution
cars$dist %>% hist()
```

```
# Alternative histogram
cars$dist %>% MASS::truehist()
```

```
# Apply arcsinh transformation
as_dist <- bestNormalize::arcsinh_x(cars$dist)
#> Registered S3 method overwritten by 'butcher':
#>   method                 from
#>   as.character.dev_topic generics
as_dist
#> Standardized asinh(x) Transformation with 50 nonmissing obs.:
#>  Relevant statistics:
#>  - mean (before standardization) = 4.230843
#>  - sd (before standardization) = 0.7710887
```

```
library(ggplot2)
ggplot() +
  geom_histogram(
    aes(x = as_dist$x.t, y = after_stat(count)),
    breaks = hist(as_dist$x.t, plot = FALSE)$breaks,
    colour = "black",
    fill   = "lightblue"
  ) +
```

```
  labs(
    title = "Histogram of xt",
    x     = "xt values",
    y     = "Count"
  ) +
  theme_minimal()
```

Consider a simple regression model (Table 3.2):

$$Y = \beta X + \epsilon$$

When both Y and X are transformed

- The coefficient estimate β represents **elasticity**: *A 1% increase in* X leads to a $\beta\%$ change in Y.

When only Y is transformed

- The coefficient estimate represents a **percentage change in** Y for a one-unit change in X.

This makes the arcsinh transformation particularly valuable for **log-linear models** where zero values exist (Figs. 3.5 and 3.6).

3.1.9 Ordered Quantile Normalization (Rank-Based Transformation)

The ordered quantile normalization (OQN) technique transforms data into a normal distribution using rank-based methods (Bartlett, 1947).

Table 3.2 Interpretation of estimated effects in recent applied economics papers

Paper	Interpretation
Azoulay et al. (2019)	Elasticity
Faber and Gaubert (2019)	Percentage
Hjort and Poulsen (2019)	Percentage
Johnson (2020)	Percentage
Beerli et al. (2021)	Percentage
Norris et al. (2021)	Percentage
Berkouwer and Dean (2022)	Percentage
Cabral et al. (2022)	Elasticity
Carranza et al. (2022)	Percentage
Mirenda et al. (2022)	Percentage

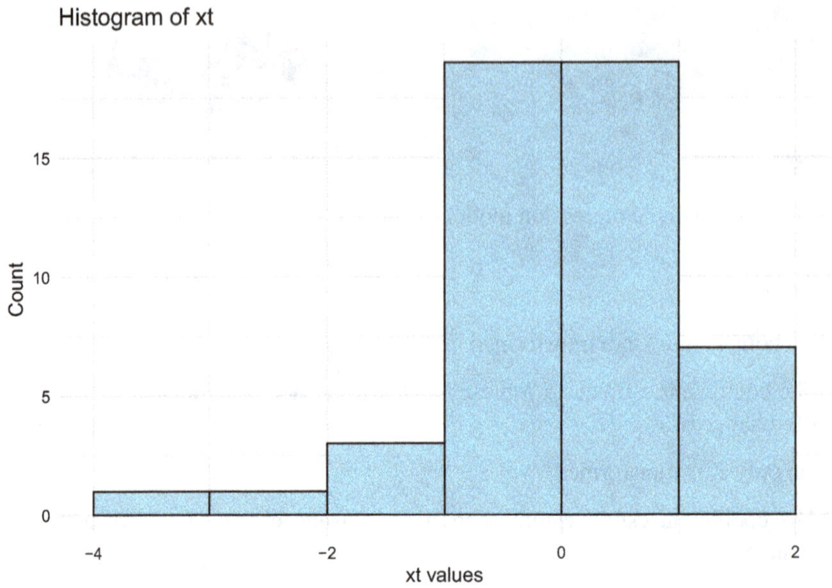

Fig. 3.5 Arcsinh transformed histogram

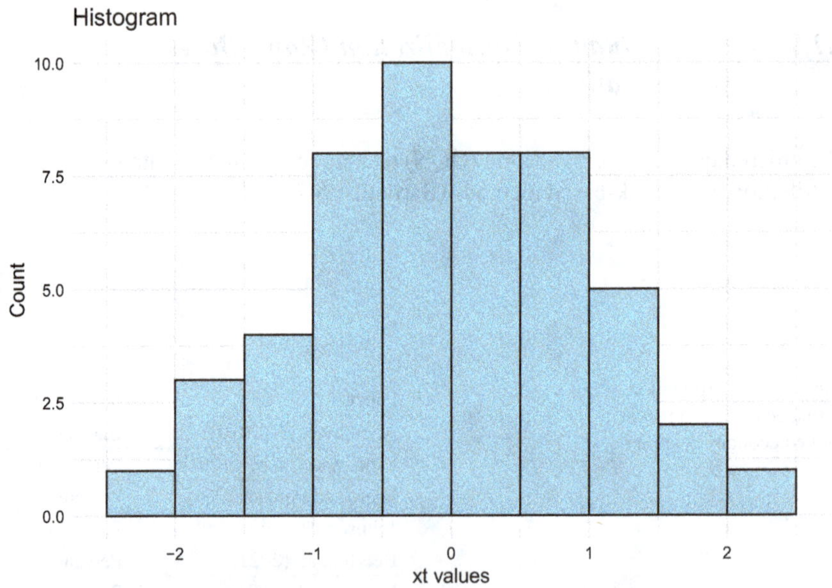

Fig. 3.6 Histogram plot

Formula

$$x_i' = \Phi^{-1}\left(\frac{\text{rank}(x_i) - 1/2}{\text{length}(x)}\right)$$

where Φ^{-1} is the inverse normal cumulative distribution function.

When to Use

- When data are **heavily skewed** or **contain extreme values**.
- When normality is required for **parametric tests**.

```
ord_dist <- bestNormalize::orderNorm(cars$dist)
#> Warning in bestNormalize::orderNorm(cars$dist): Ties in data, Normal
↪   distribution not guaranteed
ord_dist
#> orderNorm Transformation with 50 nonmissing obs and ties
#>  - 35 unique values
#>  - Original quantiles:
#>   0%  25%  50%  75% 100%
#>    2   26   36   56  120
```

```
library(ggplot2)
ggplot() +
  geom_histogram(
    aes(x = ord_dist$x.t, y = after_stat(count)),
    breaks = hist(ord_dist$x.t, plot = FALSE)$breaks,
    colour = "black",
    fill   = "lightblue"
  ) +
  labs(
    title = "Histogram",
    x     = "xt values",
    y     = "Count"
  ) +
  theme_minimal()
```

3.1.10 Lambert W x F Transformation

The Lambert W transformation is a more advanced method that normalizes data by removing skewness and heavy tails (Fig. 3.7).

When to Use

- When traditional transformations (e.g., log, Box-Cox) fail.
- When dealing with heavy-tailed distributions.

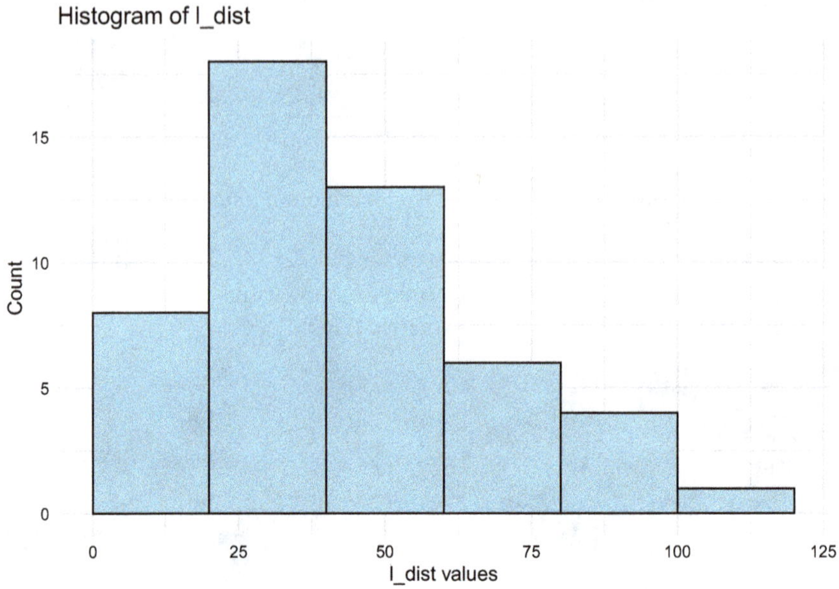

Fig. 3.7 Lambert W transformation

```r
# cars = datasets::cars
# head(cars$dist)
# cars$dist %>% hist()

# Apply Lambert W transformation
l_dist <- LambertW::Gaussianize(cars$dist)
```

```r
library(ggplot2)
ggplot() +
  geom_histogram(
    aes(x = l_dist, y = after_stat(count)),
    breaks = hist(l_dist, plot = FALSE)$breaks,
    colour = "black",
    fill   = "lightblue"
  ) +
  labs(
    title = "Histogram of l_dist",
    x     = "l_dist values",
    y     = "Count"
  ) +
  theme_minimal()
```

3.1.11 Inverse Hyperbolic Sine Transformation

The inverse hyperbolic sine (IHS) transformation is similar to the log transformation but handles zero and negative values (Johnson, 1949).

Formula

$$f(x, \theta) = \frac{\sinh^{-1}(\theta x)}{\theta} = \frac{\log\left(\theta x + (\theta^2 x^2 + 1)^{1/2}\right)}{\theta}$$

When to Use

- When data contain zeros or negative values.
- Alternative to log transformation in economic and financial modeling.

3.1.12 Box-Cox Transformation

The Box-Cox transformation is a power transformation designed to improve linearity and normality (Manly, 1976; Bickel & Doksum, 1981; Box & Cox, 1981).

Formula

$$x_i'^{\lambda} = \begin{cases} \frac{x_i^{\lambda}-1}{\lambda} & \text{if } \lambda \neq 0 \\ \log(x_i) & \text{if } \lambda = 0 \end{cases}$$

When to Use

- To fix non-linearity in the error terms of regression models.
- When data are strictly positive

```
library(MASS)
#>
#> Attaching package: 'MASS'
#> The following object is masked from 'package:dplyr':
#>
#>     select
# data(cars)
cars = datasets::cars
mod <- lm(cars$speed ~ cars$dist, data = cars)

# Check residuals
# plot(mod)
```

```
# Find optimal lambda
bc <- MASS::boxcox(mod, lambda = seq(-3, 3))
```

```
best_lambda <- bc$x[which.max(bc$y)]

# Apply transformation
mod_lambda = lm(cars$speed ^ best_lambda ~ cars$dist, data = cars)
# plot(mod_lambda)
```

For the **two-parameter Box-Cox transformation**, we use (Fig. 3.8)

$$x_i'(\lambda_1, \lambda_2) = \begin{cases} \frac{(x_i+\lambda_2)^{\lambda_1}-1}{\lambda_1} & \text{if } \lambda_1 \neq 0 \\ \log(x_i + \lambda_2) & \text{if } \lambda_1 = 0 \end{cases}$$

```
# Two-parameter Box-Cox transformation
two_bc <- geoR::boxcoxfit(cars$speed)
two_bc
#> Fitted parameters:
#>    lambda      beta   sigmasq
#>  1.028798 15.253008 31.935297
#>
#> Convergence code returned by optim: 0
# plot(two_bc)
```

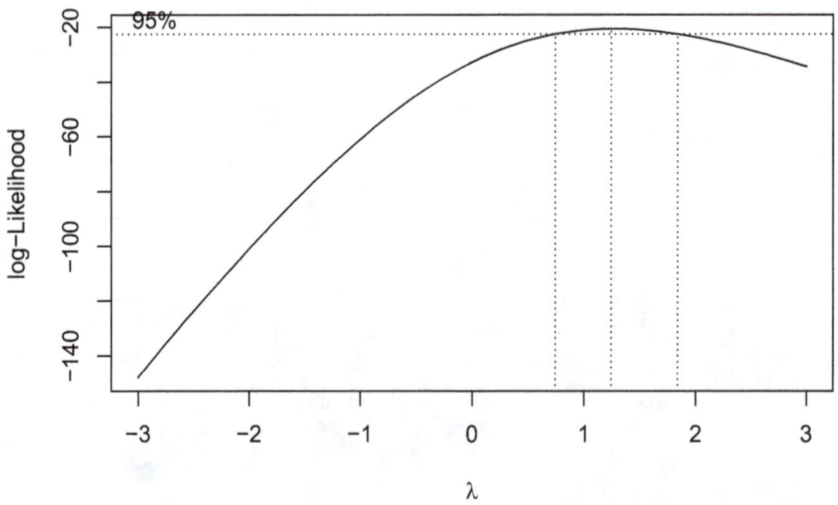

Fig. 3.8 Log-likelihood curve

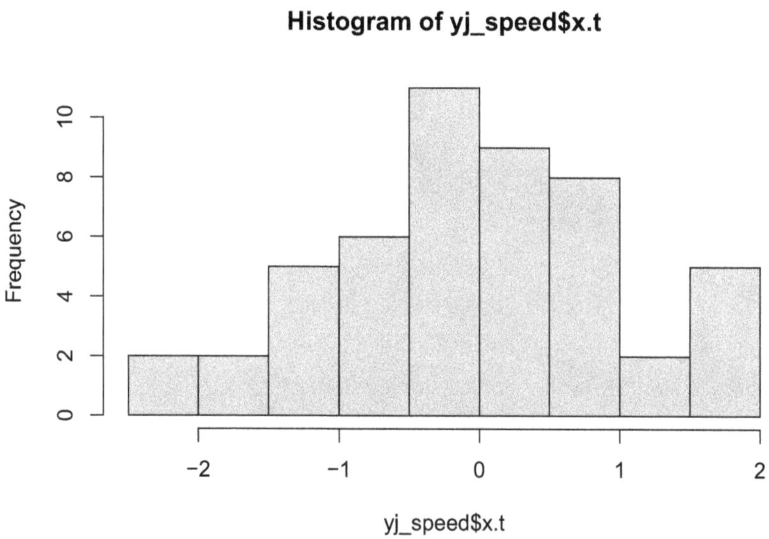

Fig. 3.9 Histogram of xt

3.1.13 Yeo-Johnson Transformation

Similar to Box-Cox (when $\lambda = 1$), but allows for negative values (Fig. 3.9).

Formula

$$x_i'^{\lambda} = \begin{cases} \frac{(x_i+1)^{\lambda}-1}{\lambda} & \text{if } \lambda \neq 0, x_i \geq 0 \\ \log(x_i + 1) & \text{if } \lambda = 0, x_i \geq 0 \\ \frac{-\left[(-x_i+1)^{2-\lambda}-1\right]}{2-\lambda} & \text{if } \lambda \neq 2, x_i < 0 \\ -\log(-x_i + 1) & \text{if } \lambda = 2, x_i < 0 \end{cases}$$

```
# data(cars)
cars = datasets::cars
yj_speed <- bestNormalize::yeojohnson(cars$speed)
```

```
hist(yj_speed$x.t)
```

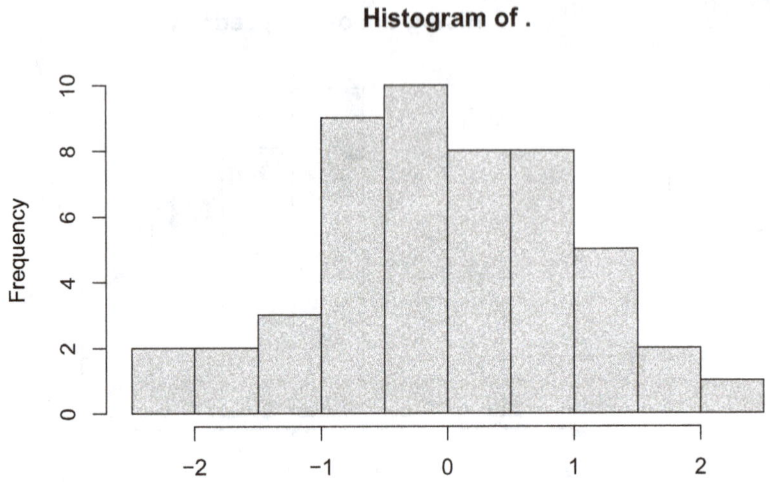

Fig. 3.10 Best transformation histogram

3.1.14 RankGauss Transformation

A rank-based transformation that maps values to a normal distribution.

When to Use

- To handle skewed data while preserving rank order.

3.1.15 Automatically Choosing the Best Transformation

The `bestNormalize` package selects the **best transformation** for a given dataset (Fig. 3.10).

```
bestdist <- bestNormalize::bestNormalize(cars$dist)
bestdist$x.t %>% hist()
```

3.2 Categorical Variables

Transforming categorical variables into numerical representations is essential for machine learning models and statistical analysis. The key objectives include

- Converting categorical data into a format suitable for numerical models.
- Improving model interpretability and performance.
- Handling high-cardinality categorical variables efficiently.

There are multiple ways to transform categorical variables, each with its advantages and use cases. The choice depends on factors like **cardinality**, **ordinality**, and **model type**.

3.2.1 One-Hot Encoding (Dummy Variables)

Creates binary indicator variables for each category.

Formula
For a categorical variable with k unique values, create k binary columns:

$$x_i' = \begin{cases} 1 & \text{if } x_i = \text{category} \\ 0 & \text{otherwise} \end{cases}$$

When to Use
- Low-cardinality categorical variables (e.g., "Red," "Blue," "Green").
- Tree-based models (e.g., Random Forest, XGBoost).
- Linear regression models (dummy variables prevent information loss).

```
library(caret)
#> Loading required package: lattice
#>
#> Attaching package: 'caret'
#> The following object is masked from 'package:purrr':
#>
#>     lift

data(iris)
dummy_vars <- dummyVars(~ Species, data = iris)
one_hot_encoded <- predict(dummy_vars, newdata = iris)
head(one_hot_encoded)
#>   Species.setosa Species.versicolor Species.virginica
#> 1              1                  0                 0
#> 2              1                  0                 0
#> 3              1                  0                 0
#> 4              1                  0                 0
#> 5              1                  0                 0
#> 6              1                  0                 0
```

3.2.2 Label Encoding

Assigns integer values to categories.

Table 3.3 Comparison of categorical encoding methods

Category	Encoded value
Red	1
Blue	2
Green	3

Formula
If a categorical variable has k unique values

$$\text{Category} \rightarrow \text{Integer}$$

Example
See Table 3.3.

When to Use

- Ordinal categorical variables (e.g., "Low," "Medium," "High").
- Neural networks (use embeddings instead of one-hot).
- Memory-efficient encoding for high-cardinality features.

```
iris$Species_encoded <- as.numeric(factor(iris$Species))
head(iris$Species_encoded)
#> [1] 1 1 1 1 1 1
```

3.2.3 *Feature Hashing (Hash Encoding)*

Maps categories to a **fixed number of hash bins**, reducing memory usage.

When to Use

- High-cardinality categorical variables (e.g., user IDs, URLs).
- Scenarios where an exact category match isn't needed.
- Sparse models (e.g., text data in NLP).

```
library(text2vec)
library(Matrix)
#>
#> Attaching package: 'Matrix'
#> The following objects are masked from 'package:tidyr':
#>
#>     expand, pack, unpack

data(iris)

# Convert the 'Species' factor to character tokens
tokens <- word_tokenizer(as.character(iris$Species))
```

```
# Create an iterator over tokens
it <- itoken(tokens, progressbar = FALSE)

# Define the hash_vectorizer with a specified hash size
(8 in this case)
vectorizer <- hash_vectorizer(hash_size = 8)

# Create a Document-Term Matrix (DTM) using the hashed features
hashed_dtm <- create_dtm(it, vectorizer)

# Inspect the first few rows of the hashed feature matrix
head(hashed_dtm)
#> 6 x 8 sparse Matrix of class "dgCMatrix"
#>
#> 1 . . . . . . 1 .
#> 2 . . . . . . 1 .
#> 3 . . . . . . 1 .
#> 4 . . . . . . 1 .
#> 5 . . . . . . 1 .
#> 6 . . . . . . 1 .
```

- **word_tokenizer**: This function splits the character vector into tokens. Since `iris$Species` is already a categorical variable with values like "`setosa`," "`versicolor`," and "`virginica`," each value becomes a token.
- **itoken**: Creates an iterator over the tokens.
- **hash_vectorizer**: Sets up a hashing vectorizer that transforms tokens into a sparse feature space of size `2^3 = 8` (because `hash_size = 8` means 2^8 bins; if you intend exactly 8 bins, you might adjust the parameter accordingly).
- **create_dtm**: Builds the document-term matrix (which in this case is analogous to a feature matrix for each observation).

3.2.4 *Binary Encoding*

Converts categories to binary representations and distributes them across multiple columns.

Example For four categories ("A," "B," "C," "D") (Table 3.4)

Table 3.4 Illustration of binary encoding for categorical variables

Category	Binary code	Encoded columns
A	00	0, 0
B	01	0, 1
C	10	1, 0
D	11	1, 1

When to Use

- High-cardinality categorical features (less memory than one-hot encoding).
- Tree-based models (preserves some ordinal information).

```
library(mltools)
#>
#> Attaching package: 'mltools'
#> The following object is masked from 'package:tidyr':
#>
#>     replace_na
library(data.table)
#>
#> Attaching package: 'data.table'
#> The following objects are masked from 'package:lubridate':
#>
#>     hour, isoweek, mday, minute, month, quarter,
#>     second, wday, week, yday, year
#> The following objects are masked from 'package:dplyr':
#>
#>     between, first, last
#> The following object is masked from 'package:purrr':
#>
#>     transpose

# Convert the Species column to a data.table and perform
one-hot encoding
binary_encoded <- one_hot(as.data.table(iris[, "Species"]))
head(binary_encoded)
#>    V1_setosa V1_versicolor V1_virginica
#>        <int>         <int>        <int>
#> 1:         1             0            0
#> 2:         1             0            0
#> 3:         1             0            0
#> 4:         1             0            0
#> 5:         1             0            0
#> 6:         1             0            0
```

3.2.5 Base-N Encoding (Generalized Binary Encoding)

Expands binary encoding to base N instead of binary.

When to Use

- Similar to binary encoding, but allows for greater flexibility.

3.2.6 Frequency Encoding

Replaces each category with its frequency (proportion) in the dataset.

Formula

$$x_i' = \frac{\text{count}(x_i)}{\text{total count}}$$

When to Use

- High-cardinality categorical variables.
- Feature engineering for boosting algorithms (e.g., LightGBM).

```
freq_encoding <- table(iris$Species) / length(iris$Species)
iris$Species_freq <-
    iris$Species %>% as.character() %>% map_dbl(~freq_encoding[.])
head(iris$Species_freq)
#> [1] 0.3333333 0.3333333 0.3333333 0.3333333 0.3333333
#> [6] 0.3333333
```

3.2.7 Target Encoding (Mean Encoding)

Encodes categories using the mean of the target variable.

Formula

$$x_i' = E[Y|X = x_i]$$

When to Use

- Predictive models with categorical features strongly correlated with the target.
- High-cardinality categorical variables.
- Risk: Can lead to data leakage (use cross-validation).

```
library(data.table)
iris_dt <- as.data.table(iris)
iris_dt[, Species_mean := mean(Sepal.Length), by = Species]
head(iris_dt$Species_mean)
#> [1] 5.006 5.006 5.006 5.006 5.006 5.006
```

Table 3.5 Example of ordinal encoding for ordered categorical variables

Category	Ordinal encoding
Low	1
Medium	2
High	3

3.2.8 *Ordinal Encoding*

Maps categories to **ordered integer values** based on logical ranking (Table 3.5).

Example

When to Use

- Ordinal variables with meaningful order (e.g., satisfaction ratings).

```
iris$Species_ordinal <-
    as.numeric(factor(iris$Species,
                      levels = c("setosa", "versicolor", "virginica")))
head(iris$Species_ordinal)
#> [1] 1 1 1 1 1 1
```

3.2.9 *Weight of Evidence Encoding*

- **Concept:**
  WoE is a method to convert categorical data into numerical values that capture the strength of the relationship between a feature (or category) and a binary outcome (like default vs. non-default).
- **The Formula:**

$$\text{WoE} = \log\left(\frac{P(X_i|Y=1)}{P(X_i|Y=0)}\right)$$

  - $(X_i|Y = 1)$: The probability (or proportion) of observing category X_i given the positive outcome (e.g., a "good" credit event).
  - $P(X_i|Y = 0)$: The probability of observing category X_i given the negative outcome (e.g., a "bad" credit event).
  - **Logarithm:** Taking the log of the ratio gives us a symmetric scale where

    A positive WoE indicates the category is more associated with the positive outcome.
    A negative WoE indicates the category is more associated with the negative outcome.

When and Why to Use WoE Encoding?

- **Logistic Regression in Credit Scoring:**
  Logistic regression models predict probabilities in terms of log-odds. WoE encoding aligns well with this because it essentially expresses how the odds of the positive outcome change with different categories. This is why it's popular in credit scoring models.
- **Interpretability:**
  The WoE transformation makes it easier to understand and interpret the relationship between each category of a variable and the outcome. Each category's WoE value tells you whether it increases or decreases the odds of a particular event (e.g., default).

Imagine you have a feature "Employment Status" with categories "Employed" and "Unemployed":

1. **Calculate Proportions:**
   - $P(\text{Employed}|Y = 1) = 0.8$ (80% of good credit cases are employed)
   - $P(\text{Employed}|Y = 0) = 0.4$ (40% of bad credit cases are employed)
2. **Compute WoE for "Employed":**

$$\text{WoE}_{\text{Employed}} = \log\left(\frac{0.8}{0.4}\right) = \log(2) \approx 0.693$$

   A positive value indicates that being employed increases the odds of a good credit outcome.
3. **Repeat for "Unemployed":**
   Suppose:
   - $P(\text{Unemployed}|Y = 1) = 0.2$
   - $P(\text{Unemployed}|Y = 0) = 0.6$

$$\text{WoE}_{\text{Unemployed}} = \log\left(\frac{0.2}{0.6}\right) = \log\left(\frac{1}{3}\right) \approx -1.099$$

   A negative value indicates that being unemployed is associated with a higher likelihood of a bad credit outcome.

Why is WoE Valuable?

- **Linear Relationship:**
  When you plug these WoE values into a logistic regression, the model essentially adds these values linearly, which fits nicely with how logistic regression models the log-odds.

- **Stability and Handling of Missing Values:**
  WoE can also help in smoothing out fluctuations in categorical data, especially when there are many levels or some levels with few observations.
- **Regulatory Acceptance:**
  In industries like banking, WoE is widely accepted because of its clear interpretability, which is crucial for compliance and transparency in credit risk modeling.

```
# Load required packages
library(dplyr)
library(knitr)

# Create a sample dataset
# We assume 100 good credit cases and 100 bad credit cases
# Good credit: 80 "Employed" and 20 "Unemployed"
# Bad credit: 40 "Employed" and 60 "Unemployed"
data <- data.frame(
  employment_status = c(rep("Employed", 80), rep("Unemployed", 20),
                        rep("Employed", 40), rep("Unemployed", 60)),
  credit = c(rep(1, 100), rep(0, 100))
)

# Calculate counts for each category
woe_table <- data %>%
  group_by(employment_status) %>%
  summarise(
    good = sum(credit == 1),
    bad = sum(credit == 0)
  ) %>%
  # Calculate the distribution for good and bad credit cases
  mutate(
    dist_good = good / sum(good),
    dist_bad = bad / sum(bad),
    WoE = log(dist_good / dist_bad)
  )

# Print the WoE table
kable(woe_table)
```

employment_status	good	bad	dist_good	dist_bad	WoE
Employed	80	40	0.8	0.4	0.6931472
Unemployed	20	60	0.2	0.6	−1.0986123

```
# Merge the WoE values into the original data
data_woe <- data %>%
  left_join(woe_table %>% dplyr::select(employment_status, WoE), by =
  ↪ "employment_status")
```

```
head(data_woe)
#>   employment_status credit       WoE
#> 1          Employed      1 0.6931472
#> 2          Employed      1 0.6931472
#> 3          Employed      1 0.6931472
#> 4          Employed      1 0.6931472
#> 5          Employed      1 0.6931472
#> 6          Employed      1 0.6931472

# Fit a logistic regression model using WoE as predictor
model <- glm(credit ~ WoE, data = data_woe, family = binomial)

# Summary of the model
summary(model)
#>
#> Call:
#> glm(formula = credit ~ WoE, family = binomial, data = data_woe)
#>
#> Coefficients:
#>              Estimate Std. Error z value Pr(>|z|)
#> (Intercept) 1.023e-12  1.552e-01   0.000        1
#> WoE         1.000e+00  1.801e-01   5.552 2.83e-08 ***
#> ---
#> Signif. codes:
#> 0 '***' 0.001 '**' 0.01 '*' 0.05 '.' 0.1 ' ' 1
#>
#> (Dispersion parameter for binomial family taken to be 1)
#>
#>     Null deviance: 277.26  on 199  degrees of freedom
#> Residual deviance: 242.74  on 198  degrees of freedom
#> AIC: 246.74
#>
#> Number of Fisher Scoring iterations: 4
```

When you fit a logistic regression using the WoE-encoded variable, the model is essentially

$$\log\left(\frac{P(Y=1)}{P(Y=0)}\right) = \beta_0 + \beta_1 \cdot \text{WoE}$$

Here, **WoE** represents the weight of evidence value for a given category.

- **Log Odds Change:**
  β_1 indicates how much the log odds of a good credit outcome change for a one-unit increase in WoE. For example, if $\beta_1 = 0.5$, then a one-unit increase in WoE is associated with an increase of 0.5 in the log odds of having a good credit outcome.

- **Odds Ratio:**
  If you exponentiate β_1, you get the odds ratio. For instance, if $\beta_1 = 0.5$, then $\exp(0.5) \approx 1.65$. This means that for each one-unit increase in WoE, the odds of having a good credit outcome are multiplied by about 1.65.

  Why is This Meaningful?

- **Direct Link to the Data:**
  The WoE value itself is a transformation of the original categorical variable that reflects the ratio of the proportions of good to bad outcomes for that category. By using WoE, you're directly incorporating this information into the model.
- **Interpretability:**
  The interpretation becomes intuitive:
  - A **positive** WoE indicates that the category is more associated with a good outcome.
  - A **negative** WoE indicates that the category is more associated with a bad outcome.

  Thus, if β_1 is positive, it suggests that as the category moves to one with a higher WoE (i.e., more favorable for a good outcome), the likelihood of a good outcome increases.

3.2.10 Helmert Encoding

Compares each category against the mean of previous categories.

When to Use
- ANOVA models and categorical regression.

3.2.11 Probability Ratio Encoding

Encodes categories using the probability ratio of the target variable.

3.2.12 Backward Difference Encoding

Compares each category against the mean of all remaining categories.

3.2.13 Leave-One-Out Encoding

Similar to target encoding, but excludes the current observation to avoid bias.

3.2.14 James-Stein Encoding

A smoothed version of target encoding, reducing overfitting.

3.2.15 M-Estimator Encoding

Uses a Bayesian prior to smooth target encoding.

3.2.16 Thermometer Encoding

Similar to one-hot encoding, but retains ordinal structure.

3.2.17 Choosing the Right Encoding Method

For a comparison of encoding methods, refer to Table 3.6.

Table 3.6 Comparison of encoding methods

Encoding method	Best for low cardinality	Best for high cardinality	Handles ordinality	Suitable for tree models	Suitable for linear models
One-hot encoding	Yes	No	No	Yes	Yes
Label encoding	Yes	Yes	Yes	No	Yes
Target encoding	Yes	Yes	No	Yes	Yes
Frequency encoding	Yes	Yes	No	Yes	Yes
Binary encoding	Yes	Yes	No	Yes	Yes

Chapter 4
Imputation (Missing Data)

Missing data is a pervasive issue in applied statistics, and this chapter offers a comprehensive treatment of its diagnosis and resolution. Beginning with a conceptual introduction, we discuss the mechanisms underlying missingness-MCAR, MAR, and MNAR-and their consequences for unbiased estimation. Later, the chapter provides practical tools for identifying patterns of missingness, including graphical and statistical diagnostics. It then reviews a spectrum of imputation methods, from simple techniques (mean substitution, regression imputation) to advanced algorithms (multiple imputation using chained equations and random forest imputation). Each method is evaluated in terms of bias, efficiency, and robustness, with detailed comparisons. Practical criteria for choosing an appropriate imputation strategy are also provided, followed by an in-depth discussion on ethical issues and transparency in reporting. Emerging trends in handling high-dimensional and MNAR data are also covered. Finally, the last section walks through a complete imputation example, ensuring readers can apply these concepts immediately.

4.1 Introduction to Missing Data

Missing data is a common problem in statistical analyses and data science, impacting the quality and reliability of insights derived from datasets. One widely used approach to address this issue is **imputation**, where missing data is replaced with *reasonable estimates*.

© The Author(s), under exclusive license to Springer Nature Switzerland AG 2025
M. Nguyen, *Advanced Modeling and Data Challenges*,
https://doi.org/10.1007/978-3-032-01719-2_4

4.1.1 Types of Imputation

Imputation can be categorized into

1. **Unit Imputation**: Replacing an entire missing observation (i.e., all features for a single data point are missing).
2. **Item Imputation**: Replacing missing values for specific variables (features) within a dataset.

While imputation offers a means to make use of incomplete datasets, it has historically been viewed skeptically. This skepticism arises from

1. Frequent **misapplication** of imputation techniques, which can introduce significant **bias** to estimates.
2. Limited **applicability**, as imputation works well only under certain assumptions about the missing data mechanism and research objectives.

Biases in imputation can arise from various factors, including

- **Imputation Method**: The chosen method can influence the results and introduce biases.
- **Missing Data Mechanism**: The nature of the missing data-whether it is Missing Completely at Random (MCAR) or Missing at Random (MAR)-affects the accuracy of imputation.
- **Proportion of Missing Data**: The amount of missing data significantly impacts the reliability of the imputation.
- **Available Information in the Dataset**: Limited information reduces the robustness of the imputed values.

4.1.2 When and Why to Use Imputation

The appropriateness of imputation depends on the nature of the missing data and the research goal:

- **Missing Data in the Outcome Variable** (y): Imputation in such cases is generally problematic, as it can distort statistical models and lead to misleading conclusions. For example, imputing outcomes in regression or classification problems can alter the underlying relationship between the dependent and independent variables.
- **Missing Data in Predictive Variables** (x): Imputation is more commonly applied here, especially for **nonrandom missing data**. Properly handled, imputation can enable the use of incomplete datasets while minimizing bias.

4.1.2.1 Objectives of Imputation

The utility of imputation methods differs substantially depending on whether the goal of the analysis is *inference/explanation* or *prediction*. Each goal has distinct priorities and tolerances for bias, variance, and assumptions about the missing data mechanism:

4.1.2.1.1 Inference/Explanation

In causal inference or explanatory analyses, the primary objective is to ensure valid statistical inference, emphasizing unbiased estimation of parameters and accurate representation of uncertainty. The treatment of missing data must align closely with the assumptions about the mechanism behind the missing data-whether it is missing completely at random (MCAR), missing at random (MAR), or missing not at random (MNAR):

- **Bias Sensitivity:** Inference analyses require that imputed data preserve the integrity of the relationships among variables. Poorly executed imputation can introduce bias, even when it addresses missingness superficially.
- **Variance and Confidence Intervals:** For inference, the quality of the standard errors, confidence intervals, and test statistics is critical. Naive imputation methods (e.g., mean imputation) often fail to appropriately reflect the uncertainty due to missingness, leading to overconfidence in parameter estimates.
- **Mechanism Considerations:** Imputation methods, such as multiple imputation (MI), attempt to generate values consistent with the observed data distribution while accounting for missing data uncertainty. However, MI's performance depends heavily on the validity of the MAR assumption. If the missingness mechanism is MNAR and not addressed adequately, the imputed data could yield biased parameter estimates, undermining the purpose of inference.

4.1.2.1.2 Prediction

In predictive modeling, the primary goal is to maximize model accuracy (e.g., minimizing mean squared error for continuous outcomes or maximizing classification accuracy). Here, the focus shifts to optimizing predictive performance rather than ensuring unbiased parameter estimates:

- **Loss of Information:** Missing data reduces the amount of usable information in a dataset. Imputation allows the model to leverage all available data, rather than excluding incomplete cases via listwise deletion, which can significantly reduce sample size and model performance.
- **Impact on Model Fit:** In predictive contexts, imputation can reduce standard errors of the predictions and stabilize model coefficients by incorporating plausible estimates for missing values.

- **Flexibility with Mechanism:** Predictive models are less sensitive to the missing data mechanism than inferential models, as long as the imputed values help reduce variability and align with patterns in the observed data. Methods like K-nearest neighbor (KNN), iterative imputation, or even machine learning models (e.g., random forests for imputation) can be valuable, regardless of strict adherence to MAR or MCAR assumptions.
- **Trade-Offs:** Overimputation, where too much noise or complexity is introduced in the imputation process, can harm prediction by introducing artifacts that degrade model generalizability.

4.1.2.1.3 Key Takeaways

The usefulness of imputation depends on whether the goal of the analysis is **inference** or **prediction**:

- **Inference/Explanation:** The primary concern is valid statistical inference, where biased estimates are unacceptable. Imputation is often of limited value for this purpose, as it may not address the underlying missing data mechanism appropriately (Rubin, 1996).
- **Prediction:** Imputation can be more useful in predictive modeling, as it reduces the loss of information from incomplete cases. By leveraging observed data, imputation can lower standard errors and improve model accuracy.

4.1.3 Importance of Missing Data Treatment in Statistical Modeling

Proper handling of missing data ensures

- **Unbiased Estimates:** Avoiding distortions in parameter estimates.
- **Accurate Standard Errors:** Ensuring valid hypothesis testing and confidence intervals.
- **Adequate Statistical Power:** Maximizing the use of available data.

Ignoring or mishandling missing data can lead to

1. **Bias:** Systematic errors in parameter estimates, especially under MAR or MNAR mechanisms.
2. **Loss of Power:** Reduced sample size leads to larger standard errors and weaker statistical significance.
3. **Misleading Conclusions:** Oversimplistic imputation methods (e.g., mean substitution) can distort relationships among variables.

4.1.4 Prevalence of Missing Data Across Domains

Missing data affects virtually all fields:

- **Business:** Nonresponses in customer surveys, incomplete sales records, and transactional errors.
- **Healthcare:** Missing data in electronic health records (EHRs) due to incomplete patient histories or inconsistent data entry.
- **Social Sciences:** Nonresponses or partial responses in large-scale surveys, leading to biased conclusions.

4.1.5 Practical Considerations for Imputation

- **Diagnostic Checks:** Always examine the patterns and mechanisms of missing data before applying imputation (Diagnosing the Missing Data Mechanism).
- **Model Selection:** Align the imputation method with the missing data mechanism and research goal.
- **Validation:** Assess the impact of imputation on results through sensitivity analyses or cross-validation.

4.2 Theoretical Foundations of Missing Data

4.2.1 Definition and Classification of Missing Data

Missing data refers to the absence of values for some variables in a dataset. The mechanisms underlying missingness significantly impact the validity of statistical analyses and the choice of handling methods. These mechanisms are classified into three categories:

- Missing Completely at Random (MCAR): The probability of missingness is independent of both observed and unobserved data. In other words, the missing data occur entirely at random and are unrelated to any values in the dataset.
- Missing at Random (MAR): The probability of missingness is related to the observed data but not to the missing data itself. This means that, after controlling for observed variables, the missingness is random.
- Missing Not at Random (MNAR): The probability of missingness depends on unobserved data or the missing values themselves. In this case, the missingness is related to the very information that is missing, making it the most challenging type to handle in analysis.

4.2.1.1 Missing Completely at Random (MCAR)

MCAR occurs when the probability of missingness is entirely random and unrelated to either observed or unobserved variables. Under this mechanism, missing data do not introduce bias in parameter estimates when ignored, although statistical efficiency is reduced due to the smaller sample size.

Mathematical Definition The missingness is independent of all data, both observed and unobserved:

$$P(Y_{\text{missing}}|Y, X) = P(Y_{\text{missing}})$$

Characteristics of MCAR
- Missingness is completely unrelated to both observed and unobserved data.
- Analyses remain unbiased even if missing data are ignored, though they may lack efficiency due to reduced sample size.
- The missing data points represent a random subset of the overall data.

Examples

- A sensor randomly fails at specific time points, unrelated to environmental or operational conditions.
- Survey participants randomly omit responses to certain questions without any systematic pattern.

Methods for Testing MCAR
1. **Little's MCAR Test:** A formal statistical test to assess whether data are MCAR. A significant result suggests deviation from MCAR.
2. **Mean Comparison Tests:**

   - T-tests or similar approaches compare observed and missing data groups on other variables. Significant differences indicate potential bias.
   - Failure to reject the null hypothesis of no difference does not confirm MCAR but suggests consistency with the MCAR assumption.

Handling MCAR
Since MCAR data introduce no bias, they can be handled using the following techniques:

1. **Complete Case Analysis (Listwise Deletion):**

   - Analyses are performed only on cases with complete data. While unbiased under MCAR, this method reduces sample size and efficiency.

2. **Universal Singular Value Thresholding (USVT):**

   - This technique is effective for MCAR data recovery but can only recover the mean structure, not the entire true distribution (Chatterjee, 2015).

3. **SoftImpute:**
   - A matrix completion method useful for some missing data problems but less effective when missingness is not MCAR (Hastie et al., 2015).
4. **Synthetic Nearest Neighbor Imputation:**
   - A robust method for imputing missing data. While primarily designed for MCAR, it can also handle certain cases of missing not at random (MNAR) (Agarwal et al., 2023). Available on GitHub: syntheticNN.

Notes
- The "missingness" on one variable can be correlated with the "missingness" on another variable without violating the MCAR assumption.
- Absence of evidence for bias (e.g., failing to reject a t-test) does not confirm that the data are MCAR.

4.2.1.2 Missing at Random (MAR)

Missing at random (MAR) occurs when missingness depends on observed variables but not the missing values themselves. This mechanism assumes that observed data provide sufficient information to explain the missingness. In other words, there is a systematic relationship between the propensity of missing values and the observed data, but not the missing data.

Mathematical Definition
The probability of missingness is conditional only on observed data:

$$P(Y_{\text{missing}}|Y, X) = P(Y_{\text{missing}}|X)$$

This implies that whether an observation is missing is unrelated to the missing values themselves but is related to the observed values of other variables.

Characteristics of MAR
- Missingness is systematically related to observed variables.
- The propensity for a data point to be missing is not related to the missing data but is related to some of the observed data.
- Analyses must account for observed data to mitigate bias.

Examples
- Women are less likely to disclose their weight, but their gender is recorded. In this case, weight is MAR.
- Missing income data is correlated with education, which is observed. For example, individuals with higher education levels might be less likely to reveal their income.

Challenges in MAR

- MAR is weaker than missing completely at random (MCAR).
- It is impossible to directly test for MAR. Evidence for MAR relies on domain expertise and indirect statistical checks rather than direct tests.

Handling MAR

Common methods for handling MAR include

- **Multiple Imputation by Chained Equations (MICE):** Iteratively imputes missing values based on observed data.
- **Maximum Likelihood Estimation:** Estimates model parameters directly while accounting for MAR assumptions.
- **Regression-Based Imputation:** Predicts missing values using observed covariates.

These methods assume that observed variables fully explain the missingness. Effective handling of MAR requires careful modeling and often domain-specific knowledge to validate the assumptions underlying the analysis.

4.2.1.3 Missing Not at Random (MNAR)

Missing not at random (MNAR) is the most complex missing data mechanism. Here, missingness depends on unobserved variables or the values of the missing data themselves. This makes MNAR particularly challenging, as ignoring this dependency introduces significant bias in analyses.

Mathematical Definition

The probability of missingness depends on the missing values:

$$P(Y_{\text{missing}}|Y, X) \neq P(Y_{\text{missing}}|X)$$

Characteristics of MNAR

- Missingness cannot be fully explained by observed data.
- The cause of missingness is directly related to the unobserved values.
- Ignoring MNAR introduces significant bias in parameter estimates, often leading to invalid conclusions.

Examples

- High-income individuals are less likely to disclose their income, and income itself is unobserved.
- Patients with severe symptoms drop out of a clinical study, leaving their health outcomes unrecorded.

Challenges in MNAR

- MNAR is the most difficult missingness mechanism to address because the missing data mechanism must be explicitly modeled.

- Identifying MNAR often requires domain knowledge and auxiliary information beyond the observed dataset.

Handling MNAR
MNAR requires explicit modeling of the missingness mechanism. Common approaches include

- **Heckman Selection Models:** These models explicitly account for the selection process leading to missing data, adjusting for potential bias (Heckman, 1976).
- **Instrumental Variables:** Variables predictive of missingness but unrelated to the outcome can be used to mitigate bias (Sun et al., 2018; Tchetgen Tchetgen & Wirth, 2017).
- **Pattern-Mixture Models**: These models separate the data into groups (patterns) based on missingness and model each group separately. They are particularly useful when the relationship between missingness and missing values is complex.
- **Sensitivity Analysis:** Examines how conclusions change under different assumptions about the missing data mechanism.
- **Use of Auxiliary Data:**
  Auxiliary data refers to external data sources or variables that can help explain the missingness mechanism.
  - **Surrogate Variables**: Adding variables that correlate with missing data can improve imputation accuracy and mitigate the MNAR challenge.
  - **Linking External Datasets**: Merging datasets from different sources can provide additional context or predictors for missingness.
  - **Applications in Business**: In marketing, customer demographics or transaction histories often serve as auxiliary data to predict missing responses in surveys.

Additionally, data collection strategies, such as follow-up surveys or targeted sampling, can help mitigate MNAR effects by collecting information that directly addresses the missingness mechanism. However, such approaches can be resource-intensive and require careful planning (Table 4.1).

Table 4.1 Types of missing data mechanisms and their implications

Mechanism	Missingness depends on	Implications	Examples
MCAR	Neither observed nor missing data	No bias; simplest to handle; decreases efficiency due to data loss	Random sensor failure
MAR	Observed data only	Requires observed data to explain missingness; common assumption in imputation methods	Gender-based missingness of weight
MNAR	Missing data itself or unobserved variables	Requires explicit modeling of the missingness mechanism; significant bias if ignored	High-income individuals not disclosing income

4.2.2 Missing Data Mechanisms

4.2.3 Relationship Between Mechanisms and Ignorability

The concept of ignorability is central to determining whether the missingness process must be explicitly modeled. Ignorability impacts the choice of methods for handling missing data and whether the missing data mechanism can be safely disregarded or must be explicitly accounted for.

4.2.3.1 Ignorable Missing Data

Missing data is **ignorable** under the following conditions:

1. The missing data mechanism is MAR or MCAR.
2. The parameters governing the missing data process are unrelated to the parameters of interest in the analysis.

In cases of ignorable missing data, there is no need to model the missingness mechanism explicitly unless you aim to improve the efficiency or precision of parameter estimates. Common imputation techniques, such as multiple imputation or maximum likelihood estimation, rely on the assumption of ignorability to produce unbiased parameter estimates.

Practical Considerations for Ignorable Missingness
Even though ignorable mechanisms simplify analysis, researchers must rigorously assess whether the missingness mechanism meets the MAR or MCAR criteria. Violations can lead to biased results, even if unintentionally overlooked.

For example, a survey on income may assume MAR if missingness is associated with respondent age (observed variable) but not income itself (unobserved variable). However, if income directly influences nonresponse, the assumption of MAR is violated.

4.2.3.2 Non-Ignorable Missing Data

Missing data is **non-ignorable** when

1. The missingness mechanism depends on the values of the missing data themselves or on unobserved variables.
2. The missing data mechanism is related to the parameters of interest, resulting in bias if the mechanism is not modeled explicitly.

This type of missingness (i.e., Missing Not at Random (MNAR)) requires modeling the missing data mechanism directly to produce unbiased estimates.

Characteristics of Non-Ignorable Missingness:

- **Dependence on Missing Values**: The likelihood of missingness is associated with the missing values themselves.
  - Example: In a study on health, individuals with more severe conditions are more likely to drop out, leading to an underrepresentation of the sickest individuals in the data.
- **Bias in Complete Case Analysis**: Analyses based solely on complete cases can lead to substantial bias.
  - Example: In income surveys, if wealthier individuals are less likely to report their income, the estimated mean income will be systematically lower than the true population mean.
- **Need for Explicit Modeling**: To address MNAR, the analyst must model the missing data mechanism. This often involves specifying relationships between observed data, missing data, and the missingness process itself.

4.2.3.3 Implications of Non-Ignorable Missingness

Non-ignorable mechanisms are often associated with sensitive or personal data:

- **Examples**:
  - Individuals with lower education levels may omit their education information.
  - Participants with controversial or stigmatized health conditions might opt out of surveys entirely.
- **Impact on Policy and Decision-Making**:
  - Biases introduced by MNAR can have serious consequences for policymaking, such as underestimating the prevalence of poverty or mischaracterizing population health needs.

By explicitly addressing non-ignorable missingness, researchers can mitigate biases and ensure that findings accurately reflect the underlying population.

4.3 Diagnosing the Missing Data Mechanism

Understanding the mechanism behind missing data is critical to choosing the appropriate methods for handling it. The three main mechanisms for missing data are **MCAR (missing completely at random)**, **MAR (missing at random)**, and **MNAR (missing not at random)**. This section discusses methods for diagnosing these mechanisms, including descriptive and inferential approaches.

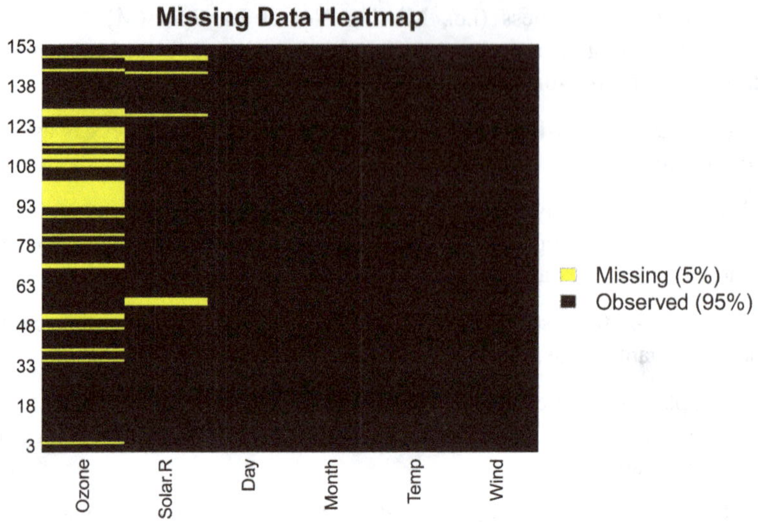

Fig. 4.1 Missing data heatmap

4.3.1 *Descriptive Methods*

4.3.1.1 Visualizing Missing Data Patterns

Visualization tools are essential for detecting patterns in missing data. Heatmaps and correlation plots can help identify systematic missingness and provide insights into the underlying mechanism (Fig. 4.1).

```
# Example: Visualizing missing data
library(Amelia)
#> Loading required package: Rcpp
#> ##
#> ## Amelia II: Multiple Imputation
#> ## (Version 1.8.3, built: 2024-11-07)
#> ## Copyright (C) 2005-2025 James Honaker, Gary King and Matthew
↪  Blackwell
#> ## Refer to http://gking.harvard.edu/amelia/ for more information
#> ##
missmap(
    airquality,
    main = "Missing Data Heatmap",
    col = c("yellow", "black"),
    legend = TRUE
)
```

- **Heatmaps**: Highlight where missingness occurs in a dataset.
- **Correlation Plots**: Show relationships between missingness indicators of different variables.

Exploring Univariate and Multivariate Missingness

- **Univariate Analysis**: Calculate the proportion of missing data for each variable.

```
# Example: Proportion of missing values
missing_proportions <- colSums(is.na(airquality)) / nrow(airquality)
print(missing_proportions)
#>      Ozone    Solar.R       Wind       Temp      Month
#> 0.24183007 0.04575163 0.00000000 0.00000000 0.00000000
#>        Day
#> 0.00000000
```

- **Multivariate Analysis**: Examine whether missingness in one variable is related to others. This can be visualized using scatterplots of observed vs. missing values (Figs. 4.2 and 4.3).

```
# Example: Missingness correlation
library(naniar)
```

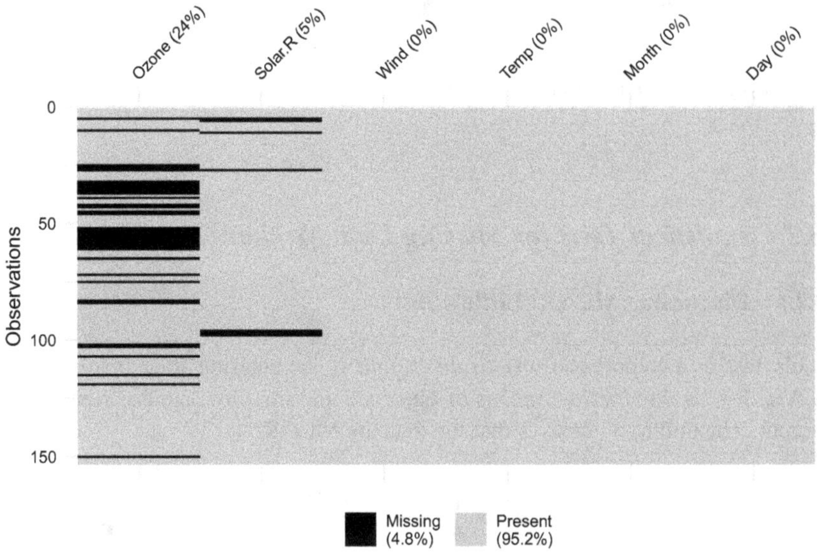

Fig. 4.2 Missing data heatmap

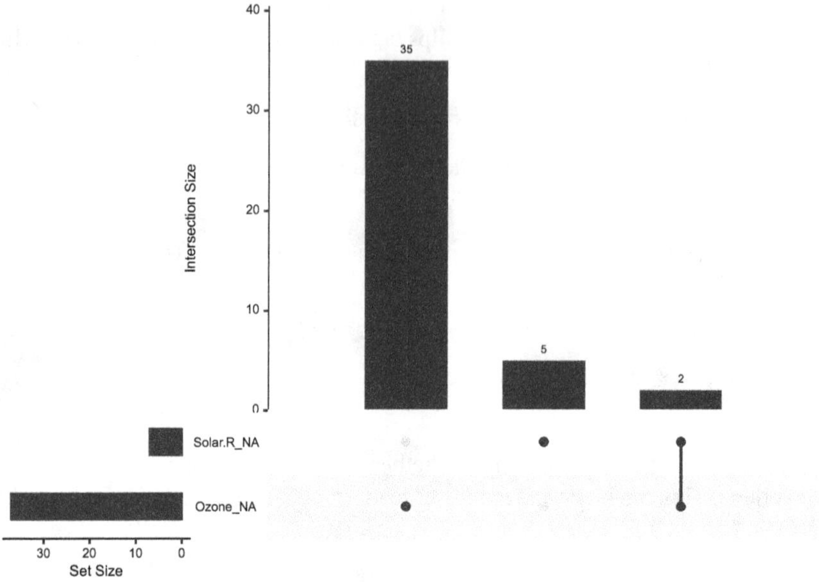

Fig. 4.3 Missing data barchart

```
vis_miss(airquality)
```

```
gg_miss_upset(airquality) # Displays a missingness upset plot
```

4.3.2 Statistical Tests for Missing Data Mechanisms

4.3.2.1 Diagnosing MCAR: Little's Test

Little's test is a hypothesis test to determine if the missing data mechanism is **MCAR**. It tests whether the means of observed and missing data are significantly different. The null hypothesis is that the data are MCAR.

$$\chi^2 = \sum_{i=1}^{n} \frac{(O_i - E_i)^2}{E_i}$$

where

- O_i= Observed frequency,
- E_i= Expected frequency under MCAR.

```
# Example: Little's test
naniar::mcar_test(airquality)
#> # A tibble: 1 x 4
#>   statistic    df p.value missing.patterns
#>       <dbl> <dbl>   <dbl>            <int>
#> 1      35.1    14 0.00142                4
misty::na.test(airquality)
#> Warning: Cases with missing on all variables were removed
#> from the analysis: NA
#>  Little's MCAR Test
#>
#>      n nIncomp nPattern  chi2 df  pval
#>    152      42        4 35.68 14 0.001
```

4.3.2.2 Diagnosing MCAR via Dummy Variables

Creating a binary indicator for missingness allows you to test whether the presence of missing data is related to observed data. For instance:

1. Create a dummy variable:
   - 1 = Missing
   - 0 = Observed
2. Conduct a chi-square test or t-test:
   - Chi-square: Compare proportions of missingness across groups.
   - T-test: Compare means of (other) observed variables with missingness indicators.

```
# Example: Chi-square test
airquality$missing_var <- as.factor(ifelse(is.na(airquality$Ozone), 1,
↪  0))
# Across groups of months
table(airquality$missing_var, airquality$Month)
#>
#>      5  6  7  8  9
#>   0 26  9 26 26 29
#>   1  5 21  5  5  1
chisq.test(table(airquality$missing_var, airquality$Month))
#>
#>  Pearson's Chi-squared test
#>
#> data:  table(airquality$missing_var, airquality$Month)
#> X-squared = 44.751, df = 4, p-value = 4.48e-09

# Example: T-test (of other variable)
t.test(Wind ~ missing_var, data = airquality)
#>
```

```
#>  Welch Two Sample t-test
#>
#> data:  Wind by missing_var
#> t = -0.60911, df = 63.646, p-value = 0.5446
#> alternative hypothesis: true difference in means between group 0 and
↪  group 1 is not equal to 0
#> 95 percent confidence interval:
#>  -1.6893132  0.8999377
#> sample estimates:
#> mean in group 0 mean in group 1
#>        9.862069       10.256757
```

4.3.3 Assessing MAR and MNAR

4.3.3.1 Sensitivity Analysis

Sensitivity analysis involves simulating different scenarios of missing data and assessing how the results change. For example, imputing missing values under different assumptions can provide insight into whether the data are MAR or MNAR.

4.3.3.2 Proxy Variables and External Data

Using proxy variables or external data sources can help assess whether missingness depends on unobserved variables (MNAR). For example, in surveys, follow-ups with non-respondents can reveal systematic differences.

4.3.3.3 Practical Challenges in Distinguishing MAR from MNAR

Distinguishing between missing at random (MAR) and missing not at random (MNAR) is a critical and challenging task in data analysis. Properly identifying the nature of the missing data has significant implications for the choice of imputation strategies, model robustness, and the validity of conclusions. While statistical tests can sometimes aid in this determination, the process often relies heavily on domain knowledge, intuition, and exploratory analysis. Below, we discuss key considerations and examples that highlight these challenges:

- **Sensitive Topics**: Missing data related to sensitive or stigmatized topics, such as income, drug use, or health conditions, are often MNAR. For example, individuals with higher incomes might deliberately choose not to report their earnings due to privacy concerns. Similarly, participants in a health survey may avoid answering questions about smoking if they perceive social disapproval. In such cases, the probability of missingness is directly related to the unobserved value itself, making MNAR likely.

- **Field-Specific Norms**: Understanding norms and typical data collection practices in a specific field can provide insights into missingness patterns. For instance, in marketing surveys, respondents may skip questions about spending habits if they consider the questions intrusive. Prior research or historical data from the same domain can help infer whether missingness is more likely MAR (e.g., random skipping due to survey fatigue) or MNAR (e.g., deliberate omission by higher spenders).
- **Analyzing Auxiliary Variables**: Leveraging auxiliary variables-those correlated with the missing variable-can help infer the missingness mechanism. For example, if missing income data strongly correlates with employment status, this suggests a MAR mechanism, as the missingness depends on observed variables. However, if missingness persists even after accounting for observable predictors, MNAR might be at play.
- **Experimental Design and Follow-Up**: In longitudinal studies, dropout rates can signal MAR or MNAR patterns. For example, if dropouts occur disproportionately among participants reporting lower satisfaction in early surveys, this indicates an MNAR mechanism. Designing follow-up surveys to specifically investigate dropout reasons can clarify missingness patterns.
- **Sensitivity Analysis**: To account for uncertainty in the missingness mechanism, researchers can conduct sensitivity analyses by comparing results under different assumptions (e.g., imputing data using both MAR and MNAR approaches). This process helps to quantify the potential impact of misclassifying the missingness mechanism on study conclusions.
- **Real-World Examples**:
  - In customer feedback surveys, higher ratings might be overrepresented due to nonresponse bias. Customers with negative experiences might be less likely to complete surveys, leading to an MNAR scenario.
  - In financial reporting, missing audit data might correlate with companies in financial distress, a classic MNAR case where the missingness depends on unobserved financial health metrics.

Summary

- **MCAR**: No pattern in missingness; use Little's test or dummy variable analysis.
- **MAR**: Missingness related to observed data; requires modeling assumptions or proxy analysis.
- **MNAR**: Missingness depends on unobserved data; requires external validation or sensitivity analysis.

4.4 Methods for Handling Missing Data

4.4.1 Basic Methods

Refer to Table 4.2 for a comparison of basic missing data handling methods.

4.4.1.1 Complete Case Analysis (Listwise Deletion)

Listwise deletion retains only cases with complete data for all features, discarding rows with any missing values.

Advantages

- Universally applicable to various statistical tests (e.g., SEM, multilevel regression).
- When data are missing completely at random (MCAR), parameter estimates and standard errors are unbiased.
- Under specific missing at random (MAR) conditions, such as when the probability of missing data depends only on independent variables, listwise deletion can still yield unbiased estimates. For instance, in the model $y = \beta_0 + \beta_1 X_1 + \beta_2 X_2 + \epsilon$, if missingness in X_1 is independent of y but depends on X_1 and X_2, the estimates remain unbiased (Little, 1992).
  - This aligns with principles of stratified sampling, which does not bias estimates.
  - In logistic regression, if missing data depend only on the dependent variable but not on independent variables, listwise deletion produces consistent slope estimates, though the intercept may be biased (Vach & Vach, 1994).
- For regression analysis, listwise deletion is more robust than maximum likelihood (ML) or multiple imputation (MI) when the MAR assumption is violated.

Disadvantages

- Results in larger standard errors compared to advanced methods.

Table 4.2 Comparison of basic missing data handling methods

Method	Advantages	Disadvantages
Listwise deletion	Simple and universally applicable; unbiased under MCAR; robust in certain MAR scenarios	Inefficient (larger standard errors); biased under MAR in many cases; discards potentially useful data
Pairwise deletion	Utilizes all available data; efficient under MCAR with low correlations; avoids discarding all cases	Biased under MAR; prone to non-positive-definite covariance matrices in small samples
Indicator method	Simple implementation; explicitly models missingness effect	Biased even under MCAR; complicates interpretation; may not reflect true underlying relationships

- If data are MAR but not MCAR, biased estimates can occur.
- In non-regression contexts, more sophisticated methods often outperform listwise deletion.

4.4.1.2 Available Case Analysis (Pairwise Deletion)

Pairwise deletion calculates estimates using all available data for each pair of variables, without requiring complete cases. It is particularly suitable for methods like linear regression, factor analysis, and SEM, which rely on correlation or covariance matrices.

Advantages
- Under MCAR, pairwise deletion produces consistent and unbiased estimates in large samples.
- Compared to listwise deletion (Glasser, 1964)
  - When variable correlations are low, pairwise deletion provides more efficient estimates.
  - When correlations are high, listwise deletion becomes more efficient.

Disadvantages
- Yields biased estimates under MAR conditions.
- In small samples, covariance matrices might not be positive definite, rendering coefficient estimation infeasible.
- Software implementation varies in how sample size is handled, potentially affecting standard errors.

Note: Carefully review software documentation to understand how sample size is treated, as this influences standard error calculations.

4.4.1.3 Indicator Method (Dummy Variable Adjustment)

Also known as the missing indicator method, this approach introduces an additional variable to indicate missingness in the dataset.

Implementation
1. Create an indicator variable:

$$D = \begin{cases} 1 & \text{if data on } X \text{ are missing} \\ 0 & \text{otherwise} \end{cases}$$

2. Modify the original variable to accommodate missingness:

$$X^* = \begin{cases} X & \text{if data are available} \\ c & \text{if data are missing} \end{cases}$$

Note: A common choice for c is the mean of X.

Interpretation

- The coefficient of D represents the difference in the expected value of Y between cases with missing data and those without.
- The coefficient of X^* reflects the effect of X on Y for cases with observed data.

Disadvantages

- Produces biased estimates of coefficients, even under MCAR conditions (Jones, 1996).
- May lead to overinterpretation of the "missingness effect," complicating model interpretation.

4.4.1.4 Advantages and Limitations of Basic Methods

4.4.2 Single Imputation Techniques

Single imputation methods replace missing data with a single value, generating a complete dataset that can be analyzed using standard techniques. While convenient, single imputation generally underestimates variability and risks biasing results.

4.4.2.1 Deterministic Methods

4.4.2.1.1 Mean, Median, Mode Imputation

This method replaces missing values with the mean, median, or mode of the observed data.

Advantages

- Simplicity and ease of implementation.
- Useful for quick exploratory data analysis.

Disadvantages

- **Bias in Variances and Relationships**: Mean imputation reduces variance and disrupts relationships among variables, leading to biased estimates of variances and covariances (Haitovsky, 1968).
- **Underestimated Standard Errors**: Results in overly optimistic conclusions and increased risk of Type I errors.

- **Dependency Structure Ignored**: Particularly problematic in high-dimensional data, as it fails to capture dependencies among features.

4.4.2.1.2 Forward and Backward Filling (Time Series Contexts)

Used in time series analysis, this method replaces missing values using the preceding (forward filling) or succeeding (backward filling) values.

Advantages
- Simple and preserves temporal ordering.
- Suitable for datasets where adjacent values are strongly correlated.

Disadvantages
- Biased if missingness spans long gaps or occurs systematically.
- Cannot capture trends or changes in the underlying process.

4.4.2.2 Statistical Prediction Models

4.4.2.2.1 Linear Regression Imputation

Missing values in a variable are imputed based on a linear regression model using observed values of other variables.

Advantages
- Preserves relationships between variables.
- More sophisticated than mean or median imputation.

Disadvantages
- Assumes linear relationships, which may not hold in all datasets.
- Fails to capture variability, leading to downwardly biased standard errors.

4.4.2.2.2 Logistic Regression for Categorical Variables

Similar to linear regression imputation but used for categorical variables. The missing category is predicted using a logistic regression model.

Advantages
- Useful for binary or multinomial categorical data.
- Preserves relationships with other variables.

Disadvantages
- Assumes the underlying logistic model is appropriate.
- Does not account for uncertainty in the imputed values.

Table 4.3 Comparison of hot deck and cold deck imputation techniques

Feature	Hot deck imputation	Cold deck imputation
Donor pool	Internal (within the dataset)	External (predefined dataset)
Selection	Random	Systematic
Variability	Retained	Reduced
Bias potential	Lower	Higher (if donor pool differs)

4.4.2.3 Nonparametric Methods

4.4.2.3.1 Hot Deck Imputation

Hot deck imputation is a method of handling missing data where missing values are replaced with observed values from "donor" cases that are similar in other characteristics. This technique has been widely used in survey data, including by organizations like the US Census Bureau, due to its flexibility and ability to maintain observed data distributions (Table 4.3).

Advantages of Hot Deck Imputation

- **Retains Observed Data Distributions**: Since missing values are imputed using actual observed data, the overall distribution remains realistic.
- **Flexible**: This method is applicable to both categorical and continuous variables.
- **Constrained Imputations**: Imputed values are always feasible, as they come from observed cases.
- **Adds Variability**: By randomly selecting donors, this method introduces variability, which can aid in robust standard error estimation.

Disadvantages of Hot Deck Imputation

- **Sensitivity to Similarity Definitions**: The quality of imputed values depends on the criteria used to define similarity between cases.
- **Computational Intensity**: Identifying similar cases and randomly selecting donors can be computationally expensive, especially for large datasets.
- **Subjectivity**: Deciding how to define "similar" can introduce subjectivity or bias.

Algorithm for Hot Deck Imputation

Let n_1 represent the number of cases with complete data on the variable Y, and n_0 represent the number of cases with missing data on Y. The steps are as follows:

1. From the n_1 cases with complete data, take a random sample (with replacement) of n_1 cases.
2. From this sampled pool, take another random sample (with replacement) of size n_0.
3. Assign the values from the sampled n_0 cases to the cases with missing data in Y.

4. Repeat this process for every variable in the dataset.
5. For multiple imputation, repeat the above four steps multiple times to create multiple imputed datasets.

Variations and Considerations

- **Skipping Step 1**: If Step 1 is skipped, the variability of imputed values is reduced. This approach might not fully account for the uncertainty in missing data, which can underestimate standard errors.
- **Defining Similarity**: A major challenge in this method is deciding what constitutes "similarity" between cases. Common approaches include matching based on distance metrics (e.g., Euclidean distance) or grouping cases by strata or clusters.

Practical Example

The US Census Bureau employs an approximate Bayesian bootstrap variation of hot deck imputation. In this approach

- Similar cases are identified based on shared characteristics or grouping variables.
- A randomly chosen value from a similar individual in the sample is used to replace the missing value.

  This method ensures imputed values are plausible while incorporating variability.
  Key Notes:

- **Good Aspects**:
  - Imputed values are constrained to observed possibilities.
  - Random selection introduces variability, helpful for multiple imputation scenarios.
- **Challenges**:
  - Defining and operationalizing "similarity" remains a critical step in applying this method effectively.

  Below is an example code snippet illustrating hot deck imputation in R:

```
library(Hmisc)
#>
#> Attaching package: 'Hmisc'
#> The following objects are masked from 'package:base':
#>
#>     format.pval, units

# Example dataset with missing values
data <- data.frame(
  ID = 1:10,
  Age = c(25, 30, NA, 40, NA, 50, 60, NA, 70, 80),
  Gender = c("M", "F", "F", "M", "M", "F", "M", "F", "M", "F")
)
```

```
# Perform Hot Deck Imputation using Hmisc::impute
data$Age_imputed <- impute(data$Age, "random")

# Display the imputed dataset
print(data)
#>    ID Age Gender Age_imputed
#> 1   1  25      M          25
#> 2   2  30      F          30
#> 3   3  NA      F          30
#> 4   4  40      M          40
#> 5   5  NA      M          80
#> 6   6  50      F          50
#> 7   7  60      M          60
#> 8   8  NA      F          60
#> 9   9  70      M          70
#> 10 10  80      F          80
```

This code randomly imputes missing values in the `Age` column based on observed data using the `Hmisc` package's `impute` function.

4.4.2.3.2 Cold Deck Imputation

Cold deck imputation is a systematic variant of hot deck imputation where the donor pool is predefined. Instead of selecting donors dynamically from within the same dataset, cold deck imputation relies on an external reference dataset, such as historical data or other high-quality external sources.

Advantages of Cold Deck Imputation

- **Utilizes High-Quality External Data**: This method is particularly useful when reliable external reference datasets are available, allowing for accurate and consistent imputations.
- **Consistency**: If the same donor pool is used across multiple datasets, imputations remain consistent, which can be advantageous in longitudinal studies or standardized processes.

Disadvantages of Cold Deck Imputation

- **Lack of Adaptability**: External data may not adequately reflect the unique characteristics or variability of the current dataset.
- **Potential for Systematic Bias**: If the donor pool is significantly different from the target dataset, imputations may introduce bias.
- **Reduces Variability**: Unlike hot deck imputation, cold deck imputation systematically selects values, which removes random variation. This can affect the estimation of standard errors and other inferential statistics.

Key Characteristics

- **Systematic Selection**: Cold deck imputation selects donor values systematically based on predefined rules or matching criteria, rather than using random sampling.
- **External Donor Pool**: Donors are typically drawn from a separate dataset or historical records.

Algorithm for Cold Deck Imputation

1. Identify an external reference dataset or predefined donor pool.
2. Define the matching criteria to find "similar" cases between the donor pool and the current dataset (e.g., based on covariates or stratification).
3. Systematically assign values from the donor pool to missing values in the current dataset based on the matching criteria.
4. Repeat the process for each variable with missing data.

Practical Considerations

- Cold deck imputation works well when external data closely resemble the target dataset. However, when there are significant differences in distributions or relationships between variables, imputations may be biased or unrealistic.
- This method is less useful for datasets without access to reliable external reference data.

Suppose we have a current dataset with missing values and a historical dataset with similar variables. The following example demonstrates how cold deck imputation can be implemented:

```
# Current dataset with missing values
current_data <- data.frame(
  ID = 1:5,
  Age = c(25, 30, NA, 45, NA),
  Gender = c("M", "F", "F", "M", "M")
)

# External reference dataset (donor pool)
reference_data <- data.frame(
  Age = c(28, 35, 42, 50),
  Gender = c("M", "F", "F", "M")
)

# Perform Cold Deck Imputation
library(dplyr)
#>
#> Attaching package: 'dplyr'
#> The following objects are masked from 'package:Hmisc':
#>
#>     src, summarize
#> The following objects are masked from 'package:stats':
#>
#>     filter, lag
#> The following objects are masked from 'package:base':
```

```
#>
#>     intersect, setdiff, setequal, union

# Define a matching function to find closest donor
impute_cold_deck <- function(missing_row, reference_data) {
  # Filter donors with the same gender
  possible_donors <- reference_data %>%
    filter(Gender == missing_row$Gender)

  # Return the mean age of matching donors as an example of systematic
  ↪ imputation
  return(mean(possible_donors$Age, na.rm = TRUE))
}

# Apply Cold Deck Imputation to the missing rows
current_data <- current_data %>%
  rowwise() %>%
  mutate(
    Age_imputed = ifelse(
      is.na(Age),
      impute_cold_deck(cur_data(), reference_data),
      Age
    )
  )
#> Warning: There was 1 warning in `mutate()`.
#> i In argument: `Age_imputed = ifelse(...)`.
#> i In row 0.
#> Caused by warning:
#> ! There was 1 warning in `filter()`.
#> i In argument: `Gender == missing_row$Gender`.
#> Caused by warning:
#> ! `cur_data()` was deprecated in dplyr 1.1.0.
#> i Please use `pick()` instead.

# Display the imputed dataset
print(current_data)
#> # A tibble: 5 x 4
#> # Rowwise:
#>      ID   Age Gender Age_imputed
#>   <int> <dbl> <chr>        <dbl>
#> 1     1    25 M             25
#> 2     2    30 F             30
#> 3     3    NA F             38.8
#> 4     4    45 M             45
#> 5     5    NA M             38.8
```

Comparison to Hot Deck Imputation

This method suits situations where external reference datasets are trusted and representative. However, careful consideration is required to ensure alignment between the donor pool and the target dataset to avoid systematic biases.

4.4.2.3.3 Random Draw from Observed Distribution

This imputation method replaces missing values by randomly sampling from the observed distribution of the variable with missing data. It is a simple, nonparametric approach that retains the variability of the original data.

Advantages

- **Preserves Variability**:
  - By randomly drawing values from the observed data, this method ensures that the imputed values reflect the inherent variability of the variable.
- **Computational Simplicity**:
  - The process is straightforward and does not require model fitting or complex calculations.

Disadvantages

- **Ignores Relationships Among Variables**:
  - Since the imputation is based solely on the observed distribution of the variable, it does not consider relationships or dependencies with other variables.
- **May Not Align with Trends**:
  - Imputed values are random and may fail to align with patterns or trends present in the data, such as time series structures or interactions.

Steps in Random Draw Imputation

1. Identify the observed (non-missing) values of the variable.
2. For each missing value, randomly sample one value from the observed distribution with or without replacement.
3. Replace the missing value with the randomly sampled value.

The following example demonstrates how to use random draw imputation to fill in missing values:

```
# Example dataset with missing values
set.seed(123)
data <- data.frame(
  ID = 1:10,
  Value = c(10, 20, NA, 30, 40, NA, 50, 60, NA, 70)
)

# Perform random draw imputation
random_draw_impute <- function(data, variable) {
  observed_values <- data[[variable]][!is.na(data[[variable]])] # Observed values
  data[[variable]][is.na(data[[variable]])] <- sample(observed_values,
                                                      sum(is.na(data[[variable]])),
                                                      replace = TRUE)
  return(data)
}
```

```
# Apply the imputation
imputed_data <- random_draw_impute(data, variable = "Value")

# Display the imputed dataset
print(imputed_data)
#>    ID Value
#> 1   1    10
#> 2   2    20
#> 3   3    70
#> 4   4    30
#> 5   5    40
#> 6   6    70
#> 7   7    50
#> 8   8    60
#> 9   9    30
#> 10 10    70
```

Considerations

- **When to Use**:
  - This method is suitable for exploratory analysis or as a quick way to handle missing data in univariate contexts.
- **Limitations**:
  - Random draws may result in values that do not fit well in the broader context of the dataset, especially in cases where the variable has strong relationships with others.

This method is a quick and computationally efficient way to address missing data but is best complemented by more sophisticated methods when relationships between variables are important (Table 4.4).

Table 4.4 Comparison of random draw and regression-based imputation methods

Feature	Random draw from observed distribution	Regression-based imputation
Complexity	Simple	Moderate to high
Preserves variability	Yes	Limited in deterministic forms
Considers relationships	No	Yes
Risk of implausible values	Low (if observed values are plausible)	Moderate to high

4.4.2.4 Semi-Parametric Methods

4.4.2.4.1 Predictive Mean Matching (PMM)

Predictive mean matching (PMM) imputes missing values by finding observed values closest in predicted value (based on a regression model) to the missing data. The donor values are then used to fill in the gaps.

Advantages
- Maintains observed variability in the data.
- Ensures imputed values are realistic since they are drawn from observed data.

Disadvantages
- Requires a suitable predictive model.
- Computationally intensive for large datasets.

Steps for PMM
1. Regress Y on X (matrix of covariates) for the n_1 (non-missing cases) to estimate coefficients $\hat{b}$ and residual variance s^2.
2. Draw from the posterior predictive distribution of residual variance:

$$s_{[1]}^2 = \frac{(n_1 - k)s^2}{\chi^2},$$

   where χ^2 is a random draw from $\chi^2_{n_1-k}$.
3. Randomly sample from the posterior distribution of $\hat{b}$:

$$b_{[1]} \sim MVN(\hat{b}, s_{[1]}^2(X'X)^{-1}).$$

4. Standardize residuals for n_1 cases:

$$e_i = \frac{y_i - \hat{b}x_i}{\sqrt{s^2(1 - k/n_1)}}.$$

5. Randomly draw a sample (with replacement) of n_0 residuals from Step 4.
6. Calculate imputed values for n_0 missing cases:

$$y_i = b_{[1]}x_i + s_{[1]}e_i.$$

7. Repeat Steps 2–6 (except Step 4) to create multiple imputations.

Notes
- PMM can handle heteroskedasticity.
- Works for multiple variables, imputing each using all others as predictors.

Example

Example from Statistics Globe

```
set.seed(1) # Seed
N  <- 100                                          # Sample size
y  <- round(runif(N,-10, 10))                       # Target variable Y
x1 <- y + round(runif(N, 0, 50))                    # Auxiliary variable 1
x2 <- round(y + 0.25 * x1 + rnorm(N,-3, 15))        # Auxiliary variable 2
x3 <- round(0.1 * x1 + rpois(N, 2))                 # Auxiliary variable 3
# (categorical variable)
x4 <- as.factor(round(0.02 * y + runif(N)))         # Auxiliary variable 4

# Insert 20% missing data in Y
y[rbinom(N, 1, 0.2) == 1] <- NA

data <- data.frame(y, x1, x2, x3, x4)              # Store data in dataset
head(data) # First 6 rows of our data
#>    y x1  x2 x3 x4
#> 1 NA 28 -10  5  0
#> 2 NA 15  -2  2  1
#> 3  1 15 -12  6  1
#> 4  8 58  22 10  1
#> 5 NA 26 -12  7  0
#> 6 NA 19  36  5  1

library("mice") # Load mice package
#>
#> Attaching package: 'mice'
#> The following object is masked from 'package:stats':
#>
#>     filter
#> The following objects are masked from 'package:base':
#>
#>     cbind, rbind

##### Impute data via predictive mean matching (single imputation)#####

imp_single <- mice::mice(data, m = 1, method = "pmm") # Impute missing
↪ values
#>
#>  iter imp variable
#>   1   1  y
#>   2   1  y
#>   3   1  y
#>   4   1  y
#>   5   1  y
data_imp_single <- mice::complete(imp_single)         # Store imputed
↪ data
# head(data_imp_single)

# Since single imputation underestiamtes stnadard errors,
# we use multiple imputaiton
```

```
##### Predictive mean matching (multiple imputation) #####

# Impute missing values multiple times
imp_multi <- mice(data, m = 5, method = "pmm")
#>
#>  iter imp variable
#>   1   1  y
#>   1   2  y
#>   1   3  y
#>   1   4  y
#>   1   5  y
#>   2   1  y
#>   2   2  y
#>   2   3  y
#>   2   4  y
#>   2   5  y
#>   3   1  y
#>   3   2  y
#>   3   3  y
#>   3   4  y
#>   3   5  y
#>   4   1  y
#>   4   2  y
#>   4   3  y
#>   4   4  y
#>   4   5  y
#>   5   1  y
#>   5   2  y
#>   5   3  y
#>   5   4  y
#>   5   5  y
data_imp_multi_all <-
    # Store multiply imputed data
    mice::complete(imp_multi,
                   "repeated",
                   include = TRUE)
#> New names:
#> * `y` -> `y...1`
#> * `x1` -> `x1...2`
#> * `x2` -> `x2...3`
#> * `x3` -> `x3...4`
#> * `x4` -> `x4...5`
#> * `y` -> `y...6`
#> * `x1` -> `x1...7`
#> * `x2` -> `x2...8`
#> * `x3` -> `x3...9`
#> * `x4` -> `x4...10`
#> * `y` -> `y...11`
#> * `x1` -> `x1...12`
#> * `x2` -> `x2...13`
#> * `x3` -> `x3...14`
#> * `x4` -> `x4...15`
#> * `y` -> `y...16`
```

```
#> * `x1` -> `x1...17`
#> * `x2` -> `x2...18`
#> * `x3` -> `x3...19`
#> * `x4` -> `x4...20`
#> * `y` -> `y...21`
#> * `x1` -> `x1...22`
#> * `x2` -> `x2...23`
#> * `x3` -> `x3...24`
#> * `x4` -> `x4...25`
#> * `y` -> `y...26`
#> * `x1` -> `x1...27`
#> * `x2` -> `x2...28`
#> * `x3` -> `x3...29`
#> * `x4` -> `x4...30`

data_imp_multi <-
    # Combine imputed Y and X1-X4 (for convenience)
    data.frame(data_imp_multi_all[, 1:6], data[, 2:5])

head(data_imp_multi)
#>   y.0 y.1 y.2 y.3 y.4 y.5 x1  x2 x3 x4
#> 1  NA  -1   6  -1  -3   3 28 -10  5  0
#> 2  NA -10  10   4   0   2 15  -2  2  1
#> 3   1   1   1   1   1   1 15 -12  6  1
#> 4   8   8   8   8   8   8 58  22 10  1
#> 5  NA   0  -1  -6   2   0 26 -12  7  0
#> 6  NA   4   0   3   3   3 19  36  5  1
```

Example from UCLA Statistical Consulting (Fig. 4.4)

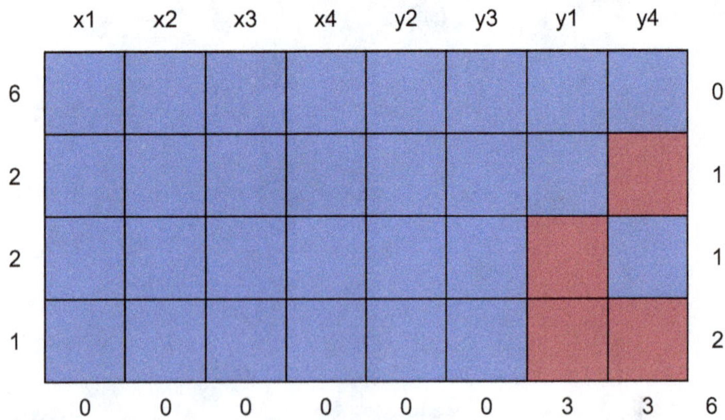

Fig. 4.4 Missing data pattern

```
library(mice)
library(VIM)
#> Loading required package: colorspace
#> Loading required package: grid
#> VIM is ready to use.
#> Suggestions and bug-reports can be submitted at:
↪  https://github.com/statistikat/VIM/issues
#>
#> Attaching package: 'VIM'
#> The following object is masked from 'package:datasets':
#>
#>     sleep
library(lattice)
library(ggplot2)
## set observations to NA
anscombe <- within(anscombe, {
    y1[1:3] <- NA
    y4[3:5] <- NA
})
## view
head(anscombe)
#>   x1 x2 x3 x4   y1   y2    y3   y4
#> 1 10 10 10  8   NA 9.14  7.46 6.58
#> 2  8  8  8  8   NA 8.14  6.77 5.76
#> 3 13 13 13  8   NA 8.74 12.74   NA
#> 4  9  9  9  8 8.81 8.77  7.11   NA
#> 5 11 11 11  8 8.33 9.26  7.81   NA
#> 6 14 14 14  8 9.96 8.10  8.84 7.04
```

```
## check missing data patterns
md.pattern(anscombe)
```

```
#>   x1 x2 x3 x4 y2 y3 y1 y4
#> 6  1  1  1  1  1  1  1  1 0
#> 2  1  1  1  1  1  1  0  1 1
#> 2  1  1  1  1  1  1  1  0 1
#> 1  1  1  1  1  1  1  0  0 2
#>    0  0  0  0  0  0  3  3 6
```

```
## Number of observations per patterns for all pairs of variables
p <- md.pairs(anscombe)
p
#> $rr
#>    x1 x2 x3 x4 y1 y2 y3 y4
#> x1 11 11 11 11  8 11 11  8
#> x2 11 11 11 11  8 11 11  8
#> x3 11 11 11 11  8 11 11  8
#> x4 11 11 11 11  8 11 11  8
#> y1  8  8  8  8  8  8  8  6
#> y2 11 11 11 11  8 11 11  8
#> y3 11 11 11 11  8 11 11  8
```

```
#> y4  8  8  8  8  6  8  8  8
#>
#> $rm
#>    x1 x2 x3 x4 y1 y2 y3 y4
#> x1  0  0  0  0  3  0  0  3
#> x2  0  0  0  0  3  0  0  3
#> x3  0  0  0  0  3  0  0  3
#> x4  0  0  0  0  3  0  0  3
#> y1  0  0  0  0  0  0  0  2
#> y2  0  0  0  0  3  0  0  3
#> y3  0  0  0  0  3  0  0  3
#> y4  0  0  0  0  2  0  0  0
#>
#> $mr
#>    x1 x2 x3 x4 y1 y2 y3 y4
#> x1  0  0  0  0  0  0  0  0
#> x2  0  0  0  0  0  0  0  0
#> x3  0  0  0  0  0  0  0  0
#> x4  0  0  0  0  0  0  0  0
#> y1  3  3  3  3  0  3  3  2
#> y2  0  0  0  0  0  0  0  0
#> y3  0  0  0  0  0  0  0  0
#> y4  3  3  3  3  2  3  3  0
#>
#> $mm
#>    x1 x2 x3 x4 y1 y2 y3 y4
#> x1  0  0  0  0  0  0  0  0
#> x2  0  0  0  0  0  0  0  0
#> x3  0  0  0  0  0  0  0  0
#> x4  0  0  0  0  0  0  0  0
#> y1  0  0  0  0  3  0  0  1
#> y2  0  0  0  0  0  0  0  0
#> y3  0  0  0  0  0  0  0  0
#> y4  0  0  0  0  1  0  0  3
```

- `rr` = number of observations where both pairs of values are observed.
- `rm` = the number of observations where both variables are missing values.
- `mr` = the number of observations where the first variable's value (e.g., the row variable) is observed and the second (or column) variable is missing.
- `mm` = the number of observations where the second variable's value (e.g., the col variable) is observed and the first (or row) variable is missing.

For a margin plot of y1 and y4, refer to Fig. 4.5

```
## Margin plot of y1 and y4
marginplot(anscombe[c(5, 8)], col = c("blue", "red", "orange"))
```

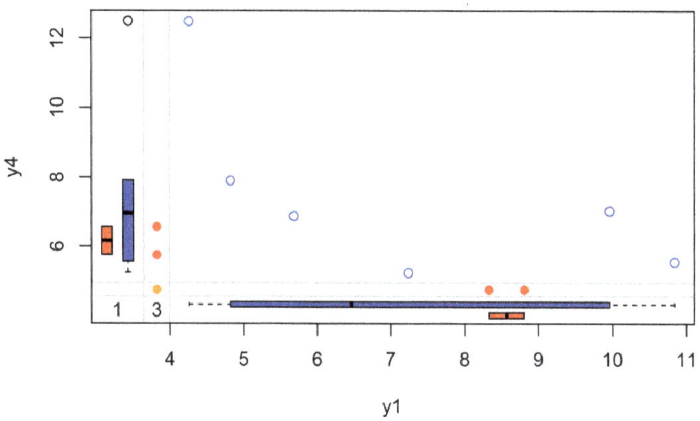

Fig. 4.5 Scatterplot of y4 versus y1 with marginal boxplots

```
## 5 imputations for all missing values
imp1 <- mice(anscombe, m = 5)
#>
#>  iter imp variable
#>   1   1  y1  y4
#>   1   2  y1  y4
#>   1   3  y1  y4
#>   1   4  y1  y4
#>   1   5  y1  y4
#>   2   1  y1  y4
#>   2   2  y1  y4
#>   2   3  y1  y4
#>   2   4  y1  y4
#>   2   5  y1  y4
#>   3   1  y1  y4
#>   3   2  y1  y4
#>   3   3  y1  y4
#>   3   4  y1  y4
#>   3   5  y1  y4
#>   4   1  y1  y4
#>   4   2  y1  y4
#>   4   3  y1  y4
#>   4   4  y1  y4
#>   4   5  y1  y4
#>   5   1  y1  y4
#>   5   2  y1  y4
#>   5   3  y1  y4
#>   5   4  y1  y4
#>   5   5  y1  y4
#> Warning: Number of logged events: 52

## linear regression for each imputed data set - 5 regression are run
fitm <- with(imp1, lm(y1 ~ y4 + x1))
summary(fitm)
```

```
#> # A tibble: 15 x 7
#>    term        estimate std.error statistic p.value  nobs
#>    <chr>          <dbl>     <dbl>     <dbl>   <dbl> <int>
#>  1 (Intercept)    7.33      2.44       3.01  0.0169    11
#>  2 y4            -0.416     0.223     -1.86  0.0996    11
#>  3 x1             0.371     0.141      2.63  0.0302    11
#>  4 (Intercept)    7.27      2.90       2.51  0.0365    11
#>  5 y4            -0.435     0.273     -1.59  0.150     11
#>  6 x1             0.387     0.160      2.41  0.0422    11
#>  7 (Intercept)    6.54      2.80       2.33  0.0479    11
#>  8 y4            -0.322     0.255     -1.26  0.243     11
#>  9 x1             0.362     0.156      2.32  0.0491    11
#> 10 (Intercept)    5.93      3.08       1.92  0.0907    11
#> 11 y4            -0.286     0.282     -1.02  0.339     11
#> 12 x1             0.418     0.176      2.37  0.0451    11
#> 13 (Intercept)    8.16      2.67       3.05  0.0158    11
#> 14 y4            -0.489     0.251     -1.95  0.0867    11
#> 15 x1             0.326     0.151      2.17  0.0622    11
#> # i 1 more variable: df.residual <dbl>

## pool coefficients and standard errors across all 5 regression models
mice::pool(fitm)
#> Class: mipo    m = 5
#>          term m   estimate       ubar           b
#> 1 (Intercept) 5  7.0445966 7.76794670 0.719350800
#> 2          y4 5 -0.3896685 0.06634920 0.006991497
#> 3          x1 5  0.3727865 0.02473847 0.001134293
#>            t dfcom       df        riv     lambda       fmi
#> 1 8.63116766     8 5.805314 0.11112601 0.10001207 0.3044313
#> 2 0.07473900     8 5.706243 0.12644909 0.11225460 0.3161877
#> 3 0.02609962     8 6.178032 0.05502168 0.05215218 0.2586992

## output parameter estimates
summary(mice::pool(fitm))
#>          term   estimate std.error statistic       df
#> 1 (Intercept)  7.0445966 2.9378849  2.397846 5.805314
#> 2          y4 -0.3896685 0.2733843 -1.425350 5.706243
#> 3          x1  0.3727865 0.1615538  2.307508 6.178032
#>      p.value
#> 1 0.05483678
#> 2 0.20638512
#> 3 0.05923999
```

4.4.2.4.2 Stochastic Imputation

Stochastic imputation is an enhancement of regression imputation that introduces randomness into the imputation process by adding a random residual to the predicted values from a regression model. This approach aims to retain the variability of the original data while reducing the bias introduced by deterministic regression imputation.

Stochastic imputation can be described as

Imputed Value = Predicted Value (from regression) + Random Residual

This method is commonly used as a foundation for multiple imputation techniques.

Advantages of Stochastic Imputation

- **Retains All the Benefits of Regression Imputation**:
  - Preserves relationships between variables in the dataset.
  - Utilizes information from observed data to inform imputations.
- **Introduces Randomness**:
  - Adds variability by including a random residual term, making imputed values more realistic and better representing the uncertainty of missing data.
- **Supports Multiple Imputation**:
  - By generating different random residuals for each iteration, it facilitates the creation of multiple plausible datasets for robust statistical analysis.

Disadvantages of Stochastic Imputation

- **Implausible Values**:
  - Depending on the random residuals, imputed values may fall outside the plausible range (e.g., negative values for variables like age or income).
- **Cannot Handle Heteroskedasticity**:
  - If the data exhibit heteroskedasticity (i.e., nonconstant variance of residuals), the randomness added by stochastic imputation may not accurately reflect the underlying variability.

Steps in Stochastic Imputation

1. Fit a regression model using cases with complete data for the variable with missing values.
2. Predict missing values using the fitted model.
3. Generate random residuals based on the distribution of residuals from the regression model.
4. Add the random residuals to the predicted values to impute missing values.

```
# Example dataset with missing values
set.seed(123)
data <- data.frame(
  X = rnorm(10, mean = 50, sd = 10),
  Y = c(100, 105, 110, NA, 120, NA, 130, 135, 140, NA)
)

# Perform stochastic imputation
```

```
stochastic_impute <- function(data, predictor, target) {
  # Subset data with complete cases
  complete_data <- data[!is.na(data[[target]]), ]

  # Fit a regression model
  model <- lm(as.formula(paste(target, "~", predictor)), data =
↪   complete_data)

  # Predict missing values
  missing_data <- data[is.na(data[[target]]), ]
  predictions <- predict(model, newdata = missing_data)

  # Add random residuals
  residual_sd <- sd(model$residuals, na.rm = TRUE)
  stochastic_values <- predictions + rnorm(length(predictions), mean = 0,
↪   sd = residual_sd)

  # Impute missing values
  data[is.na(data[[target]]), target] <- stochastic_values
  return(data)
}

# Apply stochastic imputation
imputed_data <- stochastic_impute(data, predictor = "X", target = "Y")

# Display the imputed dataset
print(imputed_data)
#>           X        Y
#> 1  44.39524 100.0000
#> 2  47.69823 105.0000
#> 3  65.58708 110.0000
#> 4  50.70508 137.0070
#> 5  51.29288 120.0000
#> 6  67.15065 114.9107
#> 7  54.60916 130.0000
#> 8  37.34939 135.0000
#> 9  43.13147 140.0000
#> 10 45.54338 127.9359
```

Notes:

- **Multiple Imputation**: Most multiple imputation methods are extensions of stochastic regression imputation. By repeating the imputation process with different random seeds, multiple datasets can be generated to account for uncertainty in the imputed values (Table 4.5).
- **Dealing with Implausible Values**: Additional constraints or transformations (e.g., truncating imputed values to a plausible range) may be necessary to address the issue of implausible values.

Table 4.5 Comparison of random draw and regression-based imputation methods

Feature	Deterministic regression imputation	Stochastic imputation
Randomness	None	Adds random residuals
Preserves variability	No	Yes
Use in multiple imputation	Limited	Well suited
Bias potential	Higher	Lower

Comparison to Deterministic Regression Imputation

```
# Income data
set.seed(1)                                      # Set seed
N <- 1000                                            # Sample size

income <-
  round(rnorm(N, 0, 500))             # Create some synthetic income data
income[income < 0] <- income[income < 0] * (-1)

x1 <- income + rnorm(N, 1000, 1500)             # Auxiliary variables
x2 <- income + rnorm(N,-5000, 2000)


# Create 10% missingness in income
income[rbinom(N, 1, 0.1) == 1] <- NA

data_inc_miss <- data.frame(income, x1, x2)
```

Single stochastic regression imputation

```
imp_inc_sri  <- mice(data_inc_miss, method = "norm.nob", m = 1)
#>
#>  iter imp variable
#>   1   1  income
#>   2   1  income
#>   3   1  income
#>   4   1  income
#>   5   1  income
data_inc_sri <- mice::complete(imp_inc_sri)
```

Single predictive mean matching

```
imp_inc_pmm  <- mice(data_inc_miss, method = "pmm", m = 1)
#>
#>  iter imp variable
#>   1   1  income
#>   2   1  income
#>   3   1  income
#>   4   1  income
#>   5   1  income
data_inc_pmm <- mice::complete(imp_inc_pmm)
```

Stochastic regression imputation contains negative values

```
data_inc_sri$income[data_inc_sri$income < 0]
#>  [1]  -23.85404  -58.37790  -61.86396  -57.47909  -21.29221
#>  [6]  -73.26549  -61.76194  -42.45942 -351.02991 -317.69090
# No values below 0
data_inc_pmm$income[data_inc_pmm$income < 0]
#> numeric(0)
```

Evidence for heteroskadastic data

```
# Heteroscedastic data

set.seed(1)                                  # Set seed
N <- 1:1000                                       # Sample size

a <- 0
b <- 1
sigma2 <- N^2
eps <- rnorm(N, mean = 0, sd = sqrt(sigma2))

y <- a + b * N + eps                              # Heteroscedastic variable
x <- 30 * N + rnorm(N[length(N)], 1000, 200) # Correlated variable

y[rbinom(N[length(N)], 1, 0.3) == 1] <- NA    # 30% missing

data_het_miss <- data.frame(y, x)
```

Single stochastic regression imputation

```
imp_het_sri  <- mice(data_het_miss, method = "norm.nob", m = 1)
#>
#>  iter imp variable
#>   1   1  y
#>   2   1  y
#>   3   1  y
#>   4   1  y
#>   5   1  y
data_het_sri <- mice::complete(imp_het_sri)
```

Single predictive mean matching

```
imp_het_pmm  <- mice(data_het_miss, method = "pmm", m = 1)
#>
#>  iter imp variable
#>   1   1  y
#>   2   1  y
#>   3   1  y
#>   4   1  y
#>   5   1  y
data_het_pmm <- mice::complete(imp_het_pmm)
```

Comparison between predictive mean matching and stochastic regression imputation (Fig. 4.6)

```
# Both plots in one graphic
par(mfrow = c(1, 2))

# Plot of observed values
plot(x[!is.na(data_het_sri$y)],
     data_het_sri$y[!is.na(data_het_sri$y)],
     main = "",
     xlab = "X",
     ylab = "Y")
# Plot of missing values
points(x[is.na(y)], data_het_sri$y[is.na(y)],
       col = "red")

# Title of plot
title("Stochastic Regression Imputation",
      line = 0.5)

# Regression line
abline(lm(y ~ x, data_het_sri),
       col = "#1b98e0", lwd = 2.5)

# Legend
legend(
  "topleft",
  c("Observed Values", "Imputed Values", "Regression Y ~ X"),
  pch = c(1, 1, NA),
  lty = c(NA, NA, 1),
  col = c("black", "red", "#1b98e0")
)

# Plot of observed values
plot(x[!is.na(data_het_pmm$y)],
     data_het_pmm$y[!is.na(data_het_pmm$y)],
     main = "",
     xlab = "X",
     ylab = "Y")


# Plot of missing values
points(x[is.na(y)], data_het_pmm$y[is.na(y)],
       col = "red")

# Title of plot
title("Predictive Mean Matching",
      line = 0.5)
abline(lm(y ~ x, data_het_pmm),
       col = "#1b98e0", lwd = 2.5)

# Legend
legend(
```

```
  "topleft",
  c("Observed Values", "Imputed Values", "Regression Y ~ X"),
  pch = c(1, 1, NA),
  lty = c(NA, NA, 1),
  col = c("black", "red", "#1b98e0")
)

mtext(
  "Imputation of Heteroskedastic Data",
  # Main title of plot
  side = 3,
  line = -1.5,
  outer = TRUE,
  cex = 2
)
```

4.4.2.5 Matrix Completion

Matrix completion is a method used to impute missing data in a feature matrix while accounting for dependence between features. This approach leverages principal components to approximate the data matrix, a process referred to as **matrix completion** (James et al., 2013, Sec 12.3).

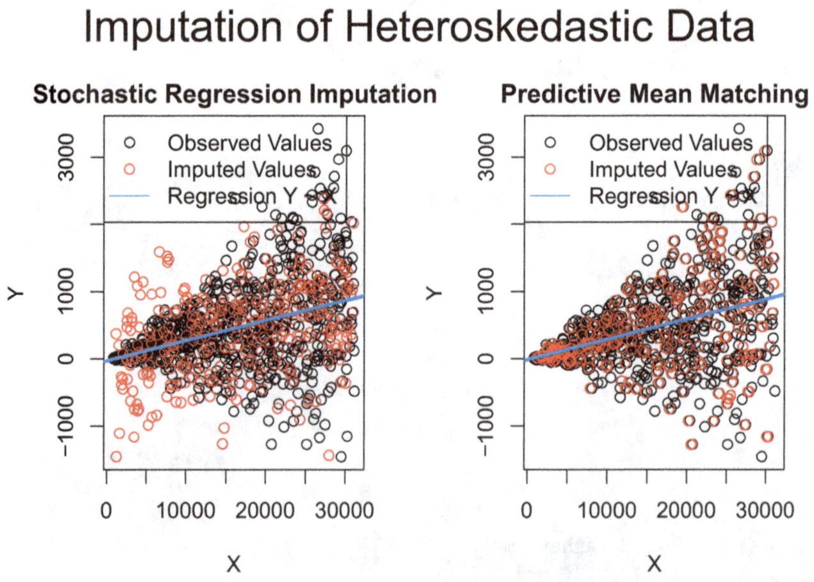

Fig. 4.6 Imputation of heteroskedastic data

Problem Setup

Consider an $n \times p$ feature matrix $\mathbf{X}$, where the element x_{ij} represents the value for the ith observation and jth feature. Some elements of $\mathbf{X}$ are missing, and we aim to impute these missing values.

The matrix $\mathbf{X}$ can be approximated using its leading principal components. Specifically, we consider M principal components that minimize the following objective:

$$\min_{\mathbf{A}\in\mathbb{R}^{n\times M},\mathbf{B}\in\mathbb{R}^{p\times M}} \left\{ \sum_{(i,j)\in\mathcal{O}} (x_{ij} - \sum_{m=1}^{M} a_{im}b_{jm})^2 \right\}$$

where $\mathcal{O}$ is the set of observed indices (i, j), which is a subset of the total $n \times p$ pairs. Here, – $\mathbf{A}$ is an $n \times M$ matrix of principal component scores. – $\mathbf{B}$ is a $p \times M$ matrix of principal component loadings.

Imputation of Missing Values

After solving the minimization problem

1. Missing observations x_{ij} can be imputed using the formula:

$$\hat{x}_{ij} = \sum_{m=1}^{M} \hat{a}_{im}\hat{b}_{jm}$$

   where $\hat{a}_{im}$ and $\hat{b}_{jm}$ are the estimated elements of $\mathbf{A}$ and $\mathbf{B}$, respectively.
2. The leading M principal component scores and loadings can be approximately recovered, as is done in complete data scenarios.

Iterative Algorithm

The eigen-decomposition used in standard principal component analysis is not applicable here because of missing values. Instead, an iterative algorithm, as described in (James et al., 2013, Alg 12.1), is employed:

1. **Initialize the Complete Matrix**: Construct an initial complete matrix $\tilde{\mathbf{X}}$ of dimension $n \times p$ where

$$\tilde{x}_{ij} = \begin{cases} x_{ij} & \text{if } (i, j) \in \mathcal{O} \\ \bar{x}_j & \text{if } (i, j) \notin \mathcal{O} \end{cases}$$

   Here, $\bar{x}_j$ is the mean of the observed values for the jth variable in the incomplete data matrix $\mathbf{X}$. $\mathcal{O}$ indexes the observed elements of $\mathbf{X}$.

2. **Iterative Steps**: Repeat the following steps until convergence:
   - **Minimize the Objective**: Solve the problem:

$$\min_{\mathbf{A}\in R^{n\times M},\mathbf{B}\in R^{p\times M}} \left\{ \sum_{(i,j)\in\mathcal{O}} (x_{ij} - \sum_{m=1}^{M} a_{im}b_{jm})^2 \right\}$$

     by computing the principal components of the current $\tilde{\mathbf{X}}$.
   - **Update Missing Values**: For each missing element $(i, j) \notin \mathcal{O}$, set

$$\tilde{x}_{ij} \leftarrow \sum_{m=1}^{M} \hat{a}_{im}\hat{b}_{jm}$$

   - **Recalculate the Objective**: Compute the objective:

$$\sum_{(i,j)\in\mathcal{O}} (x_{ij} - \sum_{m=1}^{M} \hat{a}_{im}\hat{b}_{jm})^2$$

3. **Return Imputed Values**: Once the algorithm converges, return the estimated missing entries $\tilde{x}_{ij}$ for $(i, j) \notin \mathcal{O}$.

Key Considerations

- This approach assumes that the missing data are missing at random (MAR).
- Convergence criteria for the iterative algorithm often involve achieving a threshold for the change in the objective function or limiting the number of iterations.
- The choice of M, the number of principal components, can be guided by cross-validation or other model selection techniques.

4.4.2.6 Comparison of Single Imputation Techniques

Single imputation techniques are straightforward and accessible, but they often underestimate uncertainty and fail to fully leverage relationships among variables. These limitations make them less ideal for rigorous analyses compared to multiple imputation or model-based approaches (Table 4.6).

Table 4.6 Comparison of single imputation techniques

Method	Advantages	Disadvantages
Mean, median, mode imputation	Simple, quick implementation	Biased variances and covariances; ignores relationships among variables
Forward/backward filling	Preserves temporal ordering	Biased for systematic gaps or long missing sequences
Linear regression imputation	Preserves relationships among variables	Fails to capture variability; assumes linearity
Logistic regression imputation	Handles categorical variables well	Requires appropriate model assumptions; ignores variability
PMM	Maintains variability; imputes realistic values	Computationally intensive; requires a good predictive model
Hot deck imputation	Flexible; maintains data distribution	Sensitive to donor selection; computationally demanding
Cold deck imputation	Consistent across datasets with predefined donor pools	Risk of bias if donor data are not representative
Random draw from observed	Simple; retains variability in data	Does not preserve relationships among variables; random imputation may distort trends
Matrix completion	Captures dependencies; imputes structurally consistent values	Computationally intensive; assumes principal components capture data relationships

4.4.3 Machine Learning and Modern Approaches

4.4.3.1 Tree-Based Methods

4.4.3.1.1 Random Forest Imputation (missForest)

Random forest imputation uses an iterative process where a random forest model predicts missing values for one variable at a time, treating other variables as predictors. This process continues until convergence.

- **Mathematical Framework**:
  1. For a variable X_j with missing values, treat X_j as the response variable.
  2. Fit a random forest model $f(X_{-j})$ using the other variables X_{-j} as predictors.
  3. Predict missing values $\hat{X}_j = f(X_{-j})$.
  4. Repeat for all variables with missing data until imputed values stabilize.
- **Advantages**:
  - Captures complex interactions and nonlinearities.
  - Handles mixed data types seamlessly.

- **Limitations**:
  - Computationally intensive for large datasets.
  - Sensitive to the quality of data relationships.

4.4.3.1.2 Gradient Boosting Machines (GBM)

Gradient boosting machines iteratively build models to minimize loss functions. For imputation, missing values are treated as a target variable to be predicted.

- **Mathematical Framework**: The GBM algorithm minimizes the loss function:

$$L = \sum_{i=1}^{n} \ell(y_i, f(x_i)),$$

  where ℓ is the loss function (e.g., mean squared error), y_i are observed values, and $f(x_i)$ are predictions.
- Missing values are treated as the y_i and predicted iteratively.
- **Advantages**:
  - Highly accurate predictions.
  - Captures variable importance.
- **Limitations**:
  - Overfitting risks.
  - Requires careful parameter tuning.

4.4.3.2 Neural Network-Based Imputation

4.4.3.2.1 Autoencoders

Autoencoders are unsupervised neural networks that compress and reconstruct data. Missing values are estimated during reconstruction.

- **Mathematical Framework**: An autoencoder consists of
  1. An encoder function: $h = g(Wx + b)$, which compresses the input x.
  2. A decoder function: $\hat{x} = g'(W'h + b')$, which reconstructs the data.
- The network minimizes the reconstruction loss:

$$L = \sum_{i=1}^{n} (x_i - \hat{x}_i)^2.$$

- **Advantages**:
  - Handles high-dimensional and nonlinear data.
  - Unsupervised learning.
- **Limitations**:
  - Computationally demanding.
  - Requires large datasets for effective training.

4.4.3.2.2 Generative Adversarial Networks (GANs) for Data Imputation

GANs consist of a generator and a discriminator. For imputation, the generator fills in missing values, and the discriminator evaluates the quality of the imputations.

- **Mathematical Framework**: GAN training involves optimizing

$$\min_G \max_D \mathbb{E}[\log D(x)] + \mathbb{E}[\log(1 - D(G(z)))].$$

  - $D(x)$: Discriminator's probability that x is real.
  - $G(z)$: Generator's output for latent input z.
- **Advantages**:
  - Realistic imputations that reflect underlying distributions.
  - Handles complex data types.
- **Limitations**:
  - Difficult to train and tune.
  - Computationally intensive.

4.4.3.3 Matrix Factorization and Matrix Completion

4.4.3.3.1 Singular Value Decomposition (SVD)

SVD decomposes a matrix A into three matrices:

$$A = U \Sigma V^T,$$

where U and V are orthogonal matrices and Σ contains singular values. Missing values are estimated by reconstructing A using a low-rank approximation:

$$\hat{A} = U_k \Sigma_k V_k^T.$$

- **Advantages**:
  - Captures global patterns.
  - Efficient for structured data.
- **Limitations**:
  - Assumes linear relationships.
  - Sensitive to sparsity.

4.4.3.3.2 Collaborative Filtering Approaches

Collaborative filtering uses similarities between rows (users) or columns (items) to impute missing data. For instance, the value of X_{ij} is predicted as

$$\hat{X}_{ij} = \frac{\sum_{k \in N(i)} w_{ik} X_{kj}}{\sum_{k \in N(i)} w_{ik}},$$

where w_{ik} represents similarity weights and $N(i)$ is the set of neighbors.

4.4.3.4 K-Nearest Neighbor (KNN) Imputation

KNN identifies the k nearest observations based on a distance metric and imputes missing values using a weighted average (continuous variables) or mode (categorical variables).

- **Mathematical Framework**: For a missing value x, its imputed value is

$$\hat{x} = \frac{\sum_{i=1}^{k} w_i x_i}{\sum_{i=1}^{k} w_i},$$

  where $w_i = \frac{1}{d(x,x_i)}$ and $d(x, x_i)$ is a distance metric (e.g., Euclidean or Manhattan).
- **Advantages**:
  - Simple and interpretable.
  - Nonparametric.
- **Limitations**:
  - Computationally expensive for large datasets.

Table 4.7 Advanced machine learning methods for missing data imputation

Method	Advantages	Limitations	Applications
Random forest (missForest)	Handles mixed data types, captures interactions	Computationally intensive	Mixed data types
Gradient boosting machines	High accuracy, feature importance	Sensitive to parameters	Predictive tasks
Autoencoders	Handles high-dimensional, nonlinear data	Computationally expensive	Complex datasets
GANs	Realistic imputations, complex distributions	Difficult to train, resource-intensive	Healthcare, finance
SVD	Captures global patterns, efficient	Assumes linear relationships	Recommendation systems
Collaborative filtering	Intuitive for user-item data	Struggles with sparse or new data	Recommender systems
KNN imputation	Simple, interpretable	Computationally intensive, sensitive to k	General purpose
Hybrid methods	Combines multiple strengths	Complexity in design	Flexible

4.4.3.5 Hybrid Methods

Hybrid methods combine statistical and machine learning approaches. For example, mean imputation followed by fine-tuning with machine learning models. These methods aim to leverage the strengths of multiple techniques.

4.4.3.6 Summary Table

Refer to Table 4.7 for advanced machine learning methods for missing data imputation

4.4.4 Multiple Imputation

Multiple imputation (MI) is a statistical technique for handling missing data by creating several plausible datasets through imputation, analyzing each dataset separately, and then combining the results to account for uncertainty in the imputations. MI operates under the assumption that missing data is either Missing Completely at Random (MCAR) or Missing at Random (MAR).

Unlike Single Imputation Techniques, MI reflects the uncertainty inherent in the missing data by introducing variability in the imputed values. It avoids biases introduced by ad hoc methods and produces more reliable statistical inferences.

The three fundamental steps in MI are

1. **Imputation**: Replace missing values with a set of plausible values to create multiple "completed" datasets.
2. **Analysis**: Perform the desired statistical analysis on each imputed dataset.
3. **Combination**: Combine the results using rules to account for within- and between-imputation variability.

4.4.4.1 Why Multiple Imputation Is Important

Imputed values are estimates and inherently include random error. However, when these estimates are treated as exact values in subsequent analysis, the software may overlook this additional error. This oversight results in **underestimated standard errors and overly small p-values**, leading to misleading conclusions.

Multiple imputation addresses this issue by generating multiple estimates for each missing value. These estimates differ slightly due to their random component, which reintroduces variation. This variation helps the software incorporate the uncertainty of imputed values, resulting in

- **Unbiased Parameter Estimates**
- **Accurate Standard Errors**
- **Improved p-Values**

Multiple imputation was a significant breakthrough in statistics approximately 20 years ago. It provides solutions for many missing data issues (though not all) and, when applied correctly, leads to reliable parameter estimates.

If the proportion of missing data is very small (e.g., 2–3%), the choice of imputation method is less critical.

4.4.4.2 Goals of Multiple Imputation

The primary goals of any missing data technique, including multiple imputation, are

1. **Unbiased Parameter Estimates**: Ensuring accurate regression coefficients, group means, odds ratios, etc.
2. **Accurate Standard Errors**: This leads to reliable p-values and appropriate statistical inferences.
3. **Adequate Power**: To detect meaningful and significant parameter values.

4.4.4.3 Overview of Rubin's Framework

Rubin's framework provides the theoretical foundation for MI. It uses a Bayesian model-based approach for generating imputations and a frequentist approach for evaluating the results. The central goals of Rubin's framework are to ensure that imputations:

- Retain the statistical relationships present in the data.
- Reflect the uncertainty about the true values of the missing data.

Under Rubin's framework, MI offers the following advantages:

- **Generalizability**: Unlike maximum likelihood estimation (MLE), MI can be applied to a wide range of models.
- **Statistical Properties**: When data is MAR or MCAR, MI estimates are consistent, asymptotically normal, and efficient.

Rubin also emphasized the importance of using multiple imputations, as single imputations fail to account for variability in the imputed values, leading to underestimated standard errors and overly optimistic test statistics.

4.4.4.4 Multivariate Imputation via Chained Equations (MICE)

Multivariate imputation via chained equations (MICE) is a widely used algorithm for implementing MI, particularly in datasets with mixed variable types. The steps of MICE include

1. **Initialization**: Replace missing values with initial guesses, such as the mean or median of the observed data.
2. **Iterative Imputation**:
    - For each variable with missing values, regress it on all other variables (or a subset of relevant predictors).
    - Use the regression model to predict missing values, adding a random error term drawn from the residual distribution.
3. **Convergence**: Repeat the imputation process until parameter estimates stabilize.

MICE offers flexibility in specifying regression models for each variable, accommodating continuous, categorical, and binary data.

4.4.4.5 Bayesian Ridge Regression for Imputation

Bayesian ridge regression is an advanced imputation method that incorporates prior distributions on the regression coefficients, making it particularly useful when

- Predictors are highly correlated.
- Sample sizes are small.
- Missingness is substantial.

This method treats the regression coefficients as random variables and samples from their posterior distribution, introducing variability into the imputation process.

Bayesian ridge regression is more computationally intensive than simpler methods like MICE but offers greater robustness.

4.4.4.6 Combining Results from MI (Rubin's Rules)

Once multiple datasets are imputed and analyzed, Rubin's rules are used to combine the results. The goal is to properly account for the uncertainty introduced by missing data. For a parameter of interest θ

1. **Estimate Combination**:

$$\bar{\theta} = \frac{1}{M} \sum_{m=1}^{M} \theta_m$$

   where θ_m is the estimate from the mth imputed dataset and M is the number of imputations.
2. **Variance Combination**:

$$T = \bar{W} + \left(1 + \frac{1}{M}\right) B$$

   where

   - $\bar{W}$ is the average within-imputation variance.
   - B is the between-imputation variance:

$$B = \frac{1}{M-1} \sum_{m=1}^{M} (\theta_m - \bar{\theta})^2$$

These formulas adjust the final variance to reflect uncertainty both within and across imputations.

4.4.4.6.1 Challenges

1. **Stochastic Variability**: MI results vary slightly between runs due to their reliance on random draws. To ensure reproducibility, always set a random seed.
2. **Convergence**: Iterative algorithms like MICE may struggle to converge, especially with high proportions of missing data.
3. **Assumption of MAR**: MI assumes that missing data is MAR. If data is missing not at random (MNAR), MI can produce biased results.

4.4.4.6.2 Best Practices

1. **Algorithm Selection**:
   - Use multiple imputation by chained equations (MICE) for datasets with mixed data types or when relationships between variables are complex.
   - Apply Bayesian ridge regression for small datasets or when predictors are highly correlated.
2. **Diagnostic Checks**:
   - Evaluate the quality of imputations and assess convergence using trace plots or diagnostic statistics to ensure reliable results.
3. **Data Transformations**:
   - For skewed or proportion data, consider applying log or logit transformations before imputation and inverse-transforming afterward to preserve the data's original scale.
4. **Handling Nonlinear Relationships**:
   - For nonlinear relationships or interactions, stratify imputations by the levels of the categorical variable involved to ensure accurate estimates.
5. **Number of Imputations**:
   - Use at least 20 imputations for small datasets or datasets with high missingness. This ensures robust and reliable results in downstream analyses.
6. **Avoid Rounding Imputations for Dummy Variables**:
   - Many imputation methods (e.g., Markov chain Monte Carlo [MCMC]) assume normality, even for dummy variables. While it was historically recommended to round imputed values to 0 or 1 for binary variables, research shows that this introduces bias in parameter estimates. Instead, leave imputed values as fractional, even though this may seem counterintuitive.
7. **Do Not Transform Skewed Variables Before Imputation**:
   - Transforming variables to meet normality assumptions before imputation can distort their relationships with other variables, leading to biased imputations and possibly introducing outliers. It is better to directly impute the skewed variable.
8. **Use More Imputations**:
   - Traditional advice suggests five to ten imputations are sufficient for unbiased estimates, but inconsistencies may arise in repeated analyses. [@Bodner_2008] suggests using a number of imputations equal to the percentage of missing data. As additional imputations generally do not significantly increase the computational workload, using more imputations is a prudent choice.

9. **Create Multiplicative Terms Before Imputation**:
    - When your model includes interaction or quadratic terms, generate these terms before imputing missing values. Imputing first and then generating these terms can introduce bias in their regression parameters, as highlighted by (Von Hippel, 2009).

4.5 Evaluation of Imputation Methods

4.5.1 Statistical Metrics for Assessing Imputation Quality

To evaluate the quality of imputed data, several statistical metrics are commonly used. These metrics compare the imputed values to the observed values (in cases where missingness is simulated or artificially introduced) or assess the overall impact of imputation on the quality of subsequent analyses. Key metrics include

- **Root Mean Squared Error (RMSE):** RMSE is calculated as

$$\text{RMSE} = \sqrt{\frac{1}{n}\sum_{i=1}^{n}(y_i - \hat{y}_i)^2}$$

  It measures the average magnitude of errors between the true and imputed values. Lower RMSE indicates better imputation accuracy.
- **Mean Absolute Error (MAE):** MAE measures the average absolute difference between observed and imputed values:

$$\text{MAE} = \frac{1}{n}\sum_{i=1}^{n}|y_i - \hat{y}_i|$$

  MAE provides a straightforward assessment of imputation performance and is less sensitive to outliers than RMSE.
- **Log-Likelihood and Deviance Measures:** Log-likelihood can be used to evaluate how well the imputation model fits the data. Deviance measures, based on likelihood comparisons, assess the relative goodness of fit of imputation models. These are particularly useful in evaluating methods like maximum likelihood estimation.

In practice, these metrics may be combined with graphical methods such as density plots and residual analysis to understand imputation performance more thoroughly.

4.5.2 Bias-Variance Tradeoff in Imputation

Imputation methods must balance bias and variance to achieve reliable results. Simpler methods, such as mean or mode imputation, often lead to biased parameter estimates, particularly if the missingness mechanism is nonrandom. These methods underestimate variability, shrinking standard errors and potentially leading to overconfidence in statistical inferences.

Conversely, advanced methods like Multiple Imputation or full information maximum likelihood (FIML) typically yield unbiased estimates with appropriately calibrated variances. However, these methods may increase computational complexity and require careful tuning of assumptions and parameters.

The tradeoff is summarized as follows:

- **High Bias, Low Variance:** Simpler methods (e.g., single imputation, mean imputation).
- **Low Bias, Moderate Variance:** Advanced methods (e.g., MI, FIML, Bayesian methods).

4.5.3 Sensitivity Analysis

Sensitivity analysis is crucial to assess the robustness of imputation methods under varying assumptions. Two primary areas of focus include

- **Assessing Robustness to Assumptions:** Imputation models often rely on assumptions about the missingness mechanism (see Definition and Classification of Missing Data). Sensitivity analysis involves testing how results vary when these assumptions are slightly relaxed or modified.
- **Impact on Downstream Analysis:** The quality of imputation should also be evaluated based on its influence on downstream analyses (Objectives of Imputation). For instance
  - Does the imputation affect causal inference in regression models?
  - Are the conclusions from hypothesis testing or predictive modeling robust to the imputation technique?

4.5.4 Validation Using Simulated Data and Real-World Case Studies

Validation of imputation methods is best performed through a combination of simulated data and real-world examples:

1. **Simulated Data:** – Create datasets with known missingness patterns and true values. – Apply various imputation methods and assess their performance using RMSE, MAE, and other metrics.
2. **Real-World Case Studies:**

   - Use datasets from actual studies, such as customer transaction data in marketing or financial data in portfolio analysis.
   - Evaluate the impact of imputation on actionable outcomes (e.g., market segmentation, risk assessment).

Combining these approaches ensures that methods generalize well across different contexts and data structures.

4.6 Criteria for Choosing an Effective Approach

Choosing an appropriate imputation method depends on the following criteria:

1. **Unbiased Parameter Estimates:** The technique should ensure that key estimates, such as means, variances, and regression coefficients, are unbiased, particularly in the presence of MAR or MNAR data.
2. **Adequate Power:** The method should preserve statistical power, enabling robust hypothesis testing and model estimation. This ensures that important effects are not missed due to inflated Type II error.
3. **Accurate Standard Errors:** Accurate estimation of standard errors is critical for reliable p-values and confidence intervals. Methods like single imputation often underestimate standard errors, leading to overconfident conclusions.

Preferred Methods: Multiple Imputation and Full Information Maximum Likelihood

Multiple Imputation (MI)

- MI replaces missing values with multiple plausible values drawn from a predictive distribution. It generates multiple complete datasets, analyzes each dataset, and combines the results.
- Pros: Handles uncertainty well, provides valid standard errors, and is robust under MAR.
- Cons: Computationally intensive, sensitive to model mis-specification.

Full Information Maximum Likelihood (FIML)

- FIML uses all available data to estimate parameters directly, avoiding the need to impute missing values explicitly.
- Pros: Efficient, unbiased under MAR, and computationally elegant.
- Cons: Requires correctly specified models and may be sensitive to MNAR data.

Methods to Avoid

- **Single Imputation (e.g., Mean, Mode):**
  - Leads to biased estimates and underestimates variability.
- **Listwise Deletion:**
  - Discards rows with missing data, reducing sample size and potentially introducing bias if the data is not MCAR.

Practical Considerations

- Computational efficiency and ease of implementation.
- Compatibility with downstream analysis methods.
- Alignment with the data's missingness mechanism.

4.7 Challenges and Ethical Considerations

4.7.1 Challenges in High-Dimensional Data

High-dimensional data, where the number of variables exceeds the number of observations, poses unique challenges for missing data analysis.

- **Curse of Dimensionality**: Standard imputation methods, such as mean or regression imputation, struggle with high-dimensional spaces due to sparse data distribution.
- **Regularized Methods**: Techniques such as LASSO, ridge regression, and elastic net can be used to handle high-dimensional missing data. These methods shrink model coefficients, preventing overfitting.
- **Matrix Factorization**: Methods like principal component analysis (PCA) or singular value decomposition (SVD) are often adapted to impute missing values in high-dimensional datasets by reducing the dimensionality first.

4.7.2 Missing Data in Big Data Contexts

The advent of big data introduces additional complexities for missing data handling, including computational scalability and storage constraints.

4.7.2.1 Distributed Imputation Techniques

- **MapReduce Frameworks**: Algorithms like K-nearest neighbor (KNN) imputation or multiple imputation can be adapted for distributed environments using MapReduce or similar frameworks.

- **Federated Learning**: In scenarios where data is distributed across multiple locations (e.g., in healthcare or banking), federated learning allows imputation without centralizing data, ensuring privacy.

4.7.2.2 Cloud-Based Implementations

- **Cloud-Native Algorithms**: Cloud platforms like AWS, Google Cloud, and Azure provide scalable solutions for implementing advanced imputation algorithms on large datasets.
- **AutoML Integration**: Automated machine learning (AutoML) pipelines often include missing data handling as a preprocessing step, leveraging cloud-based computational power.
- **Real-Time Imputation**: In e-commerce, cloud-based solutions enable real-time imputation for recommendation systems or fraud detection, ensuring seamless user experiences.

4.7.3 Ethical Concerns

4.7.3.1 Bias Amplification

- **Introduction of Systematic Bias**: Imputation methods can inadvertently reinforce existing biases. For example, imputing salary data based on demographic variables may propagate societal inequalities.
- **Business Implications**: In credit scoring, biased imputation of missing financial data can lead to unfair credit decisions, disproportionately affecting marginalized groups.
- **Mitigation Strategies**: Techniques such as fairness-aware machine learning and bias auditing can help identify and reduce bias introduced during imputation.

4.7.3.2 Transparency in Reporting Imputation Decisions

- **Reproducibility and Documentation**: Transparent reporting of imputation methods and assumptions is essential for reproducibility. Analysts should provide clear documentation of the imputation pipeline.
- **Stakeholder Communication**: In business settings, communicating imputation decisions to stakeholders ensures informed decision-making and trust in the results.
- **Ethical Frameworks**: Ethical guidelines, such as those provided by the European Union's GDPR or industry-specific codes, emphasize the importance of transparency in data handling.

4.8 Emerging Trends in Missing Data Handling

4.8.1 Advances in Neural Network Approaches

Neural networks have transformed the landscape of missing data imputation, offering flexible, scalable, and powerful solutions that go beyond traditional methods.

4.8.1.1 Variational Autoencoders (VAEs)

- **Overview**: Variational autoencoders (VAEs) are generative models that encode data into a latent space and reconstruct it, filling in missing values during reconstruction.
- **Advantages**:
  - Handle complex, nonlinear relationships between variables.
  - Scalable to high-dimensional datasets.
  - Generate probabilistic imputations, reflecting uncertainty.
- **Applications**:
  - In marketing, VAEs can impute missing customer behavior data while accounting for seasonal and demographic variations.
  - In finance, VAEs assist in imputing missing stock price data by modeling dependencies among assets.

4.8.1.2 GANs for Missing Data

- **Generative Adversarial Networks (GANs)**: GANs consist of a generator and a discriminator, with the generator imputing missing data and the discriminator evaluating its quality.
- **Advantages**:
  - Preserve data distributions and avoid over-smoothing.
  - Suitable for imputation in datasets with complex patterns or multimodal distributions.
- **Applications**:
  - In healthcare, GANs have been used to impute missing patient records while preserving patient privacy and data integrity.
  - In retail, GANs can model missing sales data to predict trends and optimize inventory.

4.8.2 Integration with Reinforcement Learning

Reinforcement learning (RL) is increasingly being integrated into missing data strategies, particularly in dynamic or sequential data environments.

- **Markov Decision Processes (MDPs)**: RL models missing data handling as an MDP, where actions (imputations) are optimized based on rewards (accuracy of predictions or decisions).
- **Active Imputation**:
  - RL can be used to actively query for missing data points, prioritizing those with the highest impact on downstream tasks.
  - Example: In customer churn prediction, RL can optimize the imputation of high-value customer records.
- **Applications**:
  - Financial forecasting: RL models are used to impute missing transaction data dynamically, optimizing portfolio decisions.
  - Smart cities: RL-based models handle missing sensor data to enhance real-time decision-making in traffic management.

4.8.3 Synthetic Data Generation for Missing Data

Synthetic data generation has emerged as a robust solution to address missing data, providing flexibility and privacy.

- **Data Augmentation**: Synthetic data is generated to augment datasets with missing values, reducing biases introduced by imputation.
- **Techniques**:
  - **Simulations**: Monte Carlo simulations create plausible data points based on observed distributions.
  - **Generative Models**: GANs and VAEs generate realistic synthetic data that aligns with existing patterns.
- **Applications**:
  - In fraud detection, synthetic datasets balance the impact of missing values on anomaly detection.
  - In insurance, synthetic data supports pricing models by filling in gaps from incomplete policyholder records.

4.8.4 *Federated Learning and Privacy-Preserving Imputation*

Federated learning has gained traction as a method for collaborative analysis while preserving data privacy.

- **Federated Imputation**:
  - Distributed imputation algorithms operate on decentralized data, ensuring that sensitive information remains local.
  - Example: Hospitals collaboratively impute missing patient data without sharing individual records.
- **Privacy Mechanisms**:
  - Differential privacy adds noise to imputed values, protecting individual-level data.
  - Homomorphic encryption allows computations on encrypted data, ensuring privacy throughout the imputation process.
- **Applications**:
  - Healthcare: Federated learning imputes missing diagnostic data across clinics.
  - Banking: Collaborative imputation of financial transaction data supports risk modeling while adhering to regulations.

4.8.5 *Imputation in Streaming and Online Data Environments*

The increasing use of streaming data in business and technology requires real-time imputation methods to ensure uninterrupted analysis.

- **Challenges**:
  - Imputation must occur dynamically as data streams in.
  - Low latency and high accuracy are essential to maintain real-time decision-making.
- **Techniques**:
  - **Online Learning Algorithms**: Update imputation models incrementally as new data arrives.
  - **Sliding Window Methods**: Use recent data to estimate and impute missing values in real time.
- **Applications**:
  - IoT devices: Imputation in sensor networks for smart homes or industrial monitoring ensures continuous operation despite data transmission issues.
  - Financial markets: Streaming imputation models predict and fill gaps in real-time stock price feeds to inform trading algorithms (Table 4.8).

Table 4.8 Comparison of R packages for multiple imputation of missing data

Package	Algorithm	Cont var	Cate var	Diagnostics	Complexity handling	Best use case	Limitations
missForest	Random forest	Yes	Yes	Out-of-bag error (NRMSE, PFC)	Handles complex interactions	Mixed data types with complex interactions	May overfit with small datasets
Hmisc	Additive regression, bootstrap, predictive mean matching	Yes	Yes	R^2 for imputed values	Basic to intermediate complexity	Simple datasets with low complexity	Limited to simple imputation methods
mi	Bayesian regression	Yes	Yes	Graphical diagnos-tics,convergence	Detects issues like collinearity	Datasets with irregularities	Computationally intensive for large data
MICE	Multivariate imputation via chained equations	Yes	Yes	Density plots, pooling of results	Handles variable interactions	General-purpose imputation for MAR data	Requires proper method selection for variable types
Amelia	Bootstrap-based expectation maximization (EMB)	Yes	Limited (requires normality)	Diagnostics supported	Works well with large/time series data	Time series or datasets approximating MVN	Assumes MVN, requires transformations for non-MVN

4.9 Application of Imputation

This section demonstrates how to visualize missing data and handle it using different imputation techniques.

4.9.1 Visualizing Missing Data

Visualizing missing data is an essential first step in understanding the patterns and extent of missingness in your dataset (Figs. 4.7 and 4.8).

```
library(visdat)
library(naniar)
library(ggplot2)
```

```
# Visualizing missing data
vis_miss(airquality)
```

```
# Missingness patterns using an upset plot
gg_miss_upset(airquality)
```

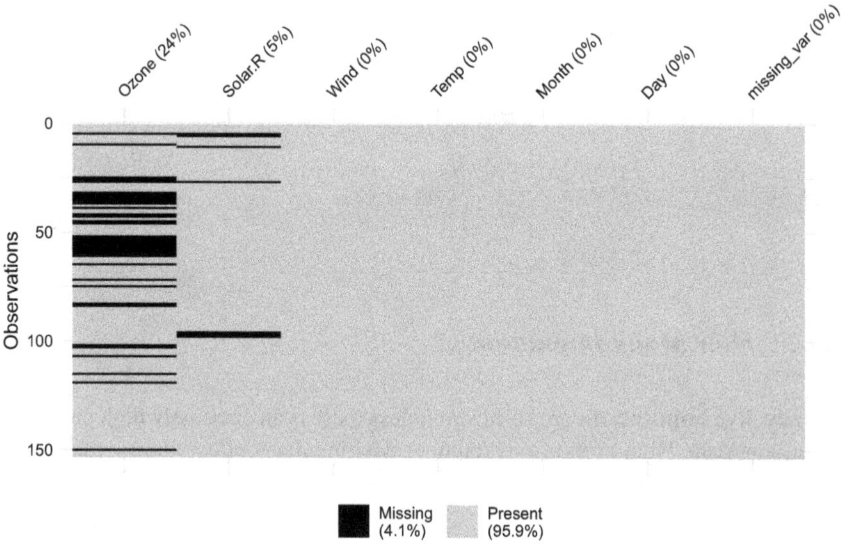

Fig. 4.7 Heatmap

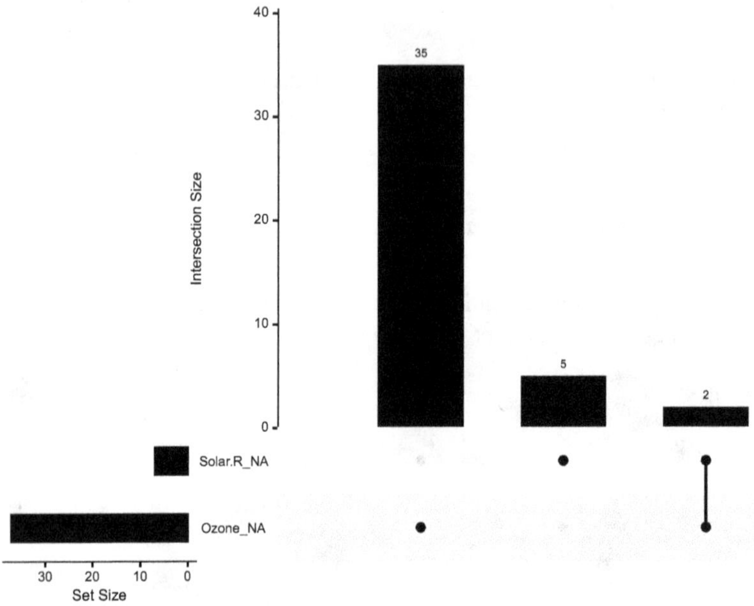

Fig. 4.8 Missingness patterns

```
# Scatter plot of missing data with faceting
ggplot(airquality, aes(x, y)) +
  geom_miss_point() +
  facet_wrap(~ group)

# Missing values by variable
gg_miss_var(data, facet = group)

# Missingness in relation to factors
gg_miss_fct(x = variable1, fct = variable2)
```

4.9.2 How Many Imputations?

Usually, five imputations are sufficient unless there is an extremely high proportion of missing data. High proportions require revisiting data collection processes.

Rubin's Rule for Relative Efficiency

According to Rubin, the relative efficiency of an estimate based on m imputations (relative to infinite imputations) is given by

$$\text{Relative Efficiency} = \left(1 + \frac{\lambda}{m}\right)^{-1}$$

where λ is the rate of missing data.

For example, with 50% missing data ($\lambda = 0.5$), the standard deviation of an estimate based on five imputations is only about 5% wider than that from infinite imputations:

$$\sqrt{1 + \frac{0.5}{5}} = 1.049$$

4.9.3 Generating Missing Data for Demonstration

```
library(missForest)
#>
#> Attaching package: 'missForest'
#> The following object is masked from 'package:VIM':
#>
#>     nrmse

# Load the data
data <- iris

# Generate 10% missing values at random
set.seed(1)
iris.mis <- prodNA(iris, noNA = 0.1)

# Remove categorical variables for numeric imputation
iris.mis.cat <- iris.mis
iris.mis <- subset(iris.mis, select = -c(Species))
```

4.9.4 Imputation with Mean, Median, and Mode

Mean, median, or mode imputation is a simple yet commonly used technique.

```
# Imputation for the entire dataset
e1071::impute(iris.mis, what = "mean")        # Replace with mean
e1071::impute(iris.mis, what = "median")      # Replace with median
```

```
# Imputation by variable
Hmisc::impute(iris.mis$Sepal.Length, mean)    # Replace with mean
Hmisc::impute(iris.mis$Sepal.Length, median)  # Replace with median
Hmisc::impute(iris.mis$Sepal.Length, 0)       # Replace with a specific
↪  value
```

Checking Accuracy
Accuracy can be checked by comparing predictions with actual values.

```
# Example data
actuals <- iris$Sepal.Width[is.na(iris.mis$Sepal.Width)]
predicteds <- rep(mean(iris$Sepal.Width, na.rm = TRUE), length(actuals))

# Using MLmetrics package
library(MLmetrics)
#>
#> Attaching package: 'MLmetrics'
#> The following object is masked from 'package:base':
#>
#>     Recall

MAE(predicteds, actuals)
#> [1] 0.2870303
MSE(predicteds, actuals)
#> [1] 0.1301598
RMSE(predicteds, actuals)
#> [1] 0.3607767
```

4.9.5 K-Nearest Neighbor (KNN) Imputation

KNN is a more sophisticated method, leveraging similar observations to fill in missing values.

```
library(DMwR2)
#> Registered S3 method overwritten by 'quantmod':
#>   method            from
#>   as.zoo.data.frame zoo
#>
#> Attaching package: 'DMwR2'
#> The following object is masked from 'package:VIM':
#>
#>     kNN
knnOutput <-
  DMwR2::knnImputation(data = iris.mis.cat,
                       meth = "median")
anyNA(knnOutput)  # Check for remaining missing values
#> [1] FALSE
```

```
actuals <- iris$Sepal.Width[is.na(iris.mis$Sepal.Width)]
predicteds <- knnOutput[is.na(iris.mis$Sepal.Width), "Sepal.Width"]
# Using MLmetrics package
library(MLmetrics)

MAE(predicteds, actuals)
#> [1] 0.2318182
MSE(predicteds, actuals)
#> [1] 0.1038636
RMSE(predicteds, actuals)
#> [1] 0.3222788
```

KNN typically improves upon mean or median imputation in terms of predictive accuracy.

4.9.6 Imputation with Decision Trees (rpart)

Decision trees, such as those implemented in `rpart`, are effective for both numeric and categorical variables.

```
library(rpart)

# Imputation for a categorical variable
class_mod <- rpart(
  Species ~ . - Sepal.Length,
  data = iris.mis.cat[!is.na(iris.mis.cat$Species), ],
  method = "class",
  na.action = na.omit
)

# Imputation for a numeric variable
anova_mod <- rpart(
  Sepal.Width ~ . - Sepal.Length,
  data = iris.mis[!is.na(iris.mis$Sepal.Width), ],
  method = "anova",
  na.action = na.omit
)

# Predictions
species_pred <- predict(class_mod,
↪  iris.mis.cat[is.na(iris.mis.cat$Species), ])
width_pred <- predict(anova_mod, iris.mis[is.na(iris.mis$Sepal.Width), ])
```

4.9.7 MICE (Multivariate Imputation via Chained Equations)

MICE assumes that the data are **Missing at Random (MAR)**. It imputes data for each variable by specifying an imputation model tailored to the variable type.

4.9.7.1 How MICE Works

For a dataset with variables $X_1, X_2, \ldots, X_k$

- If X_1 has missing data, it is regressed on the other variables.
- This process is repeated for all variables with missing data, using the previously predicted values as needed.

  By default

- **Continuous variables** use linear regression.
- **Categorical variables** use logistic regression.

4.9.7.2 Methods Available in MICE

- `pmm` **(Predictive Mean Matching)**: For numeric variables.
- `logreg` **(Logistic Regression)**: For binary variables (two levels).
- `polyreg` **(Bayesian Polytomous Regression)**: For factor variables (two levels).
- **Proportional Odds Model**: For ordered factor variables (two levels).

  To check for missing data, refer to Figs. 4.9 and 4.10

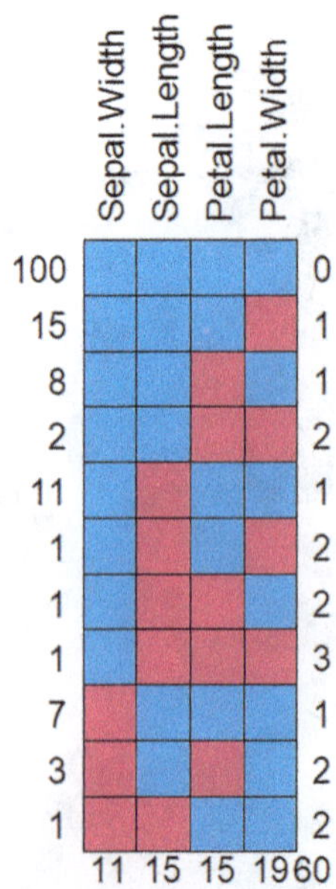

Fig. 4.9 Heatmap

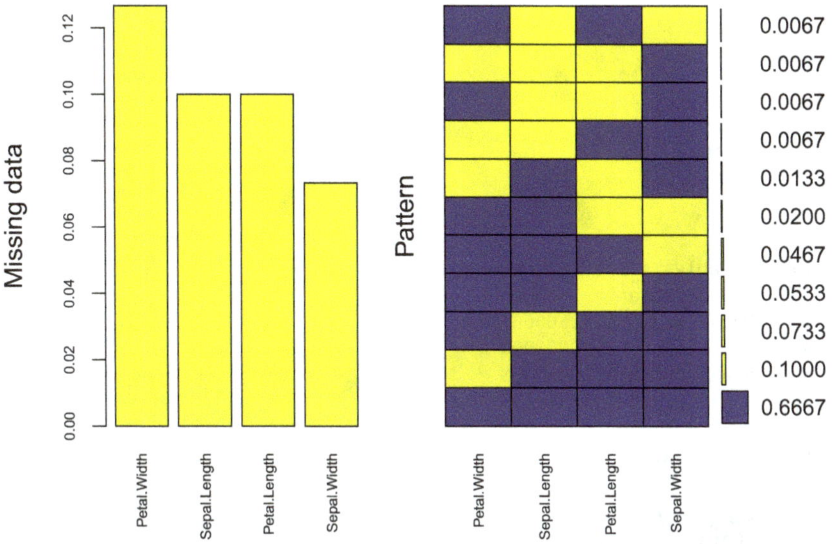

Fig. 4.10 Barplot and matrix plot

```r
# Load packages
library(mice)
library(VIM)
```

```r
# Check missing values pattern
md.pattern(iris.mis)
```

```
#>     Sepal.Width Sepal.Length Petal.Length Petal.Width
#> 100           1            1            1           1  0
#> 15            1            1            1           0  1
#> 8             1            1            0           1  1
#> 2             1            1            0           0  2
#> 11            1            0            1           1  1
#> 1             1            0            1           0  2
#> 1             1            0            0           1  2
#> 1             1            0            0           0  3
#> 7             0            1            1           1  1
#> 3             0            1            0           1  2
#> 1             0            0            1           1  2
#>              11           15           15          19 60
```

```r
# Plot missing values
aggr(
  iris.mis,
  col = c('navyblue', 'yellow'),
  numbers = TRUE,
```

```
  sortVars = TRUE,
  labels = names(iris.mis),
  cex.axis = 0.7,
  gap = 3,
  ylab = c("Missing data", "Pattern")
)
```

```
#>
#>  Variables sorted by number of missings:
#>      Variable      Count
#>   Petal.Width 0.12666667
#>  Sepal.Length 0.10000000
#>  Petal.Length 0.10000000
#>   Sepal.Width 0.07333333
```

Imputing Data

```
# Perform multiple imputation using MICE
imputed_Data <- mice(
  iris.mis,
  m = 5,               # Number of imputed datasets
  maxit = 10,          # Number of iterations
  method = 'pmm',      # Imputation method
  seed = 500           # Random seed for reproducibility
)
```

Evaluating Imputed Data (Fig. 4.11)

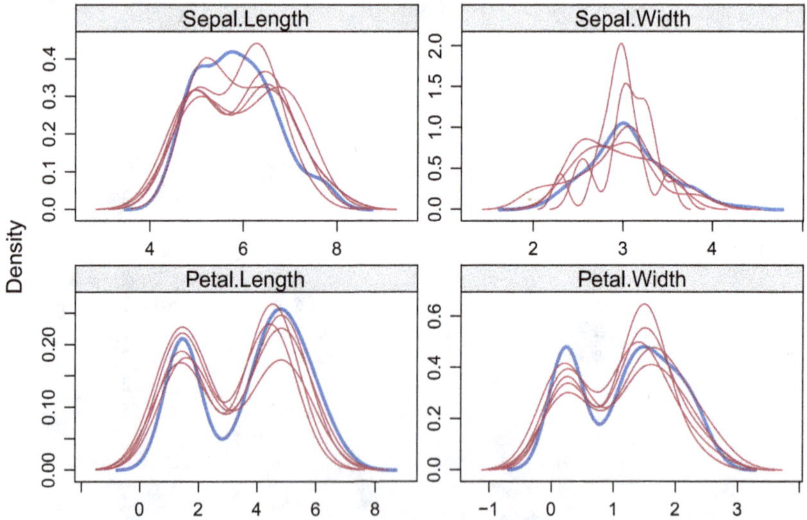

Fig. 4.11 Density plot

```
# Summary of imputed data
summary(imputed_Data)
#> Class: mids
#> Number of multiple imputations:  5
#> Imputation methods:
#> Sepal.Length  Sepal.Width Petal.Length  Petal.Width
#>        "pmm"        "pmm"        "pmm"        "pmm"
#> PredictorMatrix:
#>              Sepal.Length Sepal.Width Petal.Length
#> Sepal.Length            0           1            1
#> Sepal.Width             1           0            1
#> Petal.Length            1           1            0
#> Petal.Width             1           1            1
#>              Petal.Width
#> Sepal.Length           1
#> Sepal.Width            1
#> Petal.Length           1
#> Petal.Width            0
```

```
# Density plot: compare imputed values (red) with observed values (blue)
densityplot(imputed_Data)
```

Accessing and Using Imputed Data

```
# Access the complete datasets
completeData1 <- mice::complete(imputed_Data, 1)  # First imputed dataset
completeData2 <- mice::complete(imputed_Data, 2)  # Second imputed
↪  dataset
```

Regression Model with Imputed Dataset

```
# Fit a regression model using imputed datasets
fit <-
  with(data = imputed_Data,
       exp = lm(Sepal.Width ~ Sepal.Length + Petal.Width))

# Combine results of all 5 models
combine <- mice::pool(fit)
summary(combine)
#>           term   estimate  std.error statistic        df
#> 1  (Intercept)  1.9054698 0.33454626  5.695684 105.12438
#> 2 Sepal.Length  0.2936285 0.07011405  4.187870  88.69066
#> 3  Petal.Width -0.4742921 0.08138313 -5.827892  46.94941
#>        p.value
#> 1 1.127064e-07
#> 2 6.625536e-05
#> 3 4.915270e-07
```

4.9.8 *Amelia*

Amelia uses a **bootstrap-based expectation-maximization with bootstrapping (EMB) algorithm** for imputation, making it faster and suitable for cross-sectional and time series data.

4.9.8.1 Assumptions

- All variables must follow a **multivariate normal distribution** (MVN). Transformations may be required for non-normal data.
- Data must be **Missing at Random (MAR)**.

4.9.8.2 Comparison: Amelia vs. MICE

- **MICE** imputes on a variable-by-variable basis using separate models.
- **Amelia** uses a joint modeling approach based on MVN.
- **MICE** handles multiple data types, while **Amelia** requires variables to approximate normality.

4.9.8.3 Imputation with Amelia

```
library(Amelia)
data("iris")

# Seed 10% missing values
set.seed(123)
iris.mis <- prodNA(iris, noNA = 0.1)

# Specify columns and run Amelia
amelia_fit <- amelia(
  iris.mis,
  m = 5,                          # Number of imputations
  parallel = "multicore",         # Use multicore processing
  noms = "Species"                # Nominal variables
)
#> -- Imputation 1 --
#>
#>   1  2  3  4  5  6  7
#>
#> -- Imputation 2 --
#>
#>   1  2  3  4  5
#>
#> -- Imputation 3 --
```

```
#>
#>   1  2  3  4  5
#>
#> -- Imputation 4 --
#>
#>   1  2  3  4  5  6
#>
#> -- Imputation 5 --
#>
#>   1  2  3  4  5  6  7  8  9 10

# Access imputed outputs
# amelia_fit$imputations[[1]]
```

Amelia's workflow includes bootstrapping multiple imputations to generate robust estimates of means and variances. This process ensures flexibility and speed for large datasets.

4.9.9 missForest

The `missForest` package provides a robust nonparametric imputation method using the random forest algorithm. It is versatile, handling both continuous and categorical variables without requiring assumptions about the underlying functional forms.

Key Features of missForest:

1. **Nonparametric**: No assumptions about the functional form.
2. **Variable-Specific Models**: Builds a random forest model for each variable to impute missing values.
3. **Error Estimates**: Provides out-of-bag (OOB) imputation error estimates.
    - **NRMSE** (normalized root mean squared error) for continuous variables.
    - **PFC** (proportion of falsely classified) for categorical variables.
4. **High Control**: Offers customizable parameters like `mtry` and `ntree`.

```
library(missForest)

# Impute missing values using default parameters
iris.imp <- missForest(iris.mis)

# Check imputed values
# View the imputed dataset
# iris.imp$ximp
```

```
# Out-of-bag error estimates
iris.imp$OOBerror
#>      NRMSE        PFC
#> 0.14004144 0.02877698

# Compare imputed data with original data to calculate error
iris.err <- mixError(iris.imp$ximp, iris.mis, iris)
iris.err
#>      NRMSE        PFC
#> 0.14420833 0.09090909
```

4.9.10 Hmisc

The Hmisc package provides a suite of tools for imputing missing data, offering both simple methods (like mean or median imputation) and more advanced approaches like aregImpute.

Features of Hmisc:

1. **impute()**: Simple imputation using user-defined methods like mean, median, or a random value.
2. **aregImpute()**:
   - Combines additive regression, bootstrapping, and predictive mean matching.
   - Handles continuous and categorical variables.
   - Automatically recognizes variable types and applies appropriate methods.

Assumptions:

- Linearity in the variables being predicted.
- Fisher's optimum scoring is used for categorical variable prediction.

```
library(Hmisc)

# Impute using mean
iris.mis$imputed_SepalLength <- with(iris.mis, impute(Sepal.Length,
↪  mean))

# Impute using random value
iris.mis$imputed_SepalLength2 <- with(iris.mis, impute(Sepal.Length,
↪  'random'))

# Advanced imputation using aregImpute
impute_arg <- aregImpute(
  ~ Sepal.Length + Sepal.Width + Petal.Length + Petal.Width + Species,
  data = iris.mis,
  n.impute = 5
)
```

```
#> Iteration 1
Iteration 2
Iteration 3
Iteration 4
Iteration 5
Iteration 6
Iteration 7
Iteration 8

# Check R-squared values for predicted missing values
impute_arg
#>
#> Multiple Imputation using Bootstrap and PMM
#>
#> aregImpute(formula = ~Sepal.Length + Sepal.Width + Petal.Length +
#>     Petal.Width + Species, data = iris.mis, n.impute = 5)
#>
#> n: 150   p: 5    Imputations: 5      nk: 3
#>
#> Number of NAs:
#> Sepal.Length  Sepal.Width Petal.Length  Petal.Width
#>           17           19           12           16
#>      Species
#>           11
#>
#>              type d.f.
#> Sepal.Length    s    2
#> Sepal.Width     s    2
#> Petal.Length    s    2
#> Petal.Width     s    2
#> Species         c    2
#>
#> Transformation of Target Variables Forced to be Linear
#>
#> R-squares for Predicting Non-Missing Values for Each Variable
#> Using Last Imputations of Predictors
#> Sepal.Length  Sepal.Width Petal.Length  Petal.Width
#>        0.895        0.536        0.987        0.967
#>      Species
#>        0.984

# Access imputed values for Sepal.Length
impute_arg$imputed$Sepal.Length
#>     [,1] [,2] [,3] [,4] [,5]
#> 13   4.4  4.9  4.9  5.0  4.9
#> 14   4.8  4.4  5.0  4.5  4.5
#> 23   4.8  5.1  5.1  5.1  4.8
#> 26   5.0  4.8  4.9  4.9  5.0
#> 34   5.0  5.8  6.0  5.7  5.8
#> 39   4.4  4.9  5.0  4.5  4.6
#> 41   5.2  5.1  4.8  5.0  4.8
#> 69   5.8  6.0  6.3  6.0  6.1
#> 72   5.6  5.7  5.7  5.8  6.1
```

```
#> 89   6.1  5.7  5.7  5.6  6.9
#> 90   5.5  6.2  5.2  6.0  5.8
#> 91   5.7  6.9  6.0  6.4  6.4
#> 116  5.9  6.8  6.4  6.6  6.9
#> 118  7.9  7.9  7.9  7.9  7.9
#> 135  6.7  6.7  6.7  6.9  6.7
#> 141  7.0  6.3  5.9  6.7  7.0
#> 143  5.7  6.7  5.8  6.3  5.4
```

Note: While `missForest` often outperforms `Hmisc` in terms of accuracy, the latter is useful for datasets with simpler requirements.

4.9.11 mi

The `mi` package is a powerful tool for imputation, using Bayesian methods and providing rich diagnostics for model evaluation and convergence.

Features of mi

1. **Graphical Diagnostics**: Visualize imputation models and convergence.
2. **Bayesian Regression**: Handles separation and other issues in data.
3. **Irregularity Detection**: Automatically detects issues like high collinearity.
4. **Noise Addition**: Adds noise to address additive constraints.

```
library(mi)
#> Loading required package: Matrix
#> Loading required package: stats4
#> mi (Version 1.1, packaged: 2022-06-05 05:31:15 UTC; ben)
#> mi  Copyright (C) 2008, 2009, 2010, 2011, 2012, 2013, 2014, 2015
↪  Trustees of Columbia University
#> This program comes with ABSOLUTELY NO WARRANTY.
#> This is free software, and you are welcome to redistribute it
#> under the General Public License version 2 or later.
#> Execute RShowDoc('COPYING') for details.
#>
#> Attaching package: 'mi'
#> The following objects are masked from 'package:mice':
#>
#>     complete, pool

# Perform imputation using mi
mi_data <- mi(iris.mis, seed = 1)

# Summary of the imputation process
summary(mi_data)
#> $Sepal.Length
#> $Sepal.Length$is_missing
#> missing
#> FALSE  TRUE
```

```
#>   133    17
#>
#> $Sepal.Length$imputed
#>       Min.    1st Qu.     Median       Mean    3rd Qu.       Max.
#> -0.159508 -0.046872 -0.004186 -0.004918  0.051321  0.178300
#>
#> $Sepal.Length$observed
#>     Min.  1st Qu.   Median     Mean  3rd Qu.     Max.
#> -0.90110 -0.47329 -0.04549  0.00000  0.32120  1.23792
#>
#>
#> $Sepal.Width
#> $Sepal.Width$is_missing
#> missing
#> FALSE  TRUE
#>   131    19
#>
#> $Sepal.Width$imputed
#>     Min.  1st Qu.   Median     Mean  3rd Qu.     Max.
#> -1.67238 -0.38702 -0.07632 -0.10053  0.20766  1.12249
#>
#> $Sepal.Width$observed
#>     Min.  1st Qu.   Median     Mean  3rd Qu.     Max.
#> -1.01272 -0.30642 -0.07099  0.00000  0.39988  1.34161
#>
#>
#> $Petal.Length
#> $Petal.Length$is_missing
#> missing
#> FALSE  TRUE
#>   138    12
#>
#> $Petal.Length$imputed
#>       Min.    1st Qu.     Median       Mean    3rd Qu.       Max.
#> -0.865076 -0.602627  0.178362 -0.001594  0.410557  0.701006
#>
#> $Petal.Length$observed
#>    Min. 1st Qu.  Median    Mean 3rd Qu.    Max.
#> -0.7797 -0.6088  0.1459  0.0000  0.3880  0.9006
#>
#>
#> $Petal.Width
#> $Petal.Width$is_missing
#> missing
#> FALSE  TRUE
#>   134    16
#>
#> $Petal.Width$imputed
#>       Min.    1st Qu.     Median       Mean    3rd Qu.       Max.
#> -0.761411 -0.008254  0.205610  0.163761  0.509972  0.899349
#>
#> $Petal.Width$observed
#>     Min.  1st Qu.   Median     Mean  3rd Qu.     Max.
```

```
#> -0.69624 -0.56602  0.08503  0.00000  0.41055  0.86629
#>
#>
#> $Species
#> $Species$crosstab
#>
#>              observed imputed
#>   setosa          180      25
#>   versicolor      192       2
#>   virginica       184      17
#>
#>
#> $imputed_SepalLength
#> $imputed_SepalLength$is_missing
#> [1] "all values observed"
#>
#> $imputed_SepalLength$observed
#>    Min. 1st Qu.  Median    Mean 3rd Qu.    Max.
#> -0.9574 -0.4379  0.0000  0.0000  0.3413  1.3152
#>
#>
#> $imputed_SepalLength2
#> $imputed_SepalLength2$is_missing
#> [1] "all values observed"
#>
#> $imputed_SepalLength2$observed
#>     Min.  1st Qu.   Median     Mean  3rd Qu.     Max.
#> -0.90570 -0.48398 -0.06225  0.00000  0.35947  1.20292
```

Chapter 5
Model Specification Tests

Model specification tests are critical in econometric analysis to verify whether the assumptions underlying a model hold true. These tests help determine if the model is correctly specified, ensuring that the estimators are both reliable and efficient. A mis-specified model can lead to biased, inconsistent, or inefficient estimates, which undermines the validity of inferences drawn from the analysis.

This chapter addresses various model specification tests, including tests for:

- Nested Model Tests
- Non-Nested Model Tests
- Heteroskedasticity Tests
- Functional Form Tests
- Autocorrelation Tests
- Multicollinearity Diagnostics

Understanding these tests allows researchers to evaluate the robustness of their models and make necessary adjustments to improve model performance.

5.1 Nested Model Tests

Nested models are those where the **restricted model** is a special case of the **unrestricted model**. In other words, the restricted model can be derived from the unrestricted model by imposing constraints on certain parameters, typically setting them equal to zero. This structure allows us to formally test whether the additional variables in the unrestricted model significantly improve the model's explanatory power. The following tests help compare these models:

- Wald Test: Assesses the significance of individual coefficients or groups of coefficients.

© The Author(s), under exclusive license to Springer Nature Switzerland AG 2025

M. Nguyen, *Advanced Modeling and Data Challenges*,
https://doi.org/10.1007/978-3-032-01719-2_5

- Likelihood Ratio Test: Compares the goodness-of-fit between restricted and unrestricted models.
- F-Test: Evaluates the joint significance of multiple coefficients.
- Chow Test: Evaluates whether the coefficients of a regression model are the same across different groups or time periods.

Consider the following models:

$$\textbf{Unrestricted Model}: \quad y = \beta_0 + \beta_1 x_1 + \beta_2 x_2 + \beta_3 x_3 + \epsilon$$
$$\textbf{Restricted Model}: \quad y = \beta_0 + \beta_1 x_1 + \epsilon$$

- The **unrestricted model** includes all potential explanatory variables: x_1, x_2, and x_3.
- The **restricted model** is nested within the unrestricted model, containing a subset of variables (in this case, excluding x_2 and x_3).

Our goal is to test the null hypothesis that the restrictions are valid:

$$H_0 : \beta_2 = \beta_3 = 0 \quad \text{(restrictions are valid)}$$

against the alternative hypothesis:

$$H_1 : \text{At least one of } \beta_2, \beta_3 \neq 0 \quad \text{(restrictions are invalid)}$$

To conduct this test, we use the following methods:

5.1.1 *Wald Test*

The **Wald Test** assesses whether certain linear restrictions on the parameters of the model are valid. It is commonly used when testing the joint significance of multiple coefficients.

Consider a set of linear restrictions expressed as:

$$H_0 : R\boldsymbol{\beta} = r$$

where:

- R is a $q \times k$ restriction matrix,
- $\boldsymbol{\beta}$ is the $k \times 1$ vector of parameters,
- r is a $q \times 1$ vector representing the hypothesized values (often zeros),
- q is the number of restrictions being tested.

For example, if we want to test $H_0 : \beta_2 = \beta_3 = 0$, then:

$$R = \begin{bmatrix} 0 & 0 & 1 & 0 \\ 0 & 0 & 0 & 1 \end{bmatrix}, \quad r = \begin{bmatrix} 0 \\ 0 \end{bmatrix}$$

The Wald statistic is calculated as:

$$W = (R\hat{\boldsymbol{\beta}} - r)' \left[R \, \hat{\text{Var}}(\hat{\boldsymbol{\beta}}) \, R' \right]^{-1} (R\hat{\boldsymbol{\beta}} - r)$$

where:

- $\hat{\boldsymbol{\beta}}$ is the vector of estimated coefficients from the unrestricted model,
- $\hat{\text{Var}}(\hat{\boldsymbol{\beta}})$ is the estimated covariance matrix of $\hat{\boldsymbol{\beta}}$.

Distribution and Decision Rule

- Under H_0, the Wald statistic follows a χ^2 distribution with q degrees of freedom:

$$W \sim \chi_q^2$$

- **Decision Rule:**
  - Reject H_0 if $W > \chi^2_{q,\alpha}$, where α is the significance level.
  - A large Wald statistic indicates that the restrictions are invalid.

5.1.2 *Likelihood Ratio Test*

The **Likelihood Ratio Test** compares the goodness-of-fit between the restricted and unrestricted models. It evaluates whether the additional parameters in the unrestricted model significantly improve the likelihood of observing the data.

Same as before:

$$H_0 : \beta_2 = \beta_3 = 0 \quad \text{vs.} \quad H_1 : \text{At least one of } \beta_2, \beta_3 \neq 0$$

The LR statistic is calculated as:

$$LR = -2 \left(\ln L_R - \ln L_U \right)$$

where:

- L_R is the maximized likelihood of the restricted model,
- L_U is the maximized likelihood of the unrestricted model.

Distribution and Decision Rule

- Under H_0, the LR statistic follows a χ^2 distribution with q degrees of freedom:

$$LR \sim \chi_q^2$$

- **Decision Rule:**
  - Reject H_0 if $LR > \chi^2_{q,\alpha}$.
  - A large LR statistic suggests that the unrestricted model provides a significantly better fit.

Connection to OLS
In the case of linear regression with normally distributed errors, the LR statistic can be expressed in terms of the sum of squared residuals (SSR):

$$LR = n \ln \left(\frac{SSR_R}{SSR_U} \right)$$

where n is the sample size.

5.1.3 *F-Test (for Linear Regression)*

The **F-Test** is commonly used in linear regression to evaluate the joint significance of multiple coefficients. It compares the fit of the restricted and unrestricted models using their sum of squared residuals.

Again:

$$H_0 : \beta_2 = \beta_3 = 0 \quad \text{vs.} \quad H_1 : \text{At least one of } \beta_2, \beta_3 \neq 0$$

The F-statistic is calculated as:

$$F = \frac{(SSR_R - SSR_U)/q}{SSR_U/(n-k)}$$

where:

- SSR_R = Sum of Squared Residuals from the restricted model,
- SSR_U = Sum of Squared Residuals from the unrestricted model,
- q = Number of restrictions (here, 2),
- n = Sample size,
- k = Number of parameters in the unrestricted model.

Distribution and Decision Rule

- Under H_0, the F-statistic follows an F-distribution with $(q, n-k)$ degrees of freedom:

$$F \sim F_{q,n-k}$$

- **Decision Rule:**
  - Reject H_0 if $F > F_{q,n-k,\alpha}$.
  - A large F-statistic indicates that the restricted model fits significantly worse, suggesting that the excluded variables are important.

5.1.4 *Chow Test*

The **Chow Test** evaluates whether the coefficients of a regression model are the same across different groups or time periods. It is often used to detect **structural breaks** in the data.

Key Question
Should we run two different regressions for two groups, or can we pool the data and use a single regression?

Chow Test Procedure

1. Estimate the regression model for the **pooled data** (all observations).
2. Estimate the model separately for **Group 1** and **Group 2**.
3. Compare the **sum of squared residuals (SSR)** from these models.
   The test statistic follows an F-distribution:

$$F = \frac{(SSR_P - (SSR_1 + SSR_2))/k}{(SSR_1 + SSR_2)/(n_1 + n_2 - 2k)}$$

where:

- SSR_P = Sum of Squared Residuals for the pooled model
- SSR_1, SSR_2 = SSRs for Group 1 and Group 2 models
- k = Number of parameters
- n_1, n_2 = Number of observations in each group

Interpretation

- A **significant F-statistic** suggests structural differences between groups, implying that separate regressions are more appropriate.
- A **non-significant F-statistic** indicates no structural break, supporting the use of a pooled model.

```
# Load necessary libraries
library(car)          # For Wald Test
#> Loading required package: carData
library(lmtest)       # For Likelihood Ratio Test
#> Loading required package: zoo
#>
#> Attaching package: 'zoo'
#> The following objects are masked from 'package:base':
#>
#>     as.Date, as.Date.numeric
library(strucchange)  # For Chow Test
#> Loading required package: sandwich

# Simulated dataset
set.seed(123)
n <- 100
x1 <- rnorm(n)
x2 <- rnorm(n)
x3 <- rnorm(n)
epsilon <- rnorm(n)
y <- 2 + 1.5 * x1 + 0.5 * x2 - 0.7 * x3 + epsilon

# Creating a group variable (simulating a structural break)
group <- rep(c(0, 1), each = n / 2)  # Group 0 and Group 1

# ---------------------------------------------------------------
# Wald Test
# ---------------------------------------------------------------
unrestricted_model <- lm(y ~ x1 + x2 + x3)    # Unrestricted model
restricted_model <- lm(y ~ x1)                # Restricted model

wald_test <- linearHypothesis(unrestricted_model, c("x2 = 0", "x3 = 0"))
print(wald_test)
#>
#> Linear hypothesis test:
#> x2 = 0
#> x3 = 0
#>
#> Model 1: restricted model
#> Model 2: y ~ x1 + x2 + x3
#>
#>   Res.Df    RSS Df Sum of Sq      F    Pr(>F)
#> 1     98 182.26
#> 2     96 106.14  2    76.117 34.421 5.368e-12 ***
#> ---
#> Signif. codes:
#> 0 '***' 0.001 '**' 0.01 '*' 0.05 '.' 0.1 ' ' 1

# ---------------------------------------------------------------
# Likelihood Ratio Test
# ---------------------------------------------------------------
lr_test <- lrtest(unrestricted_model, restricted_model)
print(lr_test)
```

```
#> Likelihood ratio test
#>
#> Model 1: y ~ x1 + x2 + x3
#> Model 2: y ~ x1
#>   #Df  LogLik Df  Chisq Pr(>Chisq)
#> 1   5 -144.88
#> 2   3 -171.91 -2 54.064  1.821e-12 ***
#> ---
#> Signif. codes:
#> 0 '***' 0.001 '**' 0.01 '*' 0.05 '.' 0.1 ' ' 1

# ----------------------------------------------------------------
# F-Test (for Linear Regression)
# ----------------------------------------------------------------
SSR_U <- sum(residuals(unrestricted_model)^2)  # SSR for unrestricted
↪  model
SSR_R <- sum(residuals(restricted_model)^2)    # SSR for restricted model
q <- 2                                        # Number of restrictions
n <- length(y)                                # Sample size
k <- length(coef(unrestricted_model))         # Number of parameters in
↪  unrestricted model

# F-statistic formula
F_statistic <- ((SSR_R - SSR_U) / q) / (SSR_U / (n - k))
p_value_F <- pf(F_statistic, df1 = q, df2 = n - k, lower.tail = FALSE)

cat("F-statistic:", F_statistic, "\n")
#> F-statistic: 34.42083
cat("P-value:", p_value_F, "\n")
#> P-value: 5.367912e-12

# ----------------------------------------------------------------
# Chow Test (Proper Use of the Group Variable)
# ----------------------------------------------------------------
# Pooled model (all data)
pooled_model <- lm(y ~ x1 + x2 + x3)

# Separate models for Group 0 and Group 1
model_group0 <- lm(y[group == 0] ~ x1[group == 0] + x2[group == 0] +
↪  x3[group == 0])
model_group1 <- lm(y[group == 1] ~ x1[group == 1] + x2[group == 1] +
↪  x3[group == 1])

# Calculating SSRs
SSR_pooled <- sum(residuals(pooled_model)^2)
SSR_group0 <- sum(residuals(model_group0)^2)
SSR_group1 <- sum(residuals(model_group1)^2)

# Chow Test formula
k_chow <- length(coef(pooled_model))  # Number of parameters (including
↪  intercept)
n0 <- sum(group == 0)                 # Sample size for Group 0
n1 <- sum(group == 1)                 # Sample size for Group 1
```

```
F_chow <- ((SSR_pooled - (SSR_group0 + SSR_group1)) / k_chow) /
          ((SSR_group0 + SSR_group1) / (n0 + n1 - 2 * k_chow))

# Corresponding p-value
p_value_chow <-
    pf(
        F_chow,
        df1 = k_chow,
        df2 = (n0 + n1 - 2 * k_chow),
        lower.tail = FALSE
    )

cat("Chow Test F-statistic:", F_chow, "\n")
#> Chow Test F-statistic: 0.3551197
cat("P-value:", p_value_chow, "\n")
#> P-value: 0.8398657
```

Interpretation of the Results

1. **Wald Test**
    - **Null Hypothesis** (H_0): $\beta_2 = \beta_3 = 0$ (the coefficients for x_2 and x_3 are zero).
    - **Decision Rule:**
        - **Reject** H_0 if *p-value* < 0.05: x_2 and x_3 are jointly significant.
        - **Fail to reject** H_0 if *p-value* $<$ 0.05: x_2 and x_3 do not significantly contribute to the model.
2. **Likelihood Ratio Test (LR Test)**
    - **Null Hypothesis** (H_0): The restricted model fits the data as well as the unrestricted model.
    - **Decision Rule:**
        - **Reject** H_0 if *p-value* < 0.05: The unrestricted model fits better, indicating that x_2 and x_3 improve the model.
        - **Fail to reject** H_0 if *p-value* $\approx$ 0.05: Adding x_2 and x_3 doesn't improve the model significantly.
3. **F-Test (for Linear Regression)**
    - **Null Hypothesis** (H_0): $\beta_2 = \beta_3 = 0$ (similar to the Wald Test).
    - **Decision Rule:**
        - **Reject** H_0 if *p-value* < 0.05: The excluded variables are significant.
        - **Fail to reject** H_0 if *p-value* $\approx$ 0.05: The excluded variables are not significant.

4. **Chow Test (Using the group Variable)**
   - **Null Hypothesis** (H_0): No structural break exists; the regression coefficients are the same across **Group 0** and **Group 1**.
   - **Decision Rule:**
     - **Reject** H_0 if *p-value* < 0.05: A structural break exists—model coefficients differ significantly between the groups.
     - **Fail to reject** H_0 if *p-value* $\approx$ 0.05: No significant structural break detected; the model coefficients are stable across both groups.

5.2 Non-Nested Model Tests

Comparing models is essential to identify which specification best explains the data. While nested model comparisons rely on one model being a restricted version of another, **non-nested models** do not share such a hierarchical structure. This situation commonly arises when comparing models with:

- Different functional forms (e.g., linear vs. logarithmic relationships),
- Different sets of explanatory variables,
- Competing theoretical frameworks.

To compare non-nested models, we rely on specialized statistical tests designed to handle these complexities. The two most widely used approaches are:

- Vuong Test used to compare the fit of two non-nested models to determine which model better explains the data.
- Davidson–MacKinnon J-Test is a regression-based approach for comparing non-nested models. It evaluates which model better fits the data by incorporating the predicted values from the competing model as an additional regressor.

Consider two competing models:

- **Model A:**

$$y = \alpha_0 + \alpha_1 f(X) + \epsilon_A$$

- **Model B:**

$$y = \beta_0 + \beta_1 g(Z) + \epsilon_B$$

where:

- $f(X)$ and $g(Z)$ represent different functional forms or different sets of explanatory variables.

- The models are **non-nested** because one cannot be obtained from the other by restricting parameters.

  Our goal is to determine which model better explains the data.

5.2.1 Vuong Test

The **Vuong Test** is a likelihood-ratio-based approach for comparing non-nested models (Vuong, 1989). It is particularly useful when both models are estimated via Maximum Likelihood Estimation.

Hypotheses

- Null Hypothesis (H_0): Both models are equally close to the true data-generating process (i.e., the models have equal predictive power).
- Alternative Hypothesis (H_1):
  - Positive Test Statistic ($V > 0$): Model A is preferred.
  - Negative Test Statistic ($V < 0$): Model B is preferred.

Vuong Test Statistic

The Vuong test is based on the difference in the log-likelihood contributions of each observation under the two models. Let:

- $\ell_{A,i}$ = log-likelihood of observation i under Model A,
- $\ell_{B,i}$ = log-likelihood of observation i under Model B.

  Define the difference in log-likelihoods:

$$m_i = \ell_{A,i} - \ell_{B,i}$$

The Vuong test statistic is:

$$V = \frac{\sqrt{n}\,\bar{m}}{s_m}$$

where:

- $\bar{m} = \frac{1}{n}\sum_{i=1}^{n} m_i$ is the sample mean of the log-likelihood differences,
- $s_m = \sqrt{\frac{1}{n}\sum_{i=1}^{n}(m_i - \bar{m})^2}$ is the sample standard deviation of m_i,
- n is the sample size.

Distribution and Decision Rule

- Under H_0, the Vuong statistic asymptotically follows a **standard normal distribution**:

$$V \sim N(0, 1)$$

- **Decision Rule:**
  - If $|V| > z_{\alpha/2}$ (critical value from the standard normal distribution), **reject** H_0.
    * If $V > 0$: Prefer **Model A**.
    * If $V < 0$: Prefer **Model B**.
  - If $|V| \leq z_{\alpha/2}$: **Fail to reject** H_0; no significant difference between models.

Corrections for Model Complexity
When comparing models with different numbers of parameters, a **penalized version** of the Vuong test can be used, similar to adjusting for model complexity in criteria like AIC or BIC. The corrected statistic is:

$$V_{\text{adjusted}} = V - \frac{(k_A - k_B)\ln(n)}{2s_m\sqrt{n}}$$

where k_A and k_B are the number of parameters in Models A and B, respectively.

Limitations of the Vuong Test
- Requires models to be estimated via Maximum Likelihood.
- Sensitive to **model misspecification** and **heteroskedasticity**.
- Assumes **independent and identically distributed (i.i.d.)** errors.

5.2.2 Davidson–MacKinnon J-Test

The **Davidson–MacKinnon J-Test** provides a flexible, regression-based approach for comparing non-nested models (Davidson & MacKinnon, 1981). It evaluates whether the predictions from one model contain information not captured by the competing model. This test can be thought of as comparing models with transformed independent variables, as opposed to the next section, Comparing Models with Transformed Dependent Variables.

Hypotheses
- Null Hypothesis (H_0): The competing model does not provide additional explanatory power beyond the current model.
- Alternative Hypothesis (H_1): The competing model provides additional explanatory power.

Procedure
Consider two competing models:

- **Model A:**

$$y = \alpha_0 + \alpha_1 x + \epsilon_A$$

- **Model B:**

$$y = \beta_0 + \beta_1 \ln(x) + \epsilon_B$$

Step 1: Testing Model A Against Model B
1. **Estimate Model B** and obtain predicted values $\hat{y}_B$.
2. **Run the auxiliary regression:**

$$y = \alpha_0 + \alpha_1 x + \gamma \hat{y}_B + u$$

3. **Test the null hypothesis:**

$$H_0 : \gamma = 0$$

- If γ is significant, Model B adds explanatory power beyond Model A.
- If γ is not significant, Model A sufficiently explains the data.

Step 2: Testing Model B Against Model A
1. **Estimate Model A** and obtain predicted values $\hat{y}_A$.
2. **Run the auxiliary regression:**

$$y = \beta_0 + \beta_1 \ln(x) + \gamma \hat{y}_A + u$$

3. **Test the null hypothesis:**

$$H_0 : \gamma = 0$$

Decision Rules
- **Reject** H_0 in Step 1, Fail to Reject in Step 2: Prefer **Model B**.
- **Fail to Reject** H_0 in Step 1, Reject in Step 2: Prefer **Model A**.
- **Reject** H_0 in Both Steps: Neither model is adequate; reconsider the functional form.
- **Fail to Reject** H_0 in Both Steps: No strong evidence to prefer one model; rely on other criteria (e.g., theory, simplicity). Alternatively, R^2_{adj} can also be used to choose between the two.

Adjusted R^2
- R^2 will always increase with more variables included
- Adjusted R^2 tries to correct by penalizing inclusion of unnecessary variables.

$$R^2 = 1 - \frac{SSR/n}{SST/n}$$

$$R^2_{adj} = 1 - \frac{SSR/(n-k)}{SST/(n-1)}$$

$$= 1 - \frac{(n-1)(1-R^2)}{(n-k)}$$

- R^2_{adj} increases if and only if the t-statistic on the additional variable is greater than 1 in absolute value.
- R^2_{adj} is valid in models where there is no heteroskedasticity
- therefore it **should not** be used in determining which variables should be included in the model (the t or F-tests are more appropriate)

5.2.3 Adjusted R^2

The **coefficient of determination** (R^2) measures the proportion of the variance in the dependent variable that is explained by the model. However, a key limitation of R^2 is that it **always increases** (or at least stays the same) when additional explanatory variables are added to the model, even if those variables are not statistically significant.

To address this issue, the **adjusted** R^2 introduces a penalty for including unnecessary variables, making it a more reliable measure when comparing models with different numbers of predictors.

Formulas
Unadjusted R^2:

$$R^2 = 1 - \frac{SSR}{SST}$$

where:

- SSR = **Sum of Squared Residuals** (measures unexplained variance),
- SST = **Total Sum of Squares** (measures total variance in the dependent variable).

Adjusted R^2

$$R^2_{\text{adj}} = 1 - \frac{SSR/(n-k)}{SST/(n-1)}$$

Alternatively, it can be expressed in terms of R^2 as:

$$R^2_{\text{adj}} = 1 - \left(\frac{(1 - R^2)(n - 1)}{n - k}\right)$$

where:

- n = Number of observations,
- k = Number of estimated parameters in the model (including the intercept).

Key Insights

- **Penalty for Complexity:** Unlike R^2, the adjusted R^2 **can decrease** when irrelevant variables are added to the model because it adjusts for the number of predictors relative to the sample size.
- **Interpretation:** It represents the proportion of variance explained **after accounting for model complexity**.
- **Comparison Across Models:** Adjusted R^2 is useful for comparing models with different numbers of predictors, as it discourages overfitting.

When Does Adjusted R^2 Increase?

- The adjusted R^2 will **increase** if and only if the inclusion of a new variable **improves the model more than expected by chance**.
- **Mathematically:** This typically occurs when the **absolute value of the** t-statistic for the new variable is **greater than 1** (assuming large samples and standard model assumptions).

Limitations of Adjusted R^2

- **Sensitive to Assumptions:** Adjusted R^2 assumes **homoskedasticity** (constant variance of errors) and **no autocorrelation**. In the presence of heteroskedasticity, its interpretation may be misleading.
- **Not a Substitute for Hypothesis Testing:** It **should not** be the primary criterion for deciding which variables to include in a model.
  - Use t-tests to evaluate the significance of individual coefficients.
  - Use F-tests for assessing the joint significance of multiple variables.

5.2.4 Comparing Models with Transformed Dependent Variables

When comparing regression models with different transformations of the dependent variable, such as level and log-linear models, direct comparisons using traditional goodness-of-fit metrics like R^2 or adjusted R^2 are invalid. This is because the transformation changes the scale of the dependent variable, affecting the calculation of the Total Sum of Squares (SST), which is the denominator in R^2 calculations.

Model Specifications

1. **Level Model (Linear):**

$$y = \beta_0 + \beta_1 x_1 + \epsilon$$

2. **Log-Linear Model:**

$$\ln(y) = \beta_0 + \beta_1 x_1 + \epsilon$$

where:

- y is the dependent variable,
- x_1 is an independent variable,
- ϵ represents the error term.

Interpretation of Coefficients

- **In the Level Model:**
  The effect of x_1 on y is **constant**, regardless of the magnitude of y. Specifically, a **one-unit increase** in x_1 results in a change of β_1 units in y. This implies:

$$\Delta y = \beta_1 \cdot \Delta x_1$$

- **In the Log Model:**
  The effect of x_1 on y is **proportional** to the current level of y. A **one-unit increase** in x_1 leads to a **percentage change** in y, approximately equal to $100 \times \beta_1\%$. Specifically:

$$\Delta \ln(y) = \beta_1 \cdot \Delta x_1 \quad \Rightarrow \quad \%\Delta y \approx 100 \times \beta_1$$

  - For small values of y, the absolute change is small.
  - For large values of y, the absolute change is larger, reflecting the multiplicative nature of the model.

  Why We Cannot Compare R^2 or Adjusted R^2 Directly

- The level model explains variance in the original scale of y, while the log model explains variance in the logarithmic scale of y.
- The SST (Total Sum of Squares) differs across the models because the dependent variable is transformed, making direct comparisons of R^2 invalid.
- Adjusted R^2 does not resolve this issue because it also depends on the scale of the dependent variable.

Approach to Compare Model Fit Across Transformations

To compare models on the **same scale** as the original dependent variable (y), we need to **"un-transform"** the predictions from the log model. Here's the step-by-step procedure:

Step-by-Step Procedure

1. **Estimate the Log Model**
   Fit the log-linear model and obtain the predicted values:

$$\widehat{\ln(y)} = \hat{\beta}_0 + \hat{\beta}_1 x_1$$

2. **Un-Transform the Predictions**
   Convert the predicted values back to the original scale of y using the exponential function:

$$\hat{m} = \exp(\widehat{\ln(y)})$$

   - This transformation assumes that the errors are homoskedastic in the log model.
   - **Note:** To correct for potential bias due to Jensen's inequality, a **smearing estimator** can be applied, but for simplicity, we use the basic exponential transformation here.

3. **Fit a Regression Without an Intercept**
   Regress the actual y on the un-transformed predictions $\hat{m}$ **without an intercept**:

$$y = \alpha \hat{m} + u$$

   - The coefficient α adjusts for any scaling differences between the predicted and actual values.
   - The residual term u captures the unexplained variance.

4. **Compute the Scaled R^2**
   Calculate the **squared correlation** between the observed y and the predicted values $\hat{y}$ from the above regression:

$$R^2_{\text{scaled}} = \left(\text{Corr}(y, \hat{y})\right)^2$$

   - This **scaled** R^2 represents how well the log-transformed model predicts the original y on its natural scale.
   - Now, you can **compare** R^2_{scaled} from the log model with the **regular** R^2 from the level model.

Key Insights

- **If** R^2_{scaled} (from the log model) > R^2 (from the level model): The log model fits the data better.

- **If $R^2_{\text{scaled}} < R^2$** (from the level model): The level model provides a better fit.
- **If both are similar:** Consider other model diagnostics, theoretical justification, or model simplicity.

Caveats and Considerations

- **Heteroskedasticity:** If heteroskedasticity is present, the un-transformation may introduce bias.
- **Error Distribution:** Log-transformed models assume multiplicative errors, which may not be appropriate in all contexts.
- **Smearing Estimator (Advanced Correction):** To adjust for bias in the back-transformation, apply the smearing estimator:

$$\hat{y} = \exp(\widehat{\ln(y)}) \cdot \hat{S}$$

where $\hat{S}$ is the **mean of the exponentiated residuals** from the log model.

```
# Install and load necessary libraries
# install.packages("nonnest2") # Uncomment if not already installed
library(nonnest2)     # For Vuong Test
#> This is nonnest2 0.5-8.
#> nonnest2 has not been tested with all combinations of supported model
↪  classes.
library(lmtest)       # For J-Test

# Simulated dataset
set.seed(123)
n <- 100
x <- rnorm(n, mean = 50, sd = 10)
z <- rnorm(n, mean = 100, sd = 20)
epsilon <- rnorm(n)

# Competing models (non-nested)
# Model A: Linear relationship with x
y <- 5 + 0.3 * x + epsilon
model_A <- lm(y ~ x)

# Model B: Log-linear relationship with z
model_B <- lm(y ~ log(z))

# ---------------------------------------------------------------
# Vuong Test (Correct Function)
# ---------------------------------------------------------------
vuong_test <- vuongtest(model_A, model_B)
print(vuong_test)
#>
#> Model 1
#>  Class: lm
#>  Call: lm(formula = y ~ x)
#>
#> Model 2
```

```
#>  Class: lm
#>  Call: lm(formula = y ~ log(z))
#>
#> Variance test
#>   H0: Model 1 and Model 2 are indistinguishable
#>   H1: Model 1 and Model 2 are distinguishable
#>     w2 = 0.681,   p = 2.35e-08
#>
#> Non-nested likelihood ratio test
#>   H0: Model fits are equal for the focal population
#>   H1A: Model 1 fits better than Model 2
#>     z = 13.108,   p = <2e-16
#>   H1B: Model 2 fits better than Model 1
#>     z = 13.108,   p = 1

# -------------------------------------------------------------
# Davidson--MacKinnon J-Test
# -------------------------------------------------------------

# Step 1: Testing Model A against Model B
# Obtain fitted values from Model B
fitted_B <- fitted(model_B)

# Auxiliary regression: Add fitted_B to Model A
j_test_A_vs_B <- lm(y ~ x + fitted_B)
summary(j_test_A_vs_B)
#>
#> Call:
#> lm(formula = y ~ x + fitted_B)
#>
#> Residuals:
#>     Min      1Q  Median      3Q     Max
#> -1.8717 -0.6573 -0.1223  0.6154  2.0952
#>
#> Coefficients:
#>             Estimate Std. Error t value Pr(>|t|)
#> (Intercept) 14.70881   25.98307   0.566    0.573
#> x            0.28671    0.01048  27.358   <2e-16 ***
#> fitted_B    -0.43702    1.27500  -0.343    0.733
#> ---
#> Signif. codes:
#> 0 '***' 0.001 '**' 0.01 '*' 0.05 '.' 0.1 ' ' 1
#>
#> Residual standard error: 0.951 on 97 degrees of freedom
#> Multiple R-squared:  0.8854, Adjusted R-squared:  0.883
#> F-statistic: 374.5 on 2 and 97 DF,  p-value: < 2.2e-16

# Step 2: Testing Model B against Model A
# Obtain fitted values from Model A
fitted_A <- fitted(model_A)

# Auxiliary regression: Add fitted_A to Model B
j_test_B_vs_A <- lm(y ~ log(z) + fitted_A)
```

```
summary(j_test_B_vs_A)
#>
#> Call:
#> lm(formula = y ~ log(z) + fitted_A)
#>
#> Residuals:
#>      Min       1Q   Median       3Q      Max
#> -1.8717 -0.6573 -0.1223  0.6154  2.0952
#>
#> Coefficients:
#>             Estimate Std. Error t value Pr(>|t|)
#> (Intercept) -0.77868    2.39275  -0.325    0.746
#> log(z)       0.16829    0.49097   0.343    0.733
#> fitted_A     1.00052    0.03657  27.358   <2e-16 ***
#> ---
#> Signif. codes:
#> 0 '***' 0.001 '**' 0.01 '*' 0.05 '.' 0.1 ' ' 1
#>
#> Residual standard error: 0.951 on 97 degrees of freedom
#> Multiple R-squared:  0.8854, Adjusted R-squared:  0.883
#> F-statistic: 374.5 on 2 and 97 DF,  p-value: < 2.2e-16
```

5.3 Heteroskedasticity Tests

Heteroskedasticity occurs when the variance of the error terms (ϵ_i) in a regression model is **not constant** across observations. This violates the Classical OLS Assumption, specifically the assumption of **homoskedasticity** (Assumption A4 Homoskedasticity in the Gauss–Markov Theorem), which states:

$$\text{Var}(\epsilon_i) = \sigma^2 \quad \forall i$$

When heteroskedasticity is present:

- Ordinary Least Squares estimators remain **unbiased** but become **inefficient** (i.e., no longer Best Linear Unbiased Estimators—BLUE).
- The **standard errors** of the estimates are biased, leading to unreliable **hypothesis tests** (e.g., t-tests and F-tests).

Detecting heteroskedasticity is crucial for ensuring the validity of regression results. This section covers key tests used to identify heteroskedasticity:

- Breusch–Pagan Test
- White Test
- Goldfeld–Quandt Test
- Park Test
- Glejser Test

5.3.1 Breusch–Pagan Test

The **Breusch–Pagan (BP) Test** is one of the most widely used tests for detecting heteroskedasticity (Breusch & Pagan, 1979). It examines whether the variance of the residuals depends on the independent variables.

Hypotheses

- Null Hypothesis (H_0): Homoskedasticity ($\text{Var}(\epsilon_i) = \sigma^2$ is constant).
- Alternative Hypothesis (H_1): Heteroskedasticity exists; the variance of ϵ_i depends on the independent variables.

Procedure

1. **Estimate the original regression model:**

$$y_i = \beta_0 + \beta_1 x_{1i} + \beta_2 x_{2i} + \cdots + \beta_k x_{ki} + \epsilon_i$$

   Obtain the residuals $\hat{\epsilon}_i$ from this regression.

2. **Compute the squared residuals:**

$$\hat{\epsilon}_i^2$$

3. **Auxiliary Regression:** Regress the squared residuals on the independent variables:

$$\hat{\epsilon}_i^2 = \alpha_0 + \alpha_1 x_{1i} + \alpha_2 x_{2i} + \cdots + \alpha_k x_{ki} + u_i$$

4. **Calculate the Test Statistic:**

   The BP test statistic is:

$$\text{BP} = n \cdot R^2_{\text{aux}}$$

where:

- n is the sample size,
- R^2_{aux} is the R^2 from the auxiliary regression.

5. **Decision Rule:**

- Under H_0, the BP statistic follows a **Chi-squared distribution** with k degrees of freedom (where k is the number of independent variables):

$$\text{BP} \sim \chi^2_k$$

- **Reject** H_0 if the BP statistic exceeds the critical value from the Chi-squared distribution.

Advantages and Limitations

- **Advantage:** Simple to implement; directly tests the relationship between residual variance and regressors.
- **Limitation:** Sensitive to non-normality; less effective when heteroskedasticity is not linearly related to independent variables.

5.3.2 *White Test*

The **White Test** is a more general heteroskedasticity test that does not require specifying the form of heteroskedasticity (White, 1980). It can detect both **linear** and **nonlinear** forms.

Hypotheses

- Null Hypothesis (H_0): Homoskedasticity.
- Alternative Hypothesis (H_1): Heteroskedasticity (of any form).

Procedure

1. **Estimate the original regression model** and obtain residuals $\hat{\epsilon}_i$.
2. **Auxiliary Regression:** Regress the squared residuals on:
    - The original independent variables $(x_{1i}, x_{2i}, \ldots, x_{ki})$,
    - Their **squares** $(x_{1i}^2, x_{2i}^2, \ldots, x_{ki}^2)$,
    - Their **cross-products** (e.g., $x_{1i}x_{2i}$).

    The auxiliary regression is:

$$\hat{\epsilon}_i^2 = \alpha_0 + \alpha_1 x_{1i} + \alpha_2 x_{2i} + \cdots + \alpha_k x_{ki} + \alpha_{k+1} x_{1i}^2 + \cdots + \alpha_{2k} x_{ki}^2 + \alpha_{2k+1}(x_{1i}x_{2i}) + u_i$$

3. **Calculate the Test Statistic:**

$$\text{White} = n \cdot R_{\text{aux}}^2$$

4. **Decision Rule:**

- Under H_0, the statistic follows a **Chi-squared distribution** with degrees of freedom equal to the number of auxiliary regressors:

$$\text{White} \sim \chi_{\text{df}}^2$$

- **Reject** H_0 if the statistic exceeds the critical Chi-squared value.

Advantages and Limitations

- **Advantage:** Can detect a wide range of heteroskedasticity patterns.
- **Limitation:** May suffer from overfitting in small samples due to many auxiliary regressors.

5.3.3 Goldfeld–Quandt Test

The **Goldfeld–Quandt Test** is a simple test that detects heteroskedasticity by comparing the variance of residuals in two different subsets of the data (Goldfeld & Quandt, 1965).

Hypotheses

- Null Hypothesis (H_0): Homoskedasticity.
- Alternative Hypothesis (H_1): Heteroskedasticity; variances differ between groups.

Procedure

1. **Sort the data** based on an independent variable suspected to cause heteroskedasticity.
2. **Split the data** into three groups:
    - **Group 1:** Lower values,
    - **Group 2:** Middle values (often omitted),
    - **Group 3:** Higher values.
3. **Estimate the regression model** separately for Groups 1 and 3. Obtain the residual sum of squares (SSR_1 and SSR_2).
4. **Calculate the Test Statistic:**

$$F = \frac{SSR_2/(n_2 - k)}{SSR_1/(n_1 - k)}$$

   where:

    - n_1 and n_2 are the number of observations in Groups 1 and 3, respectively,
    - k is the number of estimated parameters.

5. **Decision Rule:**

- Under H_0, the test statistic follows an F-distribution with (n_2-k, n_1-k) degrees of freedom:

$$F \sim F_{(n_2-k, n_1-k)}$$

- **Reject** H_0 if F exceeds the critical value.

Advantages and Limitations

- **Advantage:** Simple to apply when heteroskedasticity is suspected to vary systematically with an independent variable.
- **Limitation:** Requires arbitrary splitting of data, and assumes that the error variance changes abruptly between groups.

5.3.4 *Park Test*

The **Park Test** identifies heteroskedasticity by modeling the error variance as a function of an independent variable (Park, 1966).

Hypotheses

- Null Hypothesis (H_0): Homoskedasticity.
- Alternative Hypothesis (H_1): Heteroskedasticity; variance depends on an independent variable.

Procedure

1. **Estimate the original regression** and obtain residuals $\hat{\epsilon}_i$.
2. **Transform the residuals:** Take the natural logarithm of the squared residuals:

$$\ln(\hat{\epsilon}_i^2)$$

3. **Auxiliary Regression:** Regress $\ln(\hat{\epsilon}_i^2)$ on the independent variable(s):

$$\ln(\hat{\epsilon}_i^2) = \alpha_0 + \alpha_1 \ln(x_i) + u_i$$

4. **Decision Rule:**

- Test whether $\alpha_1 = 0$ using a t-test.
- **Reject** H_0 if α_1 is statistically significant, indicating heteroskedasticity.

Advantages and Limitations

- **Advantage:** Simple to implement; works well when the variance follows a log-linear relationship.
- **Limitation:** Assumes a specific functional form for the variance, which may not hold in practice.

5.3.5 *Glejser Test*

The **Glejser Test** detects heteroskedasticity by regressing the **absolute value of residuals** on the independent variables (Glejser, 1969).

Hypotheses

- Null Hypothesis (H_0): Homoskedasticity.
- Alternative Hypothesis (H_1): Heteroskedasticity exists.

Procedure

1. **Estimate the original regression** and obtain residuals $\hat{\epsilon}_i$.
2. **Auxiliary Regression:** Regress the absolute residuals on the independent variables:

$$|\hat{\epsilon}_i| = \alpha_0 + \alpha_1 x_{1i} + \alpha_2 x_{2i} + \cdots + \alpha_k x_{ki} + u_i$$

3. **Decision Rule:**

- Test the significance of the coefficients ($\alpha_1, \alpha_2, \ldots$) using t-tests.
- **Reject** H_0 if any coefficient is statistically significant, indicating heteroskedasticity.

Advantages and Limitations

- **Advantage:** Flexible; can detect various forms of heteroskedasticity.
- **Limitation:** Sensitive to outliers since it relies on absolute residuals.

5.3.6 Summary of Heteroskedasticity Tests

Detecting heteroskedasticity is critical for ensuring the reliability of regression models. While each test has strengths and limitations, combining multiple tests can provide robust insights. Once heteroskedasticity is detected, consider using **robust standard errors** or alternative estimation techniques (e.g., Generalized Least Squares or Weighted Least Squares) to address the issue (Table 5.1).

Table 5.1 Summary of heteroskedasticity tests

Test	Type	Assumptions	Key statistic	When to use
Breusch–Pagan	Parametric	Linear relationship with predictors	χ^2	General-purpose test
White	General (non-parametric)	No functional form assumption	χ^2	Detects both linear & nonlinear forms
Goldfeld–Quandt	Group comparison	Assumes known ordering of variance	F-distribution	When heteroskedasticity varies by groups
Park	Parametric (log-linear)	Assumes log-linear variance	t-test	When variance depends on predictors
Glejser	Parametric	Based on absolute residuals	t-test	Simple test for variance dependence

```
# Install and load necessary libraries
# install.packages("lmtest")      # For Breusch--Pagan Test
# install.packages("car")         # For additional regression diagnostics
# install.packages("sandwich")    # For robust covariance estimation


library(lmtest)
library(car)
library(sandwich)

# Simulated dataset
set.seed(123)
n <- 100
x1 <- rnorm(n, mean = 50, sd = 10)
x2 <- rnorm(n, mean = 30, sd = 5)
epsilon <-
    rnorm(n, sd = x1 * 0.1)  # Heteroskedastic errors increasing with x1
y <- 5 + 0.4 * x1 - 0.3 * x2 + epsilon

# Original regression model
model <- lm(y ~ x1 + x2)

# ------------------------------------------------------------
# 1. Breusch--Pagan Test
# ------------------------------------------------------------
# Null Hypothesis: Homoskedasticity
bp_test <- bptest(model)
print(bp_test)
#>
#>  studentized Breusch-Pagan test
#>
#> data:  model
#> BP = 7.8141, df = 2, p-value = 0.0201

# ------------------------------------------------------------
# 2. White Test (using Breusch--Pagan framework with squares &
↪  interactions)
# ------------------------------------------------------------
# Create squared and interaction terms
model_white <-
    lm(residuals(model) ^ 2 ~ x1 + x2 + I(x1 ^ 2) + I(x2 ^ 2) + I(x1 *
    ↪  x2))
white_statistic <-
    summary(model_white)$r.squared * n  # White Test Statistic
df_white <-
    length(coef(model_white)) - 1              # Degrees of freedom
p_value_white <- 1 - pchisq(white_statistic, df_white)

# Display White Test result
cat("White Test Statistic:", white_statistic, "\n")
#> White Test Statistic: 11.85132
cat("Degrees of Freedom:", df_white, "\n")
#> Degrees of Freedom: 5
```

```
cat("P-value:", p_value_white, "\n")
#> P-value: 0.0368828

# -------------------------------------------------------------
# 3. Goldfeld--Quandt Test
# -------------------------------------------------------------
# Null Hypothesis: Homoskedasticity
# Sort data by x1 (suspected source of heteroskedasticity)
gq_test <-
    gqtest(model, order.by = ~ x1, fraction = 0.2)  # Omit middle 20% of
    ↪ data
print(gq_test)
#>
#>  Goldfeld-Quandt test
#>
#> data:  model
#> GQ = 1.8352, df1 = 37, df2 = 37, p-value = 0.03434
#> alternative hypothesis: variance increases from segment 1 to 2

# -------------------------------------------------------------
# 4. Park Test
# -------------------------------------------------------------
# Step 1: Get residuals and square them
residuals_squared <- residuals(model) ^ 2

# Step 2: Log-transform squared residuals
log_residuals_squared <- log(residuals_squared)

# Step 3: Regress log(residuals^2) on log(x1) (assuming variance depends
↪  on x1)
park_test <- lm(log_residuals_squared ~ log(x1))
summary(park_test)
#>
#> Call:
#> lm(formula = log_residuals_squared ~ log(x1))
#>
#> Residuals:
#>     Min      1Q  Median      3Q     Max
#> -9.3633 -1.3424  0.4218  1.6089  3.0697
#>
#> Coefficients:
#>             Estimate Std. Error t value Pr(>|t|)
#> (Intercept)  -1.6319     4.5982  -0.355    0.723
#> log(x1)       0.8903     1.1737   0.759    0.450
#>
#> Residual standard error: 2.171 on 98 degrees of freedom
#> Multiple R-squared:  0.005837,   Adjusted R-squared:  -0.004308
#> F-statistic: 0.5754 on 1 and 98 DF,  p-value: 0.4499

# -------------------------------------------------------------
# 5. Glejser Test
# -------------------------------------------------------------
# Step 1: Absolute value of residuals
```

```
abs_residuals <- abs(residuals(model))

# Step 2: Regress absolute residuals on independent variables
glejser_test <- lm(abs_residuals ~ x1 + x2)
summary(glejser_test)
#>
#> Call:
#> lm(formula = abs_residuals ~ x1 + x2)
#>
#> Residuals:
#>     Min      1Q  Median      3Q     Max
#> -4.3096 -2.2680 -0.4564  1.9554  8.3921
#>
#> Coefficients:
#>              Estimate Std. Error t value Pr(>|t|)
#> (Intercept)  0.755846   2.554842   0.296   0.7680
#> x1           0.064896   0.032852   1.975   0.0511 .
#> x2          -0.008495   0.062023  -0.137   0.8913
#> ---
#> Signif. codes:
#> 0 '***' 0.001 '**' 0.01 '*' 0.05 '.' 0.1 ' ' 1
#>
#> Residual standard error: 2.98 on 97 degrees of freedom
#> Multiple R-squared:  0.0392, Adjusted R-squared:  0.01939
#> F-statistic: 1.979 on 2 and 97 DF,  p-value: 0.1438
```

Interpretation of the Results

1. **Breusch–Pagan Test**
    - **Null Hypothesis** (H_0): Homoskedasticity (constant error variance).
    - **Alternative Hypothesis** (H_1): Heteroskedasticity exists (error variance depends on predictors).
    - **Decision Rule:**
        - **Reject** H_0 if *p-value* $< 0.05 \rightarrow$ Evidence of heteroskedasticity.
        - **Fail to reject** H_0 if *p-value* $\geq 0.05 \rightarrow$ No strong evidence of heteroskedasticity.
2. **White Test**
    - **Null Hypothesis** (H_0): Homoskedasticity.
    - **Alternative Hypothesis** (H_1): Heteroskedasticity (of any form, linear or nonlinear).
    - **Decision Rule:**
        - **Reject** H_0 if *p-value* $< 0.05 \rightarrow$ Presence of heteroskedasticity.
        - **Fail to reject** H_0 if *p-value* $\geq 0.05 \rightarrow$ Homoskedasticity likely holds.
3. **Goldfeld–Quandt Test**
    - **Null Hypothesis** (H_0): Homoskedasticity (equal variances across groups).

- **Alternative Hypothesis** (H_1): Heteroskedasticity (unequal variances between groups).
- **Decision Rule:**
  - **Reject** H_0 if *p-value* $< 0.05 \rightarrow$ Variances differ between groups, indicating heteroskedasticity.
  - **Fail to reject** H_0 if *p-value* $\geq 0.05 \rightarrow$ No significant evidence of heteroskedasticity.

4. **Park Test**
   - **Null Hypothesis** (H_0): No relationship between the variance of errors and predictor(s) (homoskedasticity).
   - **Alternative Hypothesis** (H_1): Variance of errors depends on predictor(s).
   - **Decision Rule:**
     - **Reject** H_0 if the *coefficient of* $\log(x_1)$ *is statistically significant* (*p-value* < 0.05).
     - **Fail to reject** H_0 if *p-value* ≥ 0.05.

5. **Glejser Test**
   - **Null Hypothesis** (H_0): Homoskedasticity (no relationship between absolute residuals and predictors).
   - **Alternative Hypothesis** (H_1): Heteroskedasticity exists.
   - **Decision Rule:**
     - **Reject** H_0 if any predictor is statistically significant (*p-value* < 0.05).
     - **Fail to reject** H_0 if *p-value* ≥ 0.05.

5.4 Functional Form Tests

Functional form misspecification occurs when the chosen regression model does not correctly represent the true relationship between the dependent and independent variables. This can happen due to:

- **Omitted variables** (important predictors not included),
- **Incorrect transformations** of variables (e.g., missing nonlinear relationships),
- **Incorrect interaction terms** (missing interaction effects between variables),
- **Inappropriate linearity assumptions**.

Functional form errors can lead to **biased and inconsistent** estimators, undermining the validity of statistical inferences. To detect such issues, several diagnostic tests are available.

Key Functional Form Tests

1. Ramsey RESET Test (Regression Equation Specification Error Test)
2. Harvey–Collier Test
3. Rainbow Test

Each test focuses on identifying different aspects of potential model misspecification.

5.4.1 *Ramsey RESET Test (Regression Equation Specification Error Test)*

The **Ramsey RESET Test** is one of the most widely used tests to detect functional form misspecification (Ramsey, 1969). It examines whether adding **nonlinear transformations** of the fitted values (or regressors) improves the model fit.

Hypotheses

- Null Hypothesis (H_0): The model is correctly specified.
- Alternative Hypothesis (H_1): The model suffers from omitted variables, incorrect functional form, or other specification errors.

Procedure

1. **Estimate the original regression model:**

$$y_i = \beta_0 + \beta_1 x_{1i} + \beta_2 x_{2i} + \cdots + \beta_k x_{ki} + \epsilon_i$$

2. **Obtain the fitted values:**

$$\hat{y}_i$$

3. **Augment the model with powers of the fitted values (squared, cubed, etc.):**

$$y_i = \beta_0 + \beta_1 x_{1i} + \cdots + \beta_k x_{ki} + \gamma_1 \hat{y}_i^2 + \gamma_2 \hat{y}_i^3 + u_i$$

4. **Test the joint significance** of the added terms:

$$H_0 : \gamma_1 = \gamma_2 = 0$$

5. **Compute the F-statistic:**

$$F = \frac{(SSR_{\text{restricted}} - SSR_{\text{unrestricted}})/q}{SSR_{\text{unrestricted}}/(n - k - q - 1)}$$

where:

- $SSR_{\text{restricted}}$ = Sum of Squared Residuals from the original model,
- $SSR_{\text{unrestricted}}$ = SSR from the augmented model,
- q = Number of additional terms (e.g., 2 if adding $\hat{y}^2$ and $\hat{y}^3$),
- n = Sample size,
- k = Number of predictors in the original model.

Decision Rule

- Under H_0, the F-statistic follows an F-distribution with $(q, n-k-q-1)$ degrees of freedom.
- Reject H_0 if the F-statistic exceeds the critical value, indicating functional form misspecification.

Advantages and Limitations

- **Advantage:** Simple to implement; detects omitted variables and incorrect functional forms.
- **Limitation:** Does not identify which variable or functional form is incorrect—only indicates the presence of an issue.

5.4.2 *Harvey–Collier Test*

The **Harvey–Collier Test** evaluates whether the model's residuals display systematic patterns, which would indicate functional form misspecification (Harvey & Collier, 1977). It is based on testing for a **non-zero mean** in the residuals after projection onto specific components.

Hypotheses

- Null Hypothesis (H_0): The model is correctly specified (residuals are random noise with zero mean).
- Alternative Hypothesis (H_1): The model is misspecified (residuals contain systematic patterns).

Procedure

1. **Estimate the original regression model** and obtain residuals $\hat{\epsilon}_i$.
2. **Project the residuals** onto the space spanned by a specially constructed test vector (often derived from the inverse of the design matrix in linear regression).
3. **Calculate the Harvey–Collier statistic:**

$$t = \frac{\bar{\epsilon}}{\text{SE}(\bar{\epsilon})}$$

where:

- $\bar{\epsilon}$ is the mean of the projected residuals,
- $\text{SE}(\bar{\epsilon})$ is the standard error of the mean residual.

Decision Rule

- The test statistic follows a t-distribution under H_0.
- **Reject** H_0 if the t-statistic is significantly different from zero.

Advantages and Limitations

- **Advantage:** Simple to apply and interpret; good for detecting subtle misspecifications.
- **Limitation:** Sensitive to outliers; may have reduced power in small samples.

5.4.3 Rainbow Test

The **Rainbow Test** is a general-purpose diagnostic tool for functional form misspecification (Utts, 1982). It compares the performance of the model on the **full sample** versus a **central subsample**, where the central subsample contains observations near the median of the independent variables.

Hypotheses

- Null Hypothesis (H_0): The model is correctly specified.
- Alternative Hypothesis (H_1): The model is misspecified.

Procedure

1. **Estimate the regression model** on the full dataset and record the residuals.
2. **Identify a central subsample** (e.g., observations near the median of key predictors).
3. **Estimate the model again** on the central subsample.
4. **Compare the predictive accuracy** between the full sample and subsample using an F-statistic:

$$F = \frac{(SSR_{\text{full}} - SSR_{\text{subsample}})/q}{SSR_{\text{subsample}}/(n-k-q)}$$

where q is the number of restrictions implied by using the subsample.

Decision Rule

- Under H_0, the test statistic follows an F-distribution.
- Reject H_0 if the F-statistic is significant, indicating model misspecification.

Advantages and Limitations

- **Advantage:** Robust to various forms of misspecification.
- **Limitation:** Choice of subsample may influence results; less informative about the specific nature of the misspecification.

Table 5.2 Summary of functional form tests

Test	Type	Key statistic	Purpose	When to use
Ramsey RESET Test	Augmented regression	F-test	Detects omitted variables, nonlinearities	General model specification testing
Harvey–Collier Test	Residual-based	t-test	Detects systematic patterns in residuals	Subtle misspecifications in linear models
Rainbow Test	Subsample comparison	F-test	Tests model stability across subsamples	Comparing central vs. full sample

5.4.4 Summary of Functional Form Tests

Functional form misspecification can severely distort regression results, leading to biased estimates and invalid inferences. While no single test can detect all types of misspecification, using a combination of tests provides a robust framework for model diagnostics (Table 5.2).

```
# Install and load necessary libraries
# install.packages("lmtest")       # For RESET and Harvey--Collier Test
# install.packages("car")          # For diagnostic tests
# install.packages("strucchange") # For Rainbow Test

library(lmtest)
library(car)
library(strucchange)

# Simulated dataset
set.seed(123)
n <- 100
x1 <- rnorm(n, mean = 50, sd = 10)
x2 <- rnorm(n, mean = 30, sd = 5)
epsilon <- rnorm(n)
y <- 5 + 0.4 * x1 - 0.3 * x2 + epsilon

# Original regression model
model <- lm(y ~ x1 + x2)

# ------------------------------------------------------------
# 1. Ramsey RESET Test
# ------------------------------------------------------------
# Null Hypothesis: The model is correctly specified
reset_test <-
   resettest(model, power = 2:3, type = "fitted")  # Adds ŷ^2 and ŷ^3
print(reset_test)
#>
#>  RESET test
```

```
#>
#> data:  model
#> RESET = 0.1921, df1 = 2, df2 = 95, p-value = 0.8255

# -------------------------------------------------------------
# 2. Harvey--Collier Test
# -------------------------------------------------------------
# Null Hypothesis: The model is correctly specified (residuals have zero
↪  mean)
hc_test <- harvtest(model)
print(hc_test)
#>
#>  Harvey-Collier test
#>
#> data:  model
#> HC = 0.041264, df = 96, p-value = 0.9672

# -------------------------------------------------------------
# 3. Rainbow Test
# -------------------------------------------------------------
# Null Hypothesis: The model is correctly specified
rainbow_test <- lmtest::raintest (model)
print(rainbow_test)
#>
#>  Rainbow test
#>
#> data:  model
#> Rain = 1.1857, df1 = 50, df2 = 47, p-value = 0.279
```

Interpretation of the Results

1. **Ramsey RESET Test (Regression Equation Specification Error Test)**
   - **Null Hypothesis** (H_0): The model is correctly specified.
   - **Alternative Hypothesis** (H_1): The model suffers from omitted variables, incorrect functional form, or other specification errors.
   - **Decision Rule:**
     - **Reject** H_0 if *p-value* $< 0.05 \rightarrow$ Evidence of model misspecification (e.g., missing nonlinear terms).
     - **Fail to reject** H_0 if *p-value* $\geq 0.05 \rightarrow$ No strong evidence of misspecification.

2. **Harvey–Collier Test**
   - **Null Hypothesis** (H_0): The model is correctly specified (residuals are random noise with zero mean).
   - **Alternative Hypothesis** (H_1): The model is misspecified (residuals contain systematic patterns).

- **Decision Rule:**
  - **Reject** H_0 if *p-value* $< 0.05 \rightarrow$ Model misspecification detected (non-random residual patterns).
  - **Fail to reject** H_0 if *p-value* $\geq 0.05 \rightarrow$ No evidence of misspecification.

3. **Rainbow Test**
   - **Null Hypothesis** (H_0): The model is correctly specified.
   - **Alternative Hypothesis** (H_1): The model is misspecified.
   - **Decision Rule:**
     - **Reject** H_0 if *p-value* $< 0.05 \rightarrow$ Evidence of model misspecification (model performs differently on subsets).
     - **Fail to reject** H_0 if *p-value* $\geq 0.05 \rightarrow$ Model specification appears valid.

5.5 Autocorrelation Tests

Autocorrelation (also known as **serial correlation**) occurs when the error terms (ϵ_t) in a regression model are **correlated across observations**, violating the assumption of **independence** in the classical linear regression model. This issue is particularly common in **time-series data**, where observations are ordered over time.

Consequences of Autocorrelation

- OLS estimators remain unbiased but become **inefficient**, meaning they do not have the minimum variance among all linear unbiased estimators.
- **Standard errors** are biased, leading to unreliable hypothesis tests (e.g., t-tests and F-tests).
- Potential **underestimation of standard errors**, increasing the risk of Type I errors (false positives).

5.5.1 Durbin–Watson Test

The **Durbin–Watson (DW) Test** is the most widely used test for detecting **first-order autocorrelation**, where the current error term is correlated with the previous one:

$$\epsilon_t = \rho\,\epsilon_{t-1} + u_t$$

where:

- ρ is the autocorrelation coefficient,
- u_t is a white noise error term.

Hypotheses
- Null Hypothesis (H_0): No first-order autocorrelation ($\rho = 0$).
- Alternative Hypothesis (H_1): First-order autocorrelation exists ($\rho \neq 0$).

Durbin–Watson Test Statistic
The DW statistic is calculated as:

$$DW = \frac{\sum_{t=2}^{n} (\hat{\epsilon}_t - \hat{\epsilon}_{t-1})^2}{\sum_{t=1}^{n} \hat{\epsilon}_t^2}$$

where:

- $\hat{\epsilon}_t$ are the residuals from the regression,
- n is the number of observations.

Decision Rule
- The DW statistic ranges from **0 to 4**:
  - **DW** $\approx$ **2**: No autocorrelation.
  - **DW < 2**: Positive autocorrelation.
  - **DW > 2**: Negative autocorrelation.

  For more precise interpretation:

- Use **Durbin–Watson critical values** (d_L and d_U) to form decision boundaries.
- If the test statistic falls into an **inconclusive range**, consider alternative tests like the Breusch–Godfrey test.

Advantages and Limitations
- **Advantage:** Simple to compute; specifically designed for detecting first-order autocorrelation.
- **Limitation:** Inconclusive in some cases; **invalid** when lagged dependent variables are included in the model.

5.5.2 Breusch–Godfrey Test

The **Breusch–Godfrey (BG) Test** is a more general approach that can detect **higher-order autocorrelation** (e.g., second-order, third-order) and is valid even when **lagged dependent variables** are included in the model (Breusch, 1978; Godfrey, 1978).

Hypotheses
- Null Hypothesis (H_0): No autocorrelation of any order (up to lag p).
- Alternative Hypothesis (H_1): Autocorrelation exists at some lag(s).

Procedure

1. **Estimate the original regression model** and obtain residuals $\hat{\epsilon}_t$:

$$y_t = \beta_0 + \beta_1 x_{1t} + \cdots + \beta_k x_{kt} + \epsilon_t$$

2. **Run an auxiliary regression** by regressing the residuals on the original regressors plus p lagged residuals:

$$\hat{\epsilon}_t = \alpha_0 + \alpha_1 x_{1t} + \cdots + \alpha_k x_{kt} + \rho_1 \hat{\epsilon}_{t-1} + \cdots + \rho_p \hat{\epsilon}_{t-p} + u_t$$

3. **Calculate the test statistic:**

$$\text{BG} = n \cdot R^2_{\text{aux}}$$

where:

- n is the sample size,
- R^2_{aux} is the R^2 from the auxiliary regression.

Decision Rule

- Under H_0, the BG statistic follows a **Chi-squared distribution** with p degrees of freedom:

$$\text{BG} \sim \chi^2_p$$

- **Reject** H_0 if the statistic exceeds the critical Chi-squared value, indicating the presence of autocorrelation.

Advantages and Limitations

- **Advantage:** Can detect **higher-order autocorrelation**; valid with lagged dependent variables.
- **Limitation:** More computationally intensive than the Durbin–Watson test.

5.5.3 Ljung–Box Test (or Box–Pierce Test)

The **Ljung–Box Test** is a **portmanteau test** designed to detect **autocorrelation at multiple lags** simultaneously (Box & Pierce, 1970; Ljung & Box, 1978). It is commonly used in **time-series analysis** to check residual autocorrelation after model estimation (e.g., in ARIMA models).

Hypotheses

- Null Hypothesis (H_0): No autocorrelation up to lag h.
- Alternative Hypothesis (H_1): Autocorrelation exists at one or more lags.

Ljung–Box Test Statistic
The Ljung–Box statistic is calculated as:

$$Q = n(n+2)\sum_{k=1}^{h}\frac{\hat{\rho}_k^2}{n-k}$$

where:

- n = Sample size,
- h = Number of lags tested,
- $\hat{\rho}_k$ = Sample autocorrelation at lag k.

Decision Rule
- Under H_0, the Q statistic follows a **Chi-squared distribution** with h degrees of freedom:

$$Q \sim \chi_h^2$$

- **Reject** H_0 if Q exceeds the critical value, indicating significant autocorrelation.

Advantages and Limitations
- **Advantage:** Detects autocorrelation across multiple lags simultaneously.
- **Limitation:** Less powerful for detecting specific lag structures; sensitive to model misspecification.

5.5.4 Runs Test

The **Runs Test** is a **non-parametric test** that examines the randomness of residuals. It is based on the number of **runs**—sequences of consecutive residuals with the same sign.

Hypotheses
- Null Hypothesis (H_0): Residuals are randomly distributed (no autocorrelation).
- Alternative Hypothesis (H_1): Residuals exhibit non-random patterns (indicating autocorrelation).

Procedure
1. **Classify residuals** as positive or negative.
2. **Count the number of runs:** A **run** is a sequence of consecutive positive or negative residuals.
3. **Calculate the expected number of runs under randomness:**

$$E(R) = \frac{2n_+n_-}{n} + 1$$

where:

- n_+ = Number of positive residuals,
- n_- = Number of negative residuals,
- $n = n_+ + n_-$.

4. **Compute the test statistic (Z-score):**

$$Z = \frac{R - E(R)}{\sqrt{\mathrm{Var}(R)}}$$

where $\mathrm{Var}(R)$ is the variance of the number of runs under the null hypothesis.

Decision Rule

- Under H_0, the Z-statistic follows a **standard normal distribution**:

$$Z \sim N(0, 1)$$

- **Reject** H_0 if $|Z|$ exceeds the critical value from the standard normal distribution.

Advantages and Limitations

- **Advantage:** Non-parametric; does not assume normality or linearity.
- **Limitation:** Less powerful than parametric tests; primarily useful as a supplementary diagnostic.

5.5.5 *Summary of Autocorrelation Tests*

Detecting autocorrelation is crucial for ensuring the efficiency and reliability of regression models, especially in time-series analysis. While the Durbin–Watson Test is suitable for detecting first-order autocorrelation, the Breusch–Godfrey Test and Ljung–Box Test offer more flexibility for higher-order and multi-lag dependencies. Non-parametric tests like the Runs Test serve as useful supplementary diagnostics (Table 5.3).

Table 5.3 Summary of autocorrelation tests

Test	Type	Key statistic	Detects	When to use
Durbin–Watson	Parametric	DW	First-order autocorrelation	Simple linear models without lagged dependent variables
Breusch–Godfrey	Parametric (general)	χ^2	Higher-order autocorrelation	Models with lagged dependent variables
Ljung–Box	Portmanteau (global test)	χ^2	Autocorrelation at multiple lags	Time-series models (e.g., ARIMA)
Runs Test	Non-parametric	Z-statistic	Non-random patterns in residuals	Supplementary diagnostic for randomness

```
# Install and load necessary libraries
# install.packages("lmtest")  # For Durbin--Watson and Breusch--Godfrey
↪  Tests
# install.packages("tseries") # For Runs Test
# install.packages("forecast")# For Ljung--Box Test

library(lmtest)
library(tseries)
#> Registered S3 method overwritten by 'quantmod':
#>   method            from
#>   as.zoo.data.frame zoo
library(forecast)

# Simulated time-series dataset
set.seed(123)
n <- 100
time <- 1:n
x1 <- rnorm(n, mean = 50, sd = 10)
x2 <- rnorm(n, mean = 30, sd = 5)

# Introducing autocorrelation in errors
epsilon <- arima.sim(model = list(ar = 0.6), n = n)
# AR(1) process with ρ = 0.6
y <- 5 + 0.4 * x1 - 0.3 * x2 + epsilon

# Original regression model
model <- lm(y ~ x1 + x2)

# ------------------------------------------------------------
# 1. Durbin--Watson Test
# ------------------------------------------------------------
# Null Hypothesis: No first-order autocorrelation
dw_test <- dwtest(model)
print(dw_test)
#>
#>  Durbin-Watson test
#>
#> data:  model
#> DW = 0.77291, p-value = 3.323e-10
#> alternative hypothesis: true autocorrelation is greater than 0

# ------------------------------------------------------------
# 2. Breusch--Godfrey Test
# ------------------------------------------------------------
# Null Hypothesis: No autocorrelation up to lag 2
bg_test <- bgtest(model, order = 2)  # Testing for autocorrelation up to
↪  lag 2
print(bg_test)
#>
#>  Breusch-Godfrey test for serial correlation of order
#>  up to 2
#>
#> data:  model
```

```
#> LM test = 40.314, df = 2, p-value = 1.762e-09

# -----------------------------------------------------------------
# 3. Ljung--Box Test
# -----------------------------------------------------------------
# Null Hypothesis: No autocorrelation up to lag 10
ljung_box_test <- Box.test(residuals(model), lag = 10, type =
↪ "Ljung-Box")
print(ljung_box_test)
#>
#>  Box-Ljung test
#>
#> data:  residuals(model)
#> X-squared = 50.123, df = 10, p-value = 2.534e-07

# -----------------------------------------------------------------
# 4. Runs Test (Non-parametric)
# -----------------------------------------------------------------
# Null Hypothesis: Residuals are randomly distributed
runs_test <- runs.test(as.factor(sign(residuals(model))))
print(runs_test)
#>
#>  Runs Test
#>
#> data:  as.factor(sign(residuals(model)))
#> Standard Normal = -4.2214, p-value = 2.428e-05
#> alternative hypothesis: two.sided
```

Interpretation of the Results

1. **Durbin–Watson Test**
    - **Null Hypothesis** (H_0): No first-order autocorrelation ($\rho = 0$).
    - **Alternative Hypothesis** (H_1): First-order autocorrelation exists ($\rho \neq 0$).
    - **Decision Rule:**
        - **Reject** H_0 if $DW < 1.5$ (positive autocorrelation) or $DW > 2.5$ (negative autocorrelation).
        - **Fail to reject** H_0 if $DW \approx 2$, suggesting no significant autocorrelation.

2. **Breusch–Godfrey Test**
    - **Null Hypothesis** (H_0): No autocorrelation up to lag p (here, $p = 2$).
    - **Alternative Hypothesis** (H_1): Autocorrelation exists at one or more lags.
    - **Decision Rule:**
        - **Reject** H_0 if $p\text{-value} < 0.05$, indicating significant autocorrelation.
        - **Fail to reject** H_0 if $p\text{-value} \geq 0.05$, suggesting no evidence of autocorrelation.

3. **Ljung–Box Test**
   - **Null Hypothesis** (H_0): No autocorrelation up to lag h (here, $h = 10$).
   - **Alternative Hypothesis** (H_1): Autocorrelation exists at one or more lags.
   - **Decision Rule:**
     - **Reject** H_0 if *p-value* < 0.05, indicating significant autocorrelation.
     - **Fail to reject** H_0 if *p-value* ≥ 0.05, suggesting no evidence of autocorrelation.
4. **Runs Test (Non-parametric)**
   - **Null Hypothesis** (H_0): Residuals are randomly distributed (no autocorrelation).
   - **Alternative Hypothesis** (H_1): Residuals exhibit non-random patterns (indicating autocorrelation).
   - **Decision Rule:**
     - **Reject** H_0 if *p-value* < 0.05, indicating non-randomness and potential autocorrelation.
     - **Fail to reject** H_0 if *p-value* ≥ 0.05, suggesting randomness in residuals.

5.6 Multicollinearity Diagnostics

Multicollinearity occurs when two or more independent variables in a regression model are **highly correlated**, leading to several issues:

- **Unstable coefficient estimates:** Small changes in the data can cause large fluctuations in parameter estimates.
- **Inflated standard errors:** Reduces the precision of estimated coefficients, making it difficult to determine the significance of predictors.
- **Difficulty in assessing variable importance:** It becomes challenging to isolate the effect of individual predictors on the dependent variable.

Multicollinearity does not affect the overall fit of the model (e.g., R^2 remains high), but it distorts the reliability of individual coefficient estimates.

Key Multicollinearity Diagnostics
1. Variance Inflation Factor
2. Tolerance Statistic
3. Condition Index and Eigenvalue Decomposition
4. Pairwise Correlation Matrix
5. Determinant of the Correlation Matrix

5.6.1 *Variance Inflation Factor*

The **Variance Inflation Factor (VIF)** is the most commonly used diagnostic for detecting multicollinearity. It measures how much the **variance of an estimated regression coefficient** is inflated due to multicollinearity compared to when the predictors are uncorrelated.

For each predictor X_j, the VIF is defined as:

$$\text{VIF}_j = \frac{1}{1 - R_j^2}$$

where:

- R_j^2 is the **coefficient of determination** obtained by regressing X_j on all other independent variables in the model.

Interpretation of VIF

- **VIF = 1:** No multicollinearity (perfect independence).
- **1 < VIF < 5:** Moderate correlation, typically not problematic.
- **VIF $\approx$ 5:** High correlation; consider investigating further.
- **VIF $\approx$ 10:** Severe multicollinearity; corrective action is recommended.

Procedure

1. Regress each independent variable (X_j) on the remaining predictors.
2. Compute R_j^2 for each regression.
3. Calculate $\text{VIF}_j = 1/(1 - R_j^2)$.
4. Analyze VIF values to identify problematic predictors.

Advantages and Limitations

- **Advantage:** Easy to compute and interpret.
- **Limitation:** Detects only **linear** relationships; may not capture complex multicollinearity patterns involving multiple variables simultaneously.

5.6.2 *Tolerance Statistic*

The **Tolerance Statistic** is the **reciprocal of the VIF** and measures the proportion of variance in an independent variable **not explained** by the other predictors.

$$\text{Tolerance}_j = 1 - R_j^2$$

where R_j^2 is defined as in the VIF calculation.

Interpretation of Tolerance

- **Tolerance close to 1:** Low multicollinearity.
- **Tolerance < 0.2:** Potential multicollinearity problem.
- **Tolerance < 0.1:** Severe multicollinearity.

Since **low tolerance** implies **high VIF**, both metrics provide consistent information.

Advantages and Limitations

- **Advantage:** Provides an intuitive measure of how much variance is "free" from multicollinearity.
- **Limitation:** Similar to VIF, focuses on linear dependencies.

5.6.3 Condition Index and Eigenvalue Decomposition

The **Condition Index** is a more advanced diagnostic that detects **multicollinearity involving multiple variables** simultaneously. It is based on the **eigenvalues** of the scaled independent variable matrix.

1. **Compute the scaled design matrix** $X'X$, where X is the matrix of independent variables.
2. **Perform eigenvalue decomposition** to obtain the eigenvalues $\lambda_1, \lambda_2, \ldots, \lambda_k$.
3. **Calculate the Condition Index:**

$$\mathrm{CI}_j = \sqrt{\frac{\lambda_{\max}}{\lambda_j}}$$

where:

- $\lambda_{\max}$ is the largest eigenvalue,
- λ_j is the j-th eigenvalue.

Interpretation of Condition Index

- **CI < 10:** No serious multicollinearity.
- **10 $\leq$ CI < 30:** Moderate to strong multicollinearity.
- **CI $\geq$ 30:** Severe multicollinearity.

A **high condition index** indicates near-linear dependence among variables.

Variance Decomposition Proportions

To identify which variables contribute to multicollinearity:

- Compute the **Variance Decomposition Proportions (VDPs)** for each coefficient across eigenvalues.
- If **two or more variables** have high VDPs (e.g., > 0.5) associated with a **high condition index**, this indicates severe multicollinearity.

Advantages and Limitations

- **Advantage:** Detects multicollinearity involving **multiple variables**, which VIF may miss.
- **Limitation:** Requires matrix algebra knowledge; less intuitive than VIF or tolerance.

5.6.4 Pairwise Correlation Matrix

A **Pairwise Correlation Matrix** provides a simple diagnostic by computing the correlation coefficients between each pair of independent variables.

For variables X_i and X_j, the **correlation coefficient** is:

$$\rho_{ij} = \frac{\text{Cov}(X_i, X_j)}{\sigma_{X_i}\sigma_{X_j}}$$

where:

- $\text{Cov}(X_i, X_j)$ is the covariance,
- σ_{X_i} and σ_{X_j} are standard deviations.

Interpretation of Correlation Coefficients

- $|\rho| < 0.5$: Weak correlation (unlikely to cause multicollinearity).
- $0.5 \leq |\rho| < 0.8$: Moderate correlation; monitor carefully.
- $|\rho| \geq 0.8$: Strong correlation; potential multicollinearity issue.

Advantages and Limitations

- **Advantage:** Quick and easy to compute; useful for initial screening.
- **Limitation:** Detects **only pairwise relationships**; may miss multicollinearity involving more than two variables.

5.6.5 Determinant of the Correlation Matrix

The **Determinant of the Correlation Matrix** provides a global measure of multicollinearity. A **small determinant** indicates high multicollinearity.

1. Form the **correlation matrix** R of the independent variables.
2. Compute the **determinant**:

$$\det(R)$$

Interpretation

- $\det(R) \approx 1$: No multicollinearity (perfect independence).
- $\det(R) \approx 0$: Severe multicollinearity.

A determinant **close to zero** suggests that the correlation matrix is nearly singular, indicating strong multicollinearity.

Advantages and Limitations

- **Advantage:** Provides a single summary statistic for overall multicollinearity.
- **Limitation:** Does not indicate **which** variables are causing the problem.

5.6.6 *Summary of Multicollinearity Diagnostics*

See Table 5.4.

5.6.7 *Addressing Multicollinearity*

If multicollinearity is detected, consider the following solutions:

1. **Remove or combine correlated variables:** Drop one of the correlated predictors or create an index/aggregate.

Table 5.4 Summary of multicollinearity diagnostics

Diagnostic	Type	Key metric	Threshold for concern	When to use
Variance Inflation Factor	Parametric	$VIF = \frac{1}{1-R_j^2}$	$VIF \geq 5$ (moderate), $VIF \geq 10$ (severe)	General-purpose detection
Tolerance Statistic	Parametric	$1 - R_j^2$	< 0.2 (moderate), < 0.1 (severe)	Reciprocal of VIF for variance interpretation
Condition Index	Eigenvalue-based	$\sqrt{\frac{\lambda_{\max}}{\lambda_j}}$	> 10 (moderate), > 30 (severe)	Detects multicollinearity among multiple variables
Pairwise Correlation Matrix	Correlation-based	Pearson correlation (ρ)	$\lvert\rho\rvert \geq 0.8$	Initial screening for bivariate correlations
Determinant of Correlation Matrix	Global diagnostic	$\det(R)$	≈ 0 indicates severe multicollinearity	Overall assessment of multicollinearity

2. **Principal Component Analysis:** Reduce dimensionality by transforming correlated variables into uncorrelated components.
3. Ridge Regression **(L2 regularization):** Introduces a penalty term to stabilize coefficient estimates in the presence of multicollinearity.
4. **Centering variables:** Mean-centering can help reduce multicollinearity, especially in interaction terms.

Multicollinearity can significantly distort regression estimates, leading to misleading interpretations. While VIF and Tolerance are commonly used diagnostics, advanced techniques like the Condition Index and Eigenvalue Decomposition provide deeper insights, especially when dealing with complex datasets.

```
# Install and load necessary libraries
# install.packages("car")        # For VIF calculation
# install.packages("corpcor")    # For determinant of correlation matrix


library(car)
library(corpcor)

# Simulated dataset with multicollinearity
set.seed(123)
n <- 100
x1 <- rnorm(n, mean = 50, sd = 10)
x2 <- 0.8 * x1 + rnorm(n, sd = 2)   # Highly correlated with x1
x3 <- rnorm(n, mean = 30, sd = 5)
y <- 5 + 0.4 * x1 - 0.3 * x2 + 0.2 * x3 + rnorm(n)

# Original regression model
model <- lm(y ~ x1 + x2 + x3)

# ---------------------------------------------------------------
# 1. Variance Inflation Factor (VIF)
# ---------------------------------------------------------------
# Null Hypothesis: No multicollinearity (VIF = 1)
vif_values <- vif(model)
print(vif_values)
#>        x1        x2        x3
#> 14.969143 14.929013  1.017576

# ---------------------------------------------------------------
# 2. Tolerance Statistic (Reciprocal of VIF)
# ---------------------------------------------------------------
tolerance_values <- 1 / vif_values
print(tolerance_values)
#>         x1         x2         x3
#> 0.06680409 0.06698366 0.98272742

# ---------------------------------------------------------------
# 3. Condition Index and Eigenvalue Decomposition
# ---------------------------------------------------------------
# Scaling the independent variables
```

```
X <- model.matrix(model)[,-1]  # Removing intercept
eigen_decomp <-
    eigen(cor(X))   # Eigenvalue decomposition of the correlation matrix

# Condition Index
condition_index <-
    sqrt(max(eigen_decomp$values) / eigen_decomp$values)
print(condition_index)
#> [1] 1.000000 1.435255 7.659566

# Variance Decomposition Proportions (VDP)
# Proportions calculated based on the squared coefficients
loadings <- eigen_decomp$vectors
vdp <- apply(loadings ^ 2, 2, function(x)
    x / sum(x))
print(vdp)
#>            [,1]       [,2]         [,3]
#> [1,] 0.48567837 0.01363318 5.006885e-01
#> [2,] 0.48436754 0.01638399 4.992485e-01
#> [3,] 0.02995409 0.96998283 6.307954e-05

# -------------------------------------------------------------
# 4. Pairwise Correlation Matrix
# -------------------------------------------------------------
correlation_matrix <- cor(X)
print(correlation_matrix)
#>           x1         x2         x3
#> x1  1.000000  0.9659070 -0.1291760
#> x2  0.965907  1.0000000 -0.1185042
#> x3 -0.129176 -0.1185042  1.0000000

# -------------------------------------------------------------
# 5. Determinant of the Correlation Matrix
# -------------------------------------------------------------
determinant_corr_matrix <- det(correlation_matrix)
cat("Determinant of the Correlation Matrix:",
    determinant_corr_matrix,
    "\n")
#> Determinant of the Correlation Matrix: 0.06586594
```

Interpretation of the Results

1. **Variance Inflation Factor (VIF)**

- **Formula:** $\text{VIF}_j = \frac{1}{1-R_j^2}$
- **Decision Rule:**
  - **VIF** ≈ 1: No multicollinearity.
  - $1 <$ VIF < 5: Moderate correlation, usually acceptable.
  - VIF ≥ 5: High correlation; investigate further.
  - VIF ≥ 10: Severe multicollinearity; corrective action recommended.

2. **Tolerance Statistic**

   - **Formula:** $\text{Tolerance}_j = 1 - R_j^2 = \frac{1}{\text{VIF}_j}$
   - **Decision Rule:**

     – **Tolerance** > 0.2: Low risk of multicollinearity.
     – **Tolerance** < 0.2: Possible multicollinearity problem.
     – **Tolerance** < 0.1: Severe multicollinearity.

3. **Condition Index and Eigenvalue Decomposition**

   - **Formula:** $\text{CI}_j = \sqrt{\frac{\lambda_{\max}}{\lambda_j}}$
   - **Decision Rule:**

     – **CI** < 10: No significant multicollinearity.
     – $10 \leq \text{CI} < 30$: Moderate to strong multicollinearity.
     – **CI** ≥ 30: Severe multicollinearity.

   - **Variance Decomposition Proportions (VDP):**

     – High VDP (> 0.5) associated with high CI indicates problematic variables.

4. **Pairwise Correlation Matrix**

   - **Decision Rule:**

     – $|\rho| < 0.5$: Weak correlation.
     – $0.5 \leq |\rho| < 0.8$: Moderate correlation; monitor.
     – $|\rho| \geq 0.8$: Strong correlation; potential multicollinearity issue.

5. **Determinant of the Correlation Matrix**

   - **Decision Rule:**

     – $\det(R) \approx 1$: No multicollinearity.
     – $\det(R) \approx 0$: Severe multicollinearity (near-singular matrix).

Model specification tests are essential for diagnosing and validating econometric models. They ensure that the model assumptions hold true, thereby improving the accuracy and reliability of the estimations. By systematically applying these tests, researchers can identify issues related to nested and non-nested models, heteroskedasticity, functional form, endogeneity, autocorrelation, and multicollinearity, leading to more robust and credible econometric analyses.

Chapter 6
Variable Selection

Imagine you're a detective standing before a pinboard covered in clues—some are glaringly obvious, while others might be red herrings. Your mission? To pick which pieces of evidence will crack the case. This is the essence of variable selection in statistics: deciding which variables best uncover the story behind your data. Far from a mechanical chore, it's a high-stakes balancing act blending analytical goals, domain insights, data realities, and computational feasibility.

Why Does Variable Selection Matter?

- **Focus and Clarity**: Models cluttered with unnecessary variables can obscure the real relationships or patterns in your data. By identifying the variables that truly drive your results, you sharpen your model's focus and interpretability.
- **Efficiency and Performance**: Too many variables can lead to overfitting—fitting the quirks of a single dataset rather than underlying trends. Streamlined models often run faster and generalize better.
- **Practical Constraints**: In many real-world scenarios, data collection or processing costs money, time, and effort. Prioritizing the most meaningful variables becomes not just a statistical concern, but a strategic one.

Key Influences on Variable Selection

1. **Objectives or Goals**
    - *Prediction vs. Inference*: Are you trying to forecast future outcomes or explain why certain events happen? Prediction-focused models might include as many relevant features as possible for accuracy, whereas inference-driven models often strive for parsimony and clearer relationships.
    - *Balance*: Some analyses blend both objectives, requiring careful negotiation between complexity (to maximize predictive ability) and simplicity (to maintain interpretability).

© The Author(s), under exclusive license to Springer Nature Switzerland AG 2025

M. Nguyen, *Advanced Modeling and Data Challenges*,
https://doi.org/10.1007/978-3-032-01719-2_6

2. **Previously Acquired Expertise**
   - *Domain Knowledge*: Whether you're analyzing financial trends or studying medical records, familiarity with the subject can reveal which variables are naturally linked to the phenomenon.
   - *Subtle Clues*: Experts can uncover hidden confounders—variables that outwardly seem irrelevant yet dramatically influence results.
3. **Availability and Quality of Data**
   - *Completeness*: Missing data or sparse measurements can force you to discard or transform variables. Sometimes the ideal variable simply isn't present in your dataset.
   - *Reliability*: A variable riddled with measurement errors or inconsistencies may do more harm than good.
4. **Computational Resources and Software**
   - *Toolset Capabilities*: Some statistical techniques or advanced machine learning methods thrive on large sets of variables, while others become unwieldy.
   - *Time and Memory Constraints*: Even the most sophisticated algorithms can choke on too much data if hardware resources are limited.

Selecting the right subset of variables enhances model interpretability, reduces computational cost, and prevents overfitting (Table 6.1). Broadly, variable selection methods fall into three categories:

1. **Filter Methods**: Use statistical properties of the data to select features before modeling.
   1. Information Criteria-Based Selection
      1. Akaike Information Criterion (AIC)
      2. Bayesian Information Criterion (BIC)
      3. Mallows's C Statistic
      4. Hannan-Quinn Criterion (HQC)
      5. Minimum Description Length (MDL)

Table 6.1 Throughout this chapter, let P denote the number of potential predictor variables $(X_1, X_2, \ldots, X_{P-1})$

Method category	Examples	Pros	Cons
Filter	AIC, BIC, Mutual Info, CFS	Fast, scalable, model-agnostic	Ignores feature interactions
Wrapper	Stepwise Selection, RFE	Finds optimal feature subsets	Computationally expensive
Embedded	Lasso, Decision Trees	Model-aware, balances efficiency	Selection tied to specific models

      6. Adjusted R^2
      7. Prediction Error Sum of Squares (PRESS)

   2. Univariate Selection Methods
   3. Correlation-Based Feature Selection
   4. Variance Thresholding

2. **Wrapper Methods**: Evaluate different subsets of features based on model performance.

   1. Exhaustive Search (Best Subsets Algorithm)
   2. Best Subsets Algorithm
   3. Stepwise Selection Methods

      1. Forward Selection
      2. Backward Elimination
      3. Stepwise (Both Directions) Selection

   4. Branch-and-Bound Algorithm
   5. Recursive Feature Elimination

3. **Embedded Methods**: Perform feature selection as part of the model training process.

   1. [Lasso Regression] (L1 Regularization)
   2. [Ridge Regression] (L2 Regularization)
   3. [Elastic Net] (Combination of L1 and L2)
   4. Tree-Based Feature Importance
   5. Genetic Algorithms

6.1 Filter Methods (Statistical Criteria, Model-Agnostic)

6.1.1 Information Criteria-Based Selection

6.1.1.1 Mallows's C Statistic

The C_p statistic (Mallows, 1973, *Technometrics*, 15, 661-675) (Mallows, 1995) is a criterion used to evaluate the predictive ability of a fitted model. It balances model complexity and goodness-of-fit.

For a model with p parameters, let $\hat{Y}_{ip}$ be the predicted value of Y_i. The **total standardized mean square error of prediction** is:

$$\Gamma_p = \frac{\sum_{i=1}^{n} E(\hat{Y}_{ip} - E(Y_i))^2}{\sigma^2}$$

Expanding Γ_p:

$$\Gamma_p = \frac{\sum_{i=1}^{n}[E(\hat{Y}_{ip}) - E(Y_i)]^2 + \sum_{i=1}^{n} \text{Var}(\hat{Y}_{ip})}{\sigma^2}$$

- The **first term** in the numerator represents the squared bias.
- The **second term** represents the prediction variance.

Key Insights

1. **Bias-Variance Tradeoff**:
   - The **bias** decreases as more variables are added to the model.
   - If the **full model** ($p = P$) is assumed to be the true model, $E(\hat{Y}_{ip}) - E(Y_i) = 0$, implying no bias.
   - The **prediction variance** increases as more variables are added: $\sum \text{Var}(\hat{Y}_{ip}) = p\sigma^2$.
   - Therefore, the optimal model balances bias and variance by minimizing Γ_p.
2. **Estimating** Γ_p: Since Γ_p depends on unknown parameters (e.g., β), we use an estimate:

$$C_p = \frac{SSE_p}{\hat{\sigma}^2} - (n - 2p)$$

   - SSE_p: Sum of squared errors for the model with p predictors.
   - $\hat{\sigma}^2$: Mean-squared error (MSE) of the full model with all $P - 1$ predictors.
3. **Properties of** C_p:
   - As more variables are added, SSE_p decreases, but the penalty term $2p$ increases.
   - When there is no bias, $E(C_p) \approx p$. Hence, good models have C_p values close to p.
4. **Model Selection Criteria**:
   - **Prediction-focused models**: Consider models with $C_p \leq p$.
   - **Parameter estimation-focused models**: Consider models with $C_p \leq 2p - (P - 1)$ to avoid excess bias.

```
# Simulated data
set.seed(123)
n <- 100
x1 <- rnorm(n)
x2 <- rnorm(n)
x3 <- rnorm(n)
y <- 5 + 3*x1 - 2*x2 + rnorm(n, sd=2)

# Full model and candidate models
full_model <- lm(y ~ x1 + x2 + x3)
```

```
model_1 <- lm(y ~ x1)
model_2 <- lm(y ~ x1 + x2)

# Extract SSE and calculate Cp
calculate_cp <- function(model, full_model_sse, full_model_mse, n) {
  sse <- sum(residuals(model)^2)
  p <- length(coefficients(model))
  cp <- (sse / full_model_mse) - (n - 2 * p)
  return(cp)
}

# Full model statistics
full_model_sse <- sum(residuals(full_model)^2)
full_model_mse <- mean(residuals(full_model)^2)

# Cp values for each model
cp_1 <- calculate_cp(model_1, full_model_sse, full_model_mse, n)
cp_2 <- calculate_cp(model_2, full_model_sse, full_model_mse, n)

# Display results
cat("C_p values:\n")
#> C_p values:
cat("Model 1 (y ~ x1):", round(cp_1, 2), "\n")
#> Model 1 (y ~ x1): 83.64
cat("Model 2 (y ~ x1 + x2):", round(cp_2, 2), "\n")
#> Model 2 (y ~ x1 + x2): 6.27
```

For Mallows's C_p criterion, **lower values** are preferred. Specifically:

1. **Ideal Value**: When the model is a good fit and has the correct number of predictors, C_p should be close to the number of predictors p plus 1 (i.e., $p + 1$).
2. **Model Comparison**: Among competing models, you generally prefer the one with the smallest C_p, as long as it is close to $p + 1$.
3. **Overfitting Indicator**: If C_p is significantly lower than $p + 1$, it may suggest overfitting.
4. **Underfitting Indicator**: If C_p is much higher than $p + 1$, it suggests that the model is underfitting the data and missing important predictors.

6.1.1.2 Akaike Information Criterion (AIC)

The **Akaike Information Criterion (AIC)** is a widely used model selection metric that evaluates the trade-off between model fit and complexity. It was introduced by Hirotugu Akaike and is rooted in information theory, measuring the relative quality of statistical models for a given dataset.

For a model with p parameters, the AIC is given by:

$$AIC = n \ln\left(\frac{SSE_p}{n}\right) + 2p$$

where:

- n is the number of observations.
- SSE_p is the sum of squared errors for a model with p parameters.

Key Insights

1. **Components of AIC**:
   - The **first term** ($n \ln(SSE_p/n)$): Reflects the goodness-of-fit of the model. It decreases as SSE_p decreases, meaning the model better explains the data.
   - The **second term** ($2p$): Represents a penalty for model complexity. It increases with the number of parameters to discourage overfitting.
2. **Model Selection Principle**:
   - Smaller AIC values indicate a better balance between fit and complexity.
   - Adding parameters generally reduces SSE_p, but increases the penalty term ($2p$). If AIC increases when a parameter is added, that parameter is likely unnecessary.
3. **Tradeoff**:
   - AIC emphasizes the trade-off between:
     - **Precision of fit**: Reducing the error in explaining the data.
     - **Parsimony**: Avoiding unnecessary parameters to maintain model simplicity.
4. **Comparative Criterion**:
   - AIC does not provide an absolute measure of model quality; instead, it compares relative performance. The model with the lowest AIC is preferred.

```
# Simulated data
set.seed(123)
n <- 100
x1 <- rnorm(n)
x2 <- rnorm(n)
x3 <- rnorm(n)
y <- 5 + 3*x1 - 2*x2 + rnorm(n, sd=2)

# Candidate models
model_1 <- lm(y ~ x1)
model_2 <- lm(y ~ x1 + x2)
model_3 <- lm(y ~ x1 + x2 + x3)

# Function to manually compute AIC
calculate_aic <- function(model, n) {
  sse <- sum(residuals(model)^2)
  p <- length(coefficients(model))
  aic <- n * log(sse / n) + 2 * p
```

```
  return(aic)
}

# Calculate AIC for each model
aic_1 <- calculate_aic(model_1, n)
aic_2 <- calculate_aic(model_2, n)
aic_3 <- calculate_aic(model_3, n)

# Display results
cat("AIC values:\n")
#> AIC values:
cat("Model 1 (y ~ x1):", round(aic_1, 2), "\n")
#> Model 1 (y ~ x1): 207.17
cat("Model 2 (y ~ x1 + x2):", round(aic_2, 2), "\n")
#> Model 2 (y ~ x1 + x2): 150.87
cat("Model 3 (y ~ x1 + x2 + x3):", round(aic_3, 2), "\n")
#> Model 3 (y ~ x1 + x2 + x3): 152.59
```

Interpretation

- Compare the AIC values across models:
  - A smaller AIC indicates a model with a better balance between fit and complexity.
  - If the AIC increases when moving from one model to another (e.g., from Model 2 to Model 3), the additional parameter(s) in the larger model may not be justified.

Advantages

- Simple to compute and widely applicable.
- Penalizes overfitting, encouraging parsimonious models.

Limitations

- Assumes that the model errors are normally distributed and independent.
- Does not evaluate absolute model fit, only relative performance.
- Sensitive to sample size; for smaller samples, consider using the corrected version, AICc.

Corrected AIC (AICc)
For small sample sizes ($n/p \leq 40$), the corrected AIC, $AICc$, adjusts for the sample size:

$$AICc = AIC + \frac{2p(p+1)}{n-p-1}$$

This adjustment prevents over-penalizing models with more parameters when n is small.

6.1.1.3 Bayesian Information Criterion (BIC)

The **Bayesian Information Criterion (BIC)**, also known as Schwarz Criterion, is another popular metric for model selection. It extends the concept of AIC by introducing a stronger penalty for model complexity, particularly when the number of observations is large. BIC is grounded in Bayesian probability theory and provides a framework for selecting the most plausible model among a set of candidates.

For a model with p parameters, the BIC is defined as:

$$BIC = n \ln\left(\frac{SSE_p}{n}\right) + p \ln(n)$$

where:

- n is the number of observations.
- SSE_p is the sum of squared errors for the model with p parameters.
- p is the number of parameters, including the intercept.

Key Insights

1. **Components of BIC**:
   - The **first term** ($n \ln(SSE_p/n)$): Measures the goodness-of-fit, similar to AIC. It decreases as SSE_p decreases, indicating a better fit to the data.
   - The **second term** ($p \ln(n)$): Penalizes model complexity. Unlike AIC's penalty ($2p$), the penalty in BIC increases with $\ln(n)$, making it more sensitive to the number of observations.
2. **Model Selection Principle**:
   - Smaller BIC values indicate a better model.
   - Adding parameters reduces SSE_p, but the penalty term $p \ln(n)$ grows more rapidly than AIC's $2p$ for large n. This makes BIC more conservative than AIC in selecting models with additional parameters.
3. **Trade-off**:
   - Like AIC, BIC balances:
     - **Precision of fit**: Capturing the underlying structure in the data.
     - **Parsimony**: Avoiding overfitting by discouraging unnecessary parameters.
   - BIC tends to favor simpler models compared to AIC, particularly when n is large.
4. **Comparative Criterion**:
   - BIC, like AIC, is used to compare models. The model with the smallest BIC is preferred.

```r
# Function to manually compute BIC
calculate_bic <- function(model, n) {
  sse <- sum(residuals(model)^2)
  p <- length(coefficients(model))
  bic <- n * log(sse / n) + p * log(n)
  return(bic)
}

# Calculate BIC for each model
bic_1 <- calculate_bic(model_1, n)
bic_2 <- calculate_bic(model_2, n)
bic_3 <- calculate_bic(model_3, n)

# Display results
cat("BIC values:\n")
#> BIC values:
cat("Model 1 (y ~ x1):", round(bic_1, 2), "\n")
#> Model 1 (y ~ x1): 212.38
cat("Model 2 (y ~ x1 + x2):", round(bic_2, 2), "\n")
#> Model 2 (y ~ x1 + x2): 158.68
cat("Model 3 (y ~ x1 + x2 + x3):", round(bic_3, 2), "\n")
#> Model 3 (y ~ x1 + x2 + x3): 163.01
```

Interpretation

- Compare the BIC values across models:
  - Smaller BIC values suggest a better model.
  - If BIC increases when moving to a larger model, the added complexity may not justify the reduction in SSE_p.

Comparison with AIC

Criterion	Penalty term	Sensitivity to sample size	Preferred model selection
AIC	$2p$	Less sensitive	More parameters
BIC	$p \ln(n)$	More sensitive	Simpler models

- BIC generally prefers simpler models than AIC, especially when n is large.
- In small datasets, AIC may perform better because BIC's penalty grows significantly with $\ln(n)$.

Advantages

- Strong penalty for complexity makes it robust against overfitting.
- Incorporates sample size explicitly, favoring simpler models as n grows.
- Easy to compute and interpret.

Limitations

- Assumes that model errors are normally distributed and independent.
- May underfit in smaller datasets where the penalty term dominates.
- Like AIC, BIC is not an absolute measure of model quality but a relative one.

6.1.1.4 Hannan–Quinn Criterion (HQC)

The **Hannan–Quinn Criterion (HQC)** is a statistical metric for model selection, similar to AIC and BIC. It evaluates the trade-off between model fit and complexity, offering a middle ground between the conservative penalty of BIC and the less stringent penalty of AIC. HQC is especially useful in time-series modeling and situations where large datasets are involved.

The HQC for a model with p parameters is defined as:

$$HQC = n \ln\left(\frac{SSE_p}{n}\right) + 2p \ln(\ln(n))$$

where:

- n: Number of observations.
- SSE_p: Sum of Squared Errors for the model with p predictors.
- p: Number of parameters, including the intercept.

Key Insights

1. **Components**:
   - The **first term** ($n \ln(SSE_p/n)$): Measures the goodness-of-fit, similar to AIC and BIC. Smaller SSE indicates a better fit.
   - The **second term** ($2p \ln(\ln(n))$): Penalizes model complexity. The penalty grows logarithmically with the sample size, similar to BIC but less severe.
2. **Model Selection Principle**:
   - Smaller HQC values indicate a better balance between model fit and complexity.
   - Models with lower HQC are preferred.
3. **Penalty Comparison**:
   - HQC's penalty lies between that of AIC and BIC:
     - AIC: $2p$ (less conservative, favors complex models).
     - BIC: $p \ln(n)$ (more conservative, favors simpler models).
     - HQC: $2p \ln(\ln(n))$ (balances AIC and BIC).
4. **Use Case**:
   - HQC is particularly suited for large datasets or time-series models where overfitting is a concern, but BIC may overly penalize model complexity.

```r
# Simulated data
set.seed(123)
n <- 100
x1 <- rnorm(n)
x2 <- rnorm(n)
x3 <- rnorm(n)
x4 <- rnorm(n)
y <- 5 + 3*x1 - 2*x2 + x3 + rnorm(n, sd=2)

# Prepare models
data <- data.frame(y, x1, x2, x3, x4)
model_1 <- lm(y ~ x1, data = data)
model_2 <- lm(y ~ x1 + x2, data = data)
model_3 <- lm(y ~ x1 + x2 + x3, data = data)

# Function to calculate HQC
calculate_hqc <- function(model, n) {
  sse <- sum(residuals(model)^2)
  p <- length(coefficients(model))
  hqc <- n * log(sse / n) + 2 * p * log(log(n))
  return(hqc)
}

# Calculate HQC for each model
hqc_1 <- calculate_hqc(model_1, n)
hqc_2 <- calculate_hqc(model_2, n)
hqc_3 <- calculate_hqc(model_3, n)

# Display results
cat("HQC values:\n")
#> HQC values:
cat("Model 1 (y ~ x1):", round(hqc_1, 2), "\n")
#> Model 1 (y ~ x1): 226.86
cat("Model 2 (y ~ x1 + x2):", round(hqc_2, 2), "\n")
#> Model 2 (y ~ x1 + x2): 156.44
cat("Model 3 (y ~ x1 + x2 + x3):", round(hqc_3, 2), "\n")
#> Model 3 (y ~ x1 + x2 + x3): 141.62
```

Interpretation

1. **Comparing HQC Values**:
   - Smaller HQC values indicate a better balance between goodness-of-fit and parsimony.
   - Select the model with the smallest HQC.
2. **Tradeoffs**:
   - HQC balances fit and complexity more conservatively than AIC but less so than BIC.
   - It is particularly useful when overfitting is a concern but avoiding overly simplistic models is also important.

Comparison with Other Criteria

Criterion	Penalty Term	Sensitivity to sample size	Preferred model selection
AIC	$2p$	Low	Favors more complex models
BIC	$p\ln(n)$	High	Favors simpler models
HQC	$2p\ln(\ln(n))$	Moderate	Balances fit and parsimony

Advantages

- Less sensitive to sample size than BIC, avoiding excessive penalization for large datasets.
- Provides a balanced approach to model selection, reducing the risk of overfitting while avoiding overly simplistic models.
- Particularly useful in time-series analysis.

Limitations

- Like AIC and BIC, assumes that model errors are normally distributed and independent.
- HQC is not as widely implemented in statistical software, requiring custom calculations.

Practical Considerations

- **When to use HQC?**
  - When both AIC and BIC provide conflicting recommendations.
  - For large datasets or time-series models where BIC may overly penalize complexity.
- **When to use AIC or BIC?**
  - AIC for smaller datasets or when the goal is prediction.
  - BIC for large datasets or when parsimony is critical.

6.1.1.5 Minimum Description Length (MDL)

The **Minimum Description Length (MDL)** principle is a model selection framework rooted in information theory. It balances model complexity and goodness-of-fit by seeking the model that minimizes the total length of encoding the data and the model itself. MDL is a generalization of other model selection criteria like AIC and BIC but offers a more theoretical foundation.

Theoretical Foundation
MDL is based on the idea that the best model is the one that compresses the data most efficiently. It represents a trade-off between:

1. **Model Complexity**: The cost of describing the model, including the number of parameters.
2. **Data Fit**: The cost of describing the data given the model.

   The total description length is expressed as:

$$L(M, D) = L(M) + L(D|M)$$

where:

- $L(M)$: The length of encoding the model (complexity of the model).
- $L(D|M)$: The length of encoding the data given the model (fit to the data).

Key Insights

1. **Model Complexity** ($L(M)$):
   - More complex models require longer descriptions, as they involve more parameters.
   - Simpler models are favored unless the added complexity significantly improves the fit.
2. **Data Fit** ($L(D|M)$):
   - Measures how well the model explains the data.
   - Poorly fitting models require more bits to describe the residual error.
3. **Trade-off**:
   - MDL balances these two components, selecting the model that minimizes the total description length.

Connection to Other Criteria

- MDL is closely related to BIC. In fact, the BIC criterion can be derived as an approximation of MDL for certain statistical models:

$$BIC = n \ln(SSE_p/n) + p \ln(n)$$

- However, MDL is more flexible and does not rely on specific assumptions about the error distribution.

Practical Use Cases

- **Time-Series Modeling**: MDL is particularly effective for selecting models in time-series data, where overfitting is common.
- **Machine Learning**: MDL is used in regularization techniques and decision tree pruning to prevent overfitting.
- **Signal Processing**: In applications such as compression and coding, MDL directly guides optimal model selection.

```r
# Simulated data
set.seed(123)
n <- 100
x1 <- rnorm(n)
x2 <- rnorm(n)
x3 <- rnorm(n)
y <- 5 + 3*x1 - 2*x2 + x3 + rnorm(n, sd=2)

# Prepare models
data <- data.frame(y, x1, x2, x3)
model_1 <- lm(y ~ x1, data = data)
model_2 <- lm(y ~ x1 + x2, data = data)
model_3 <- lm(y ~ x1 + x2 + x3, data = data)

# Function to calculate MDL
calculate_mdl <- function(model, n) {
  sse <- sum(residuals(model)^2)
  p <- length(coefficients(model))
  mdl <- p * log(n) + n * log(sse / n)
  return(mdl)
}

# Calculate MDL for each model
mdl_1 <- calculate_mdl(model_1, n)
mdl_2 <- calculate_mdl(model_2, n)
mdl_3 <- calculate_mdl(model_3, n)

# Display results
cat("MDL values:\n")
#> MDL values:
cat("Model 1 (y ~ x1):", round(mdl_1, 2), "\n")
#> Model 1 (y ~ x1): 219.87
cat("Model 2 (y ~ x1 + x2):", round(mdl_2, 2), "\n")
#> Model 2 (y ~ x1 + x2): 173.42
cat("Model 3 (y ~ x1 + x2 + x3):", round(mdl_3, 2), "\n")
#> Model 3 (y ~ x1 + x2 + x3): 163.01
```

Interpretation

1. **Choosing the Best Model**:
   - The model with the smallest MDL value is preferred, as it achieves the best trade-off between fit and complexity.
2. **Practical Implications**:
   - MDL discourages overfitting by penalizing complex models that do not significantly improve data fit.

Advantages

- Theoretically grounded in information theory.
- Offers a natural framework for balancing complexity and fit.
- Flexible and can be applied across various modeling frameworks.

Limitations

- Computationally intensive, especially for nonlinear models.
- Requires careful formulation of $L(M)$ and $L(D I M)$ for non-standard models.
- Less common in standard statistical software compared to AIC or BIC.

6.1.1.6 Prediction Error Sum of Squares (PRESS)

The **Prediction Error Sum of Squares (PRESS)** statistic measures the predictive ability of a model by evaluating how well it performs on data not used in fitting the model. PRESS is particularly useful for assessing model validity and identifying overfitting.

The PRESS statistic for a model with p parameters is defined as:

$$PRESS_p = \sum_{i=1}^{n}(Y_i - \hat{Y}_{i(i)})^2$$

where:

- $\hat{Y}_{i(i)}$ is the prediction of the i-th response when the i-th observation is **omitted** during model fitting.
- Y_i is the observed response for the i-th observation.

Key Insights

1. **Leave-One-Out Cross-Validation (LOOCV)**:
    - PRESS is computed by excluding each observation one at a time and predicting its response using the remaining data.
    - This process evaluates the model's generalizability and reduces overfitting.
2. **Model Selection Principle**:
    - Smaller values of $PRESS_p$ indicate better predictive performance.
    - A small $PRESS_p$ suggests that the model captures the underlying structure of the data without overfitting.
3. **Computational Complexity**:
    - Computing $\hat{Y}_{i(i)}$ for each observation can be computationally intensive for models with large p or datasets with many observations.
    - Alternative approximations (e.g., using leverage values) can simplify computations.

```r
# Function to compute PRESS
calculate_press <- function(model) {
  residuals <- residuals(model)
  h <- lm.influence(model)$hat # leverage values
  press <- sum((residuals / (1 - h))^2) # PRESS formula using leverage
  return(press)
}

# Calculate PRESS for each model
press_1 <- calculate_press(model_1)
press_2 <- calculate_press(model_2)
press_3 <- calculate_press(model_3)

# Display results
cat("PRESS values:\n")
#> PRESS values:
cat("Model 1 (y ~ x1):", round(press_1, 2), "\n")
#> Model 1 (y ~ x1): 854.36
cat("Model 2 (y ~ x1 + x2):", round(press_2, 2), "\n")
#> Model 2 (y ~ x1 + x2): 524.56
cat("Model 3 (y ~ x1 + x2 + x3):", round(press_3, 2), "\n")
#> Model 3 (y ~ x1 + x2 + x3): 460
```

Interpretation

- Compare the PRESS values across models:
  - Models with smaller PRESS values are preferred as they exhibit better predictive ability.
  - A large PRESS value indicates potential overfitting or poor model generalizability.

Advantages

- Provides an unbiased measure of predictive performance.
- Helps identify overfitting by simulating the model's performance on unseen data.

Limitations

- Computationally intensive for large datasets or models with many predictors.
- Sensitive to influential observations; high-leverage points can disproportionately affect results.

Alternative Approaches To address the computational challenges of PRESS, alternative methods can be employed:

- **Approximation using leverage values**: As shown in the example, leverage values simplify the calculation of $\hat{Y}_{i(i)}$.
- **K-Fold Cross-Validation**: Dividing the dataset into k folds reduces computational burden compared to LOOCV while still providing robust estimates.

6.1.2 Univariate Selection Methods

Univariate selection methods evaluate individual variables in isolation to determine their relationship with the target variable. These methods are often categorized under filter methods, as they do not involve any predictive model but instead rely on statistical significance and information-theoretic measures.

Univariate selection is particularly useful for:

- **Preprocessing large datasets** by eliminating irrelevant features.
- **Reducing dimensionality** before applying more complex feature selection techniques.
- **Improving interpretability** by identifying the most relevant features.

The two main categories of univariate selection methods are:

1. **Statistical Tests**: Evaluate the significance of relationships between individual features and the target variable.
2. **Information-Theoretic Measures**: Assess the dependency between variables based on information gain and mutual information.

6.1.2.1 Statistical Tests

Statistical tests assess the significance of relationships between individual predictors and the target variable. The choice of test depends on the type of data:

Test	Used for	Example use case
Chi-Square test	Categorical predictors vs. Categorical target	Checking if gender affects purchase behavior
ANOVA (Analysis of variance)	Continuous predictors vs. Categorical target	Testing if different income groups have varying spending habits
Correlation coefficients	Continuous predictors vs. Continuous target	Measuring the relationship between advertising spend and sales

Check out [Descriptive Statistics] and [Basic Statistical Inference] for more details.

6.1.2.2 Information-Theoretic Measures

Information-theoretic measures assess variable relevance based on how much information they provide about the target.

6.1.2.2.1 Information Gain

Information Gain (IG) measures the reduction in uncertainty about the target variable when a predictor is known. It is used extensively in decision trees.

Formula

$$IG = H(Y) - H(Y|X)$$

where:

- $H(Y)$ = Entropy of target variable
- $H(Y|X)$ = Conditional entropy of target given predictor

  A higher IG indicates a more informative variable.

```
# Load Library
library(FSelector)

# Example: Computing Information Gain
data(iris)
information.gain(Species ~ ., iris)
#>              attr_importance
#> Sepal.Length       0.4521286
#> Sepal.Width        0.2672750
#> Petal.Length       0.9402853
#> Petal.Width        0.9554360
```

6.1.2.2.2 Mutual Information

Mutual Information (MI) quantifies how much knowing one variable reduces uncertainty about another. Unlike correlation, it captures both linear and nonlinear relationships.

- **Pros**: Captures nonlinear relationships, robust to outliers.
- **Cons**: More computationally intensive than correlation.

Formula

$$MI(X, Y) = \sum_{x,y} P(x, y) \log \frac{P(x, y)}{P(x)P(y)}$$

where:

- $P(x, y)$ = Joint probability distribution of X and Y.
- $P(x)$, $P(y)$ = Marginal probability distributions.

```
# Load Library
library(infotheo)

# Compute Mutual Information Between Two Features
set.seed(123)
X <- sample(1:5, 100, replace=TRUE)
Y <- sample(1:5, 100, replace=TRUE)

mutinformation(X, Y)
#> [1] 0.06852247
```

- Since X and Y are independently sampled, we expect them to have no mutual dependence.
- The MI value should be close to 0, indicating that knowing X provides almost no information about Y, and vice versa.
- If the MI value is significantly greater than 0, it could be due to random fluctuations, especially in small samples.

Method	Type	Suitable for	Pros	Cons
Chi-Square test	Statistical Test	Categorical vs. Categorical	Simple, interpretable	Requires large sample sizes
ANOVA	Statistical test	Continuous vs. Categorical	Handles multiple groups	Assumes normality
Correlation	Statistical Test	Continuous vs. Continuous	Easy to compute	Only captures linear relations
Information gain	Information-based	Any variable type	Good for decision trees	Requires computation of entropy
Mutual information	Information-based	Any variable type	Captures non-linear dependencies	More computationally expensive

6.1.3 Correlation-Based Feature Selection

Evaluates features based on their correlation with the target and redundancy with other features. Check out [Descriptive Statistics] and [Basic Statistical Inference] for more details.

6.1.4 Variance Thresholding

Variance Thresholding is a simple yet effective filter method used for feature selection. It removes features with **low variance**, assuming that low-variance features contribute little to model prediction. This technique is particularly useful when:

- **Handling high-dimensional datasets** where many features contain little useful information.
- **Reducing computational complexity** by removing uninformative features.
- **Avoiding overfitting** by eliminating features that are nearly constant across samples.

This method is most effective when dealing with **binary features** (e.g., categorical variables encoded as 0s and 1s) and **numerical features** with low variance.

Variance measures the **spread** of a feature's values. A feature with **low variance** contains nearly the same value for all observations, making it less useful for predictive modeling.

For a feature X, variance is calculated as:

$$Var(X) = \frac{1}{n}\sum_{i=1}^{n}(X_i - \bar{X})^2$$

where:

- X_i is an individual observation,
- $\bar{X}$ is the mean of X,
- n is the number of observations.

Example: Features with Low and High Variance

Feature	Sample values	Variance
Low variance	0, 0, 0, 1, 0, 0, 0, 0, 0, 0	0.04
High variance	3, 7, 1, 9, 2, 8, 6, 4, 5, 0	7.25

6.1.4.1 Identifying Low-Variance Features

A **variance threshold** is set to remove features below a certain variance level. The default threshold is **0**, which removes features with a single constant value across all samples.

```r
# Load necessary library
library(caret)
#> Loading required package: ggplot2
#> Loading required package: lattice

# Generate synthetic dataset
set.seed(123)
data <- data.frame(
  Feature1 = c(rep(0, 50), rep(1, 50)), # Low variance
  Feature2 = rnorm(100, mean=10, sd=1), # High variance
  Feature3 = runif(100, min=5, max=15), # Moderate variance
  Feature4 = c(rep(3, 95), rep(4, 5))  # Almost constant
)


# Compute Variance of Features
variances <- apply(data, 2, stats::var)
print(variances)
#>  Feature1  Feature2  Feature3  Feature4
#> 0.2525253 0.8332328 8.6631461 0.0479798

# Set threshold and remove low-variance features
threshold <- 0.1
selected_features <- names(variances[variances > threshold])
filtered_data <- data[, selected_features]

print(selected_features)  # Remaining features after filtering
#> [1] "Feature1" "Feature2" "Feature3"
```

6.1.4.2 Handling Binary Categorical Features

For **binary features** (0/1 values), variance is computed as:

$$Var(X) = p(1 - p)$$

where p is the proportion of **ones**.

- If $p \approx 0$ or $p \approx 1$, variance is **low**, meaning the feature is almost constant.
- If $p = 0.5$, variance is **highest**, meaning equal distribution of 0s and 1s.

```r
# Binary feature dataset
binary_data <- data.frame(
  Feature_A = c(rep(0, 98), rep(1, 2)), # Low variance (almost all 0s)
  Feature_B = sample(0:1, 100, replace=TRUE) # Higher variance
)

# Compute variance for binary features
binary_variances <- apply(binary_data, 2, stats::var)
print(binary_variances)
```

```
#> Feature_A Feature_B
#> 0.01979798 0.24757576

# Apply threshold (removing features with variance < 0.01)
threshold <- 0.01
filtered_binary <- binary_data[, binary_variances > threshold]
print(colnames(filtered_binary))
#> [1] "Feature_A" "Feature_B"
```

Aspect	Pros	Cons
Efficiency	Fast and computationally cheap	May remove useful features
Interpretability	Simple to understand and implement	Ignores correlation with target
Applicability	Works well for removing constant or near-constant features	Not useful for detecting redundant but high-variance features

6.2 Wrapper Methods (Model-Based Subset Evaluation)

6.2.1 Best Subsets Algorithm

The **Best Subsets Algorithm** is a systematic method for selecting the best combination of predictors. Unlike exhaustive search, which evaluates all possible subsets of predictors, this algorithm efficiently narrows the search space while guaranteeing the identification of the best subset for each size.

- The algorithm is based on the **"leap and bounds"** method introduced by (Furnival & Wilson, 2000).
- It combines:
  - **Comparison of SSE**: Evaluates models by their Sum of Squared Errors.
  - **Control over sequence**: Optimizes the order in which subset models are computed.
- **Guarantees**: Finds the best m subset models within each subset size while reducing computational burden compared to evaluating all possible subsets.

Key Features

1. **Subset Comparison**:
   - The algorithm ranks subsets based on a criterion such as R^2, adjusted R^2, AIC, or BIC.
   - It evaluates models of varying sizes, starting from 1 predictor to p predictors.

2. **Efficiency**:
    - By leveraging "leap and bounds," the algorithm avoids evaluating subsets unlikely to yield the best results.
    - This reduces the computational cost significantly compared to evaluating 2^p subsets in exhaustive search.
3. **Output**:
    - Produces the best subsets for each model size, which can be compared using criteria like AIC, BIC, or PRESS.

```
# Load the leaps package
library("leaps")

# Simulated data
set.seed(123)
n <- 100
x1 <- rnorm(n)
x2 <- rnorm(n)
x3 <- rnorm(n)
x4 <- rnorm(n)
y <- 5 + 3*x1 - 2*x2 + x3 + rnorm(n, sd=2)

# Prepare data for best subsets model
data <- data.frame(y, x1, x2, x3, x4)

# Perform best subsets model
best_subsets <- regsubsets(y ~ ., data = data, nvmax = 4)

# Summarize results
best_summary <- summary(best_subsets)

# Display model selection metrics
cat("Best Subsets Summary:\n")
#> Best Subsets Summary:
cat("Adjusted R^2:\n", best_summary$adjr2, "\n")
#> Adjusted R^2:
#>  0.3651194 0.69237 0.7400424 0.7374724
cat("Cp:\n", best_summary$cp, "\n")
#> Cp:
#>  140.9971 19.66463 3.060205 5
cat("BIC:\n", best_summary$bic, "\n")
#> BIC:
#>  -37.23673 -106.1111 -119.3802 -114.8383

# Visualize results
plot(best_subsets, scale = "adjr2")
title("Best Subsets: Adjusted R^2")
```

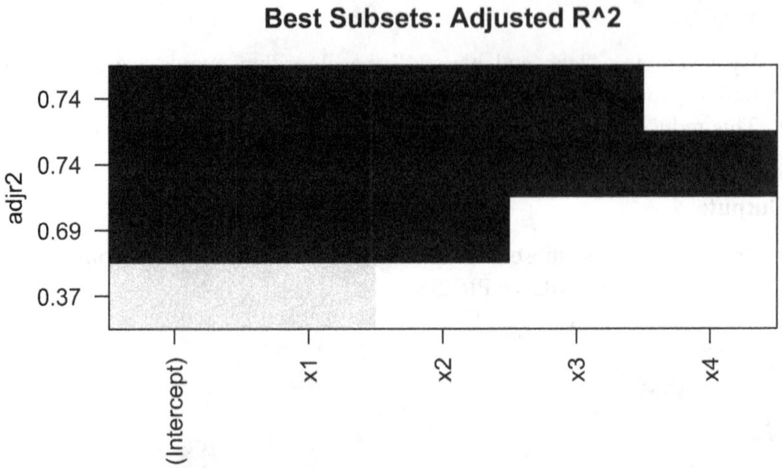

Interpretation

1. **Model Size**:
   - Examine the metrics for models with 1, 2, 3, and 4 predictors.
   - Choose the model size that optimizes your preferred metric (e.g., maximized adjusted R^2, minimized BIC).
2. **Model Comparison**:
   - For small datasets, adjusted R^2 is often a reliable criterion.
   - For larger datasets, use BIC or AIC to avoid overfitting.
3. **Efficiency**:
   - The algorithm evaluates far fewer models than an exhaustive search while still guaranteeing optimal results for each subset size.

Advantages

- Computationally efficient compared to evaluating all possible subsets.
- Guarantees the best subsets for each model size.
- Flexibility to use different selection criteria (e.g., R^2, AIC, BIC).

Limitations

- May become computationally intensive for very large p (e.g., hundreds of predictors).
- Assumes linear relationships among predictors and the outcome.

Practical Considerations

- **When to use Best Subsets?**
  - For datasets with a moderate number of predictors ($p \leq 20$).
  - When you need an optimal solution for each subset size.

- **Alternatives**:
  - **Stepwise Selection**: Faster but less reliable.
  - **Regularization Techniques**: LASSO and Ridge regression handle large p and collinearity effectively.

6.2.2 *Stepwise Selection Methods*

Stepwise selection procedures are iterative methods for selecting predictor variables. These techniques balance model simplicity and predictive accuracy by systematically adding or removing variables based on predefined criteria.

Notes

- Computer implementations often replace exact F-values with "significance" levels:
  - **SLE**: Significance level to enter.
  - **SLS**: Significance level to stay.
- These thresholds serve as guides rather than strict tests of significance.

Balancing SLE and SLS

- Large **SLE** values: May include too many variables, risking overfitting.
- Small **SLE** values: May exclude important variables, leading to underfitting and overestimation of σ^2.
- A reasonable range for SLE is **0.05 to 0.5**.
- Practical advice:
  - If **SLE > SLS**, cycling may occur (adding and removing the same variable repeatedly). To fix this, set $SLS = SLE/2$ (Bendel & Afifi, 1977).
  - If **SLE < SLS**, the procedure becomes conservative, retaining variables with minimal contribution.

6.2.2.1 Forward Selection

Forward Selection starts with an empty model (only the intercept) and sequentially adds predictors. At each step, the variable that most improves the model fit (based on criteria like R^2, AIC, or F-statistic) is added. The process stops when no variable improves the model significantly.

Steps

1. Begin with the null model: $Y = \beta_0$.
2. Evaluate each predictor's contribution to the model (e.g., using F-statistic or AIC).

3. Add the predictor with the most significant improvement to the model.
4. Repeat until no remaining variable significantly improves the model.

6.2.2.2 Backward Elimination

Backward Elimination starts with the full model, containing all predictors, and sequentially removes the least significant predictor (based on criteria like p-value or F-statistic). The process stops when all remaining predictors meet the significance threshold.

Steps

1. Begin with the full model: $Y = \beta_0 + \beta_1 X_1 + \cdots + \beta_p X_p$.
2. Identify the predictor with the smallest contribution to the model (e.g., highest p-value).
3. Remove the least significant predictor.
4. Repeat until all remaining predictors are statistically significant.

6.2.2.3 Stepwise (Both Directions) Selection

Stepwise Selection combines forward selection and backward elimination. At each step, it evaluates whether to add or remove predictors based on predefined criteria. This iterative process ensures that variables added in earlier steps can be removed later if they no longer contribute significantly.

Steps

1. Start with the null model or a user-specified initial model.
2. Evaluate all predictors not in the model for inclusion (forward step).
3. Evaluate all predictors currently in the model for removal (backward step).
4. Repeat steps 2 and 3 until no further addition or removal improves the model.

6.2.2.4 Comparison of Methods

Method	Starting point	Adds predictors	Removes predictors	Pros	Cons
Forward selection	Null model (no terms)	Yes	No	Simple, fast, useful for small models	Cannot remove irrelevant predictors
Backward elimination	Full model (all terms)	No	Yes	Removes redundant variables	May exclude important variables early
Stepwise selection	User-defined or null	Yes	Yes	Combines flexibility of both methods	More computationally intensive

```
# Simulated Data
set.seed(123)
n <- 100
x1 <- rnorm(n)
x2 <- rnorm(n)
x3 <- rnorm(n)
x4 <- rnorm(n)
y <- 5 + 3 * x1 - 2 * x2 + x3 + rnorm(n, sd = 2)

# Prepare data
data <- data.frame(y, x1, x2, x3, x4)

# Null and Full Models
null_model <- lm(y ~ 1, data = data)
full_model <- lm(y ~ ., data = data)

# Forward Selection
forward_model <- step(
    null_model,
    scope = list(lower = null_model, upper = full_model),
    direction = "forward"
)
#> Start:  AIC=269.2
#> y ~ 1
#>
#>        Df Sum of Sq     RSS    AIC
#> + x1    1    537.56  909.32 224.75
#> + x2    1    523.27  923.62 226.31
#> <none>              1446.88 269.20
#> + x3    1     23.56 1423.32 269.56
#> + x4    1      8.11 1438.78 270.64
#>
#> Step:  AIC=224.75
#> y ~ x1
#>
#>        Df Sum of Sq    RSS    AIC
#> + x2    1    473.21 436.11 153.27
#> + x3    1     62.65 846.67 219.61
#> <none>              909.32 224.75
#> + x4    1      3.34 905.98 226.38
#>
#> Step:  AIC=153.27
#> y ~ x1 + x2
#>
#>        Df Sum of Sq    RSS    AIC
#> + x3    1    71.382 364.73 137.40
#> <none>              436.11 153.27
#> + x4    1     0.847 435.27 155.08
#>
#> Step:  AIC=137.4
#> y ~ x1 + x2 + x3
#>
#>        Df Sum of Sq    RSS    AIC
```

```
#> <none>                364.73 137.40
#> + x4    1      0.231 364.50 139.34

# Backward Elimination
backward_model <- step(full_model, direction = "backward")
#> Start:  AIC=139.34
#> y ~ x1 + x2 + x3 + x4
#>
#>        Df Sum of Sq    RSS    AIC
#> - x4    1      0.23 364.73 137.40
#> <none>              364.50 139.34
#> - x3    1     70.77 435.27 155.08
#> - x2    1    480.14 844.64 221.37
#> - x1    1    525.72 890.22 226.63
#>
#> Step:  AIC=137.4
#> y ~ x1 + x2 + x3
#>
#>        Df Sum of Sq    RSS    AIC
#> <none>              364.73 137.40
#> - x3    1     71.38 436.11 153.27
#> - x2    1    481.94 846.67 219.61
#> - x1    1    528.02 892.75 224.91

# Stepwise Selection (Both Directions)
stepwise_model <- step(
    null_model,
    scope = list(lower = null_model, upper = full_model),
    direction = "both"
)
#> Start:  AIC=269.2
#> y ~ 1
#>
#>        Df Sum of Sq     RSS    AIC
#> + x1    1    537.56  909.32 224.75
#> + x2    1    523.27  923.62 226.31
#> <none>              1446.88 269.20
#> + x3    1     23.56 1423.32 269.56
#> + x4    1      8.11 1438.78 270.64
#>
#> Step:  AIC=224.75
#> y ~ x1
#>
#>        Df Sum of Sq     RSS    AIC
#> + x2    1    473.21  436.11 153.27
#> + x3    1     62.65  846.67 219.61
#> <none>               909.32 224.75
#> + x4    1      3.34  905.98 226.38
#> - x1    1    537.56 1446.88 269.20
#>
#> Step:  AIC=153.27
#> y ~ x1 + x2
#>
```

```
#>          Df Sum of Sq    RSS    AIC
#> + x3      1     71.38 364.73 137.40
#> <none>                436.11 153.27
#> + x4      1      0.85 435.27 155.08
#> - x2      1    473.21 909.32 224.75
#> - x1      1    487.50 923.62 226.31
#>
#> Step:  AIC=137.4
#> y ~ x1 + x2 + x3
#>
#>          Df Sum of Sq    RSS    AIC
#> <none>                364.73 137.40
#> + x4      1      0.23 364.50 139.34
#> - x3      1     71.38 436.11 153.27
#> - x2      1    481.94 846.67 219.61
#> - x1      1    528.02 892.75 224.91

# Summarize Results
cat("Forward Selection:\n")
#> Forward Selection:
print(summary(forward_model))
#>
#> Call:
#> lm(formula = y ~ x1 + x2 + x3, data = data)
#>
#> Residuals:
#>     Min      1Q  Median      3Q     Max
#> -4.9675 -1.1364  0.1726  1.3983  4.9332
#>
#> Coefficients:
#>             Estimate Std. Error t value Pr(>|t|)
#> (Intercept)   5.2332     0.1990  26.300  < 2e-16 ***
#> x1            2.5541     0.2167  11.789  < 2e-16 ***
#> x2           -2.2852     0.2029 -11.263  < 2e-16 ***
#> x3            0.9018     0.2080   4.335  3.6e-05 ***
#> ---
#> Signif. codes:
#> 0 '***' 0.001 '**' 0.01 '*' 0.05 '.' 0.1 ' ' 1
#>
#> Residual standard error: 1.949 on 96 degrees of freedom
#> Multiple R-squared:  0.7479, Adjusted R-squared:   0.74
#> F-statistic: 94.94 on 3 and 96 DF,  p-value: < 2.2e-16

cat("\nBackward Elimination:\n")
#>
#> Backward Elimination:
print(summary(backward_model))
#>
#> Call:
#> lm(formula = y ~ x1 + x2 + x3, data = data)
#>
#> Residuals:
#>     Min      1Q  Median      3Q     Max
```

```
#> -4.9675 -1.1364  0.1726  1.3983  4.9332
#>
#> Coefficients:
#>             Estimate Std. Error t value Pr(>|t|)
#> (Intercept)   5.2332     0.1990  26.300  < 2e-16 ***
#> x1            2.5541     0.2167  11.789  < 2e-16 ***
#> x2           -2.2852     0.2029 -11.263  < 2e-16 ***
#> x3            0.9018     0.2080   4.335  3.6e-05 ***
#> ---
#> Signif. codes:
#> 0 '***' 0.001 '**' 0.01 '*' 0.05 '.' 0.1 ' ' 1
#>
#> Residual standard error: 1.949 on 96 degrees of freedom
#> Multiple R-squared:  0.7479, Adjusted R-squared:   0.74
#> F-statistic: 94.94 on 3 and 96 DF,  p-value: < 2.2e-16

cat("\nStepwise Selection:\n")
#>
#> Stepwise Selection:
print(summary(stepwise_model))
#>
#> Call:
#> lm(formula = y ~ x1 + x2 + x3, data = data)
#>
#> Residuals:
#>     Min      1Q  Median      3Q     Max
#> -4.9675 -1.1364  0.1726  1.3983  4.9332
#>
#> Coefficients:
#>             Estimate Std. Error t value Pr(>|t|)
#> (Intercept)   5.2332     0.1990  26.300  < 2e-16 ***
#> x1            2.5541     0.2167  11.789  < 2e-16 ***
#> x2           -2.2852     0.2029 -11.263  < 2e-16 ***
#> x3            0.9018     0.2080   4.335  3.6e-05 ***
#> ---
#> Signif. codes:
#> 0 '***' 0.001 '**' 0.01 '*' 0.05 '.' 0.1 ' ' 1
#>
#> Residual standard error: 1.949 on 96 degrees of freedom
#> Multiple R-squared:  0.7479, Adjusted R-squared:   0.74
#> F-statistic: 94.94 on 3 and 96 DF,  p-value: < 2.2e-16
```

Interpretation of Results

1. **Forward Selection**:
   - Starts with no predictors and sequentially adds variables.
   - Variables are chosen based on their contribution to improving model fit (e.g., reducing SSE, increasing R^2).
2. **Backward Elimination**:
   - Begins with all predictors and removes the least significant one at each step.

- Stops when all remaining predictors meet the significance criterion.

3. **Stepwise Selection**:
   - Combines forward selection and backward elimination.
   - At each step, evaluates whether to add or remove variables based on predefined criteria.

Practical Considerations

- **SLE and SLS Tuning**:
  - Choose thresholds carefully to balance model simplicity and predictive performance.
  - For most applications, set $SLS = SLE/2$ to prevent cycling.
- **Order of Entry**:
  - Stepwise selection is unaffected by the order of variable entry. Results depend on the data and significance criteria.
- **Automated Procedures**:
  - Forward selection is simpler but less robust than forward stepwise.
  - Backward elimination works well when starting with all predictors.

Advantages

- Automates variable selection, reducing manual effort.
- Balances model fit and parsimony using predefined criteria.
- Flexible: Works with different metrics (e.g., SSE, R^2, AIC, BIC).

Limitations

- Can be sensitive to significance thresholds (SLE, SLS).
- Risk of excluding important variables in datasets with multicollinearity.
- May overfit or underfit if SLE/SLS thresholds are poorly chosen.

Practical Use Cases

- **Forward Stepwise**:
  - When starting with minimal knowledge of predictor importance.
- **Backward Elimination**:
  - When starting with many predictors and needing to reduce model complexity.
- **Stepwise (Both Directions)**:
  - For a balanced approach that adapts as variables are added or removed.

6.2.3 Branch-and-Bound Algorithm

The **Branch-and-Bound Algorithm** is a systematic optimization method used for solving subset selection problems (Furnival & Wilson, 2000). It identifies the best subsets of predictors by exploring the solution space efficiently, avoiding the need to evaluate all possible subsets.

The algorithm is particularly suited for problems with a large number of potential predictors. It systematically evaluates subsets of predictors, using bounds to prune the search space and reduce computational effort.

- **Branching**: Divides the solution space into smaller subsets (branches).
- **Bounding**: Calculates bounds on the best possible solution within each branch to decide whether to explore further or discard.

Key Features

1. **Subset Selection**:
    - Used to identify the best subset of predictors.
    - Evaluates subsets based on criteria like R^2, adjusted R^2 (Fig. 6.1), AIC, BIC, or Mallows's C_p.
2. **Efficiency**:
    - Avoids exhaustive search, which evaluates 2^p subsets for p predictors.
    - Reduces the computational burden by eliminating branches that cannot contain the optimal solution.

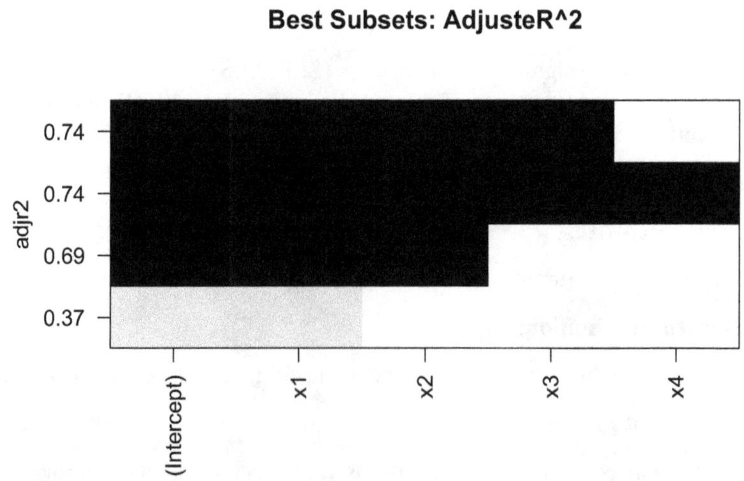

Fig. 6.1 Best subset adjusted R2

3. **Guarantee**:
   - Finds the globally optimal subset for the specified criterion.

Algorithm Steps

1. **Initialization**:
   - Start with the full set of predictors.
   - Define the criterion for evaluation (e.g., adjusted R^2, AIC, BIC).
2. **Branching**:
   - Divide the predictors into smaller subsets (branches) systematically.
3. **Bounding**:
   - Compute bounds for the criterion in each branch.
   - If the bound indicates that the branch cannot improve the current best solution, discard it.
4. **Pruning**:
   - Skip evaluating subsets within discarded branches.
5. **Stopping**:
   - The algorithm terminates when all branches are either evaluated or pruned.

```
# Load the leaps package
library("leaps")

# Simulated data
set.seed(123)
n <- 100
x1 <- rnorm(n)
x2 <- rnorm(n)
x3 <- rnorm(n)
x4 <- rnorm(n)
y <- 5 + 3 * x1 - 2 * x2 + x3 + rnorm(n, sd = 2)

# Prepare data for subset selection
data <- data.frame(y, x1, x2, x3, x4)

# Perform best subset selection using branch-and-bound
best_subsets <-
    regsubsets(y ~ .,
               data = data,
               nvmax = 4,
               method = "seqrep")

# Summarize results
best_summary <- summary(best_subsets)
```

```
# Display results
cat("Best Subsets Summary:\n")
#> Best Subsets Summary:
print(best_summary)
#> Subset selection object
#> Call: regsubsets.formula(y ~ ., data = data, nvmax = 4, method =
↪   "seqrep")
#> 4 Variables  (and intercept)
#>     Forced in Forced out
#> x1      FALSE      FALSE
#> x2      FALSE      FALSE
#> x3      FALSE      FALSE
#> x4      FALSE      FALSE
#> 1 subsets of each size up to 4
#> Selection Algorithm: 'sequential replacement'
#>          x1  x2  x3  x4
#> 1  ( 1 ) "*" " " " " " "
#> 2  ( 1 ) "*" "*" " " " "
#> 3  ( 1 ) "*" "*" "*" " "
#> 4  ( 1 ) "*" "*" "*" "*"
```

```
# Visualize best subsets
plot(best_subsets, scale = "adjr2")
title("Best Subsets: AdjusteR^2")
```

Interpretation

1. **Optimal Subsets**:
   - The algorithm identifies the best subset of predictors for each model size.
   - Evaluate models based on metrics like adjusted R^2, BIC, or AIC.
2. **Visualization**:
   - The plot of adjusted R^2 helps identify the model size with the best trade-off between fit and complexity.
3. **Efficiency**:
   - The Branch-and-Bound Algorithm achieves optimal results without evaluating all 2^p subsets.

Advantages

- Guarantees the best subset for a given criterion.
- Reduces computational cost compared to exhaustive search.
- Handles moderate-sized problems efficiently.

Limitations

- Computationally intensive for very large datasets or many predictors ($p > 20$).
- Requires a clearly defined evaluation criterion.

Practical Considerations

- **When to use Branch-and-Bound?**
  - When p (number of predictors) is moderate ($p \leq 20$).
  - When global optimality for subset selection is essential.
- **Alternatives**:
  - For larger p, consider heuristic methods like stepwise selection or regularization techniques (e.g., LASSO, Ridge).
- **Applications**:
  - Branch-and-Bound is widely used in fields like statistics, operations research, and machine learning where optimal subset selection is crucial.

6.2.4 Recursive Feature Elimination

Recursive Feature Elimination (RFE) is a feature selection method that systematically removes predictors from a model to identify the most relevant subset. RFE is commonly used in machine learning and regression tasks to improve model interpretability and performance by eliminating irrelevant or redundant features.
Theoretical Foundation

1. **Objective**:
   - Select the subset of predictors that maximizes the performance of the model (e.g., minimizes prediction error or maximizes explained variance).
2. **Approach**:
   - RFE is a backward selection method that recursively removes the least important predictors based on their contribution to the model's performance.
3. **Key Features**:
   - **Feature Ranking**: Predictors are ranked based on their importance (e.g., weights in a linear model, coefficients in regression, or feature importance scores in tree-based models).
   - **Recursive Elimination**: At each step, the least important predictor is removed, and the model is refit with the remaining predictors.
4. **Evaluation Criterion**:
   - Model performance is evaluated using metrics such as R^2, AIC, BIC, or cross-validation scores (Fig. 6.2).

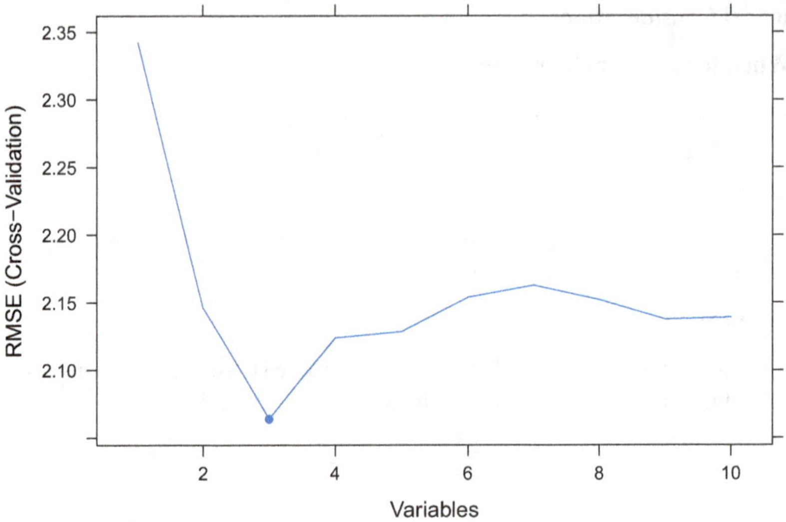

Fig. 6.2 RMSE CV

Steps in RFE

1. **Initialize**:
   - Start with the full set of predictors.
2. **Rank Features**:
   - Train the model and compute feature importance scores (e.g., coefficients, weights, or feature importance).
3. **Eliminate Features**:
   - Remove the least important feature(s) based on the ranking.
4. **Refit Model**:
   - Refit the model with the reduced set of predictors.
5. **Repeat**:
   - Continue the process until the desired number of features is reached.

```
# Install and load caret package
if (!requireNamespace("caret", quietly = TRUE)) install.packages("caret")
library(caret)

# Simulated data
set.seed(123)
n <- 100
p <- 10
X <- matrix(rnorm(n * p), nrow = n, ncol = p)
y <- 5 + rowSums(X[, 1:3]) + rnorm(n, sd = 2)  # Only first 3 variables
↪  are relevant
```

```
# Convert to a data frame
data <- as.data.frame(X)
data$y <- y

# Define control for RFE
control <- rfeControl(functions = lmFuncs, # Use linear model for
↪  evaluation
                      method = "cv",        # Use cross-validation
                      number = 10)          # Number of folds

# Perform RFE
set.seed(123)
rfe_result <- rfe(data[, -ncol(data)], data$y,
                  sizes = c(1:10),          # Subset sizes to evaluate
                  rfeControl = control)

# Display RFE results
cat("Selected Predictors:\n")
#> Selected Predictors:
print(predictors(rfe_result))
#> [1] "V1" "V2" "V3"

cat("\nModel Performance:\n")
#>
#> Model Performance:
print(rfe_result$results)
#>    Variables     RMSE  Rsquared      MAE    RMSESD
#> 1          1 2.342391 0.2118951 1.840089 0.5798450
#> 2          2 2.146397 0.3549917 1.726721 0.5579553
#> 3          3 2.063923 0.4209751 1.662022 0.5362445
#> 4          4 2.124035 0.3574168 1.698954 0.5469199
#> 5          5 2.128605 0.3526426 1.684169 0.5375931
#> 6          6 2.153916 0.3226917 1.712140 0.5036119
#> 7          7 2.162787 0.3223240 1.716382 0.5089243
#> 8          8 2.152186 0.3222999 1.698816 0.5064040
#> 9          9 2.137741 0.3288444 1.687136 0.5016526
#> 10        10 2.139102 0.3290236 1.684362 0.5127830
#>    RsquaredSD     MAESD
#> 1   0.2043123 0.4254393
#> 2   0.2434206 0.4490311
#> 3   0.2425727 0.4217594
#> 4   0.2261644 0.4212677
#> 5   0.2469891 0.4163273
#> 6   0.2236060 0.4058412
#> 7   0.2347677 0.3940646
#> 8   0.2419215 0.3826101
#> 9   0.2436893 0.3684675
#> 10  0.2457675 0.3715176
```

```
# Plot performance
plot(rfe_result, type = "l")
```

Interpretation

1. **Selected Predictors**:
   - The algorithm identifies the most relevant predictors based on their impact on model performance.
2. **Model Performance**:
   - The cross-validation results indicate how the model performs with different subset sizes, helping to select the optimal number of features.
3. **Feature Ranking**:
   - RFE ranks features by their importance, providing insights into which predictors are most influential.

Advantages

- **Improved Interpretability**: Reduces the number of predictors, simplifying the model.
- **Performance Optimization**: Eliminates irrelevant or redundant features, improving predictive accuracy.
- **Flexibility**: Works with various model types (e.g., linear models, tree-based methods, SVMs).

Limitations

- **Computationally Intensive**: Repeatedly trains models, which can be slow for large datasets or complex models.
- **Model Dependency**: Feature importance rankings depend on the underlying model, which may introduce bias.

Practical Considerations

- **When to use RFE?**
  - For high-dimensional datasets where feature selection is critical for model interpretability and performance.
  - In machine learning workflows where feature importance scores are available.
- **Extensions**:
  - Combine RFE with regularization methods (e.g., LASSO, Ridge) for additional feature selection.
  - Use advanced models (e.g., Random Forest, Gradient Boosting) for feature ranking.

6.3 Embedded Methods (Integrated into Model Training)

6.3.1 *Regularization-Based Selection*

- **Lasso (L1 Regularization)**: Shrinks some coefficients to zero, performing automatic selection.
- **Ridge (L2 Regularization)**: Shrinks coefficients but does not eliminate variables.
- **Elastic Net**: Combines L1 and L2 penalties for better feature selection.

6.3.2 *Tree-Based Feature Importance*

Decision trees and ensemble methods (Random Forests, Gradient Boosting) rank features based on their contribution to predictions.

6.3.3 *Genetic Algorithms*

Genetic Algorithms (GAs) are inspired by the principles of natural selection and genetics. They are metaheuristic optimization techniques that iteratively evolve a population of solutions to find an optimal or near-optimal subset of predictors for regression or classification tasks.

Theoretical Foundation

1. **Objective**:
   - Select a subset of predictors that optimizes a predefined fitness function (e.g., R^2, AIC, BIC, or prediction error).
2. **Key Concepts**:
   - **Population**: A collection of candidate solutions (subsets of predictors).
   - **Chromosome**: Represents a solution as a binary vector (e.g., 1 if a variable is included, `0` otherwise).
   - **Fitness Function**: Evaluates the quality of a solution based on the selected variables.
   - **Crossover**: Combines two solutions to create a new solution.
   - **Mutation**: Randomly alters a solution to maintain diversity in the population.
   - **Selection**: Chooses solutions with higher fitness to create the next generation.
3. **Advantages**:
   - Explores a wide solution space effectively.
   - Escapes local optima by introducing randomness.

Steps in Genetic Algorithms

1. **Initialization**:
   - Generate an initial population of candidate solutions randomly.
2. **Evaluation**:
   - Compute the fitness function for each solution in the population.
3. **Selection**:
   - Select solutions with higher fitness values to be parents for the next generation.
4. **Crossover**:
   - Combine pairs of parent solutions to create offspring.
5. **Mutation**:
   - Randomly modify some solutions to introduce variability.
6. **Replacement**:
   - Replace the old population with the new generation.
7. **Stopping Criteria**:
   - Terminate the algorithm after a fixed number of generations or when the improvement in fitness is below a threshold.

```
# Install and load GA package
if (!requireNamespace("GA", quietly = TRUE))
    install.packages("GA")
library(GA)
#> Loading required package: foreach
#> Loading required package: iterators
#> Package 'GA' version 3.2.4
#> Type 'citation("GA")' for citing this R package in publications.
#>
#> Attaching package: 'GA'
#> The following object is masked from 'package:utils':
#>
#>     de


# Simulated data
set.seed(123)
n <- 100
p <- 10
X <- matrix(rnorm(n * p), nrow = n, ncol = p)

# Only first 3 variables are relevant
y <-
```

```
    5 + rowSums(X[, 1:3]) + rnorm(n, sd = 2)

# Convert to a data frame
data <- as.data.frame(X)
data$y <- y

# Define the fitness function
fitness_function <- function(binary_vector) {
    selected_vars <- which(binary_vector == 1)
    if (length(selected_vars) == 0)
        return(-Inf) # Penalize empty subsets
    model <- lm(y ~ ., data = data[, c(selected_vars, ncol(data))])
    - AIC(model)  # Return negative AIC (minimization problem)
}

# Run Genetic Algorithm
set.seed(123)
ga_result <-
    ga(
        type = "binary",
        # Binary encoding for variable selection
        fitness = fitness_function,
        nBits = p,
        # Number of predictors
        popSize = 50,
        # Population size
        maxiter = 100,
        # Maximum number of generations
        run = 10,
        # Stop if no improvement in 10 generations
        seed = 123
    )

# Extract the best solution
best_solution <- ga_result@solution[1, ]
selected_vars <- which(best_solution == 1)
cat("Selected Variables (Column Indices):\n", selected_vars, "\n")
#> Selected Variables (Column Indices):
#>  1 2 3 4

# Fit the final model with selected variables
final_model <-
    lm(y ~ ., data = data[, c(selected_vars, ncol(data))])
cat("Final Model Summary:\n")
#> Final Model Summary:
print(summary(final_model))
#>
#> Call:
#> lm(formula = y ~ ., data = data[, c(selected_vars, ncol(data))])
#>
#> Residuals:
#>     Min      1Q  Median      3Q     Max
#> -5.4013 -1.3823  0.0151  1.0796  5.1537
```

```
#>
#> Coefficients:
#>              Estimate Std. Error t value Pr(>|t|)
#> (Intercept)    5.2726     0.2101  25.096  < 2e-16 ***
#> V1             1.1477     0.2290   5.012 2.49e-06 ***
#> V2             0.9469     0.2144   4.416 2.66e-05 ***
#> V3             0.6864     0.2199   3.121  0.00239 **
#> V4             0.3881     0.1997   1.943  0.05496 .
#> ---
#> Signif. codes:
#> 0 '***' 0.001 '**' 0.01 '*' 0.05 '.' 0.1 ' ' 1
#>
#> Residual standard error: 2.058 on 95 degrees of freedom
#> Multiple R-squared:  0.3574, Adjusted R-squared:  0.3304
#> F-statistic: 13.21 on 4 and 95 DF,  p-value: 1.352e-08
```

Interpretation

1. **Selected Variables**:
   - The genetic algorithm identifies the subset of predictors that minimizes the fitness function (negative AIC in this example).
2. **Model Performance**:
   - The final model can be evaluated using metrics such as adjusted R^2, prediction error, or cross-validation.
3. **Convergence**:
   - The algorithm evolves toward better solutions over generations, as indicated by improvements in the best fitness value.

Advantages

- Can handle high-dimensional datasets and complex fitness functions.
- Avoids getting trapped in local optima by introducing randomness.
- Flexible: Can optimize any user-defined fitness function.

Limitations

- Computationally intensive for large datasets or many predictors.
- Requires careful tuning of hyperparameters (e.g., population size, mutation rate).
- May not guarantee a globally optimal solution.

Practical Considerations

- **When to use Genetic Algorithms?**
  - For complex variable selection problems where traditional methods (e.g., stepwise selection) are insufficient.
  - When the fitness function is nonlinear or involves interactions among predictors.

- **Tuning Tips**:
  - Adjust population size and mutation rate based on dataset size and complexity.
  - Use parallel computing to speed up the evaluation of fitness functions.

6.4 Summary Table

See Table 6.2.

Table 6.2 Comparison of variable selection methods

Method	Type	Approach	Key criterion	Notes
Mallows's C Statistic	Information Criterion	Subset Selection	Model complexity vs Fit	Balances fit and simplicity.
Akaike Information Criterion (AIC)	Information Criterion	Model selection	Minimizes AIC	Penalizes model complexity.
Bayesian Information Criterion (BIC)	Information Criterion	Model selection	Minimizes BIC	Stronger penalty for complexity.
Hannan-Quinn Criterion (HQC)	Information Criterion	Model selection	Minimizes HQC	Combines AIC and BIC features.
Minimum Description Length (MDL)	Information Criterion	Model selection	Data + model encoding Costs	Focuses on encoding efficiency.
Prediction Error Sum of Squares (PRESS)	Error-based	Cross-validation	Minimizes prediction error	Measures predictive accuracy.
Best Subsets Algorithm	Exhaustive search	Subset Selection	Best fit across subsets	Considers all variable combinations.
Forward selection	Stepwise	Add Variables	Significance testing	Adds variables one at a time.
Backward elimination	Stepwise	Remove Variables	Significance Testing	Removes variables iteratively.
Stepwise (both directions)	Stepwise	Add/remove Variables	Significance Testing	Combines forward and backward methods.
Branch-and-Bound Algorithm	Optimized Search	Subset selection	Efficient subset search	Avoids exhaustive search.
Recursive feature elimination (RFE)	Iterative optimization	Feature removal	Model performance	Removes least important predictors.
Genetic Algorithms	Heuristic Search	Evolutionary Process	Fitness function	Mimics natural selection for subsets.

Chapter 7
Hypothesis Testing

Hypothesis testing is one of the cornerstones of statistical inference, used widely across disciplines such as economics, finance, psychology, and more. Researchers employ hypothesis testing to draw conclusions about population parameters based on sample data. Central to this process is the concept of the *p-value*, which helps quantify how unlikely the observed data (or more extreme data) would be if the null hypothesis were true.

However, as data collection has become easier and cheaper—especially in the age of big data—there is a growing awareness that *large sample sizes (large n)* can inflate the likelihood of finding statistically significant, but practically negligible, effects. Moreover, this can lead to "p-value hacking," where researchers run numerous tests or adopt flexible analytical approaches until they find a (sometimes minuscule) effect that achieves a conventional significance level (often $p < .05$).

7.1 Null Hypothesis Significance Testing

Null Hypothesis Significance Testing (NHST) is the foundation of statistical inference. It provides a structured approach to evaluating whether observed data provide sufficient evidence to reject a null hypothesis (H_0) in favor of an alternative hypothesis (H_a).

NHST follows these key steps:

1. **Define Hypotheses**
    - The null hypothesis (H_0) represents the default assumption (e.g., no effect, no difference).
    - The alternative hypothesis (H_a) represents the competing claim (e.g., a nonzero effect, a relationship between variables).

© The Author(s), under exclusive license to Springer Nature Switzerland AG 2025

M. Nguyen, *Advanced Modeling and Data Challenges*,
https://doi.org/10.1007/978-3-032-01719-2_7

2. **Select a Test Statistic**
   - The test statistic (e.g., T, W, F) quantifies evidence against H_0.
   - It follows a known distribution under H_0 (e.g., normal, Chi-square, F-distribution).
3. **Decision Rule & p-value**
   - If the test statistic **exceeds a critical value** or the **p-value is below** α, we reject H_0.
   - Otherwise, we fail to reject H_0, meaning the evidence is insufficient to rule it out.

7.1.1 Error Types in Hypothesis Testing

In hypothesis testing, we may **incorrectly** reject or fail to reject the null hypothesis, leading to two types of errors:

- **Type I Error (False Positive)**:
  - **Rejecting** H_0 when it is actually true.
  - Example: Concluding an effect exists when it does not.
- **Type II Error (False Negative)**:
  - **Failing to reject** H_0 when it is actually false.
  - Example: Missing a real effect because the test lacked power.

  The **power of a test** is the probability of correctly rejecting H_0 when it is false:

$$\text{Power} = 1 - P(\text{Type II Error})$$

A **higher power** (typically $\geq$0.8) reduces Type II errors and increases the likelihood of detecting true effects.

7.1.2 Hypothesis Testing Framework

Hypothesis tests can be **two-sided** or **one-sided**, depending on the research question.

7.1.2.1 Two-Sided Test

In a two-sided test, we examine whether a parameter is significantly different from a hypothesized value (usually zero):

$$H_0 : \beta_j = 0$$
$$H_1 : \beta_j \neq 0$$

Under the null hypothesis, and assuming standard ordinary least squares assumptions (A1–A3a, A5), the asymptotic distribution of the OLS estimator is:

$$\sqrt{n}\hat{\beta}_j \sim N(0, \text{Avar}(\sqrt{n}\hat{\beta}_j))$$

where Avar(·) denotes the asymptotic variance.

7.1.2.2 One-Sided Test

For a one-sided hypothesis test, the null hypothesis includes a range of values, and we test against a directional alternative:

$$H_0 : \beta_j \geq 0$$
$$H_1 : \beta_j < 0$$

The "hardest" null value to reject is $\beta_j = 0$. Under this specific null, the estimator follows the same asymptotic distribution:

$$\sqrt{n}\hat{\beta}_j \sim N(0, \text{Avar}(\sqrt{n}\hat{\beta}_j))$$

7.1.3 Interpreting Hypothesis Testing Results

When conducting hypothesis tests, it is essential to distinguish between **population parameters** and **sample estimates**:

- Hypotheses are always written in terms of the population parameter (β), not the sample estimate ($\hat{\beta}$).
- Some disciplines use different notations:
  - β: Standardized coefficient (useful for comparing relative effects, scale-free).
  - **b**: Unstandardized coefficient (more interpretable in practical applications, e.g., policy decisions).

The relationship between these coefficients is:

$$\beta_j = \mathbf{b}_j \frac{s_{x_j}}{s_y}$$

where s_{x_j} and s_y are the standard deviations of the independent and dependent variables.

7.1.4 Understanding p-Values

The p-value is the probability, under the assumption that H_0 is true, of observing a test statistic at least as extreme as the one computed from the sample data. Formally,

$$p\text{-value} = P(\text{Test Statistic} \geq \text{observed value} \mid H_0 \text{ is true})$$

Interpretation

- A small p-value indicates that *if* H_0 were true, seeing the observed data (or something more extreme) would be unlikely.
- By convention, if $p < \alpha$ (often 0.05), the result is deemed "statistically significant," and we reject H_0.
- **Important Caveat**: "Statistically significant" is *not* the same as "practically significant" or "economically significant." A difference can be statistically significant yet trivial in magnitude, with negligible real-world implications.

Misconceptions

- The p-value is *not* the probability that H_0 is true or false.
- A p-value above 0.05 does *not* prove that there is "no effect." It simply suggests that the data do not provide sufficient evidence (at the chosen significance level) to reject H_0.
- A p-value below 0.05 does *not* prove that an effect is "real" or large. It indicates that the data are unusual enough under H_0 that we decide to reject H_0, given our chosen threshold.

7.1.5 The Role of Sample Size

A critical factor influencing the outcome of hypothesis tests is *sample size* (n).
Increasing Power with Large n

- **Statistical Power**: The probability of correctly rejecting H_0 when H_0 is false. Large sample sizes increase statistical power, making it easier to detect even tiny deviations from H_0.

- **Implication**: If the true effect size in the population is very small (e.g., a 0.2% difference in average returns between two trading strategies), a study with a large enough n might still find it statistically significant (p-value <0.05).

Tendency Toward Over-Sensitivity
As n grows, the standard errors decrease. Thus, even minuscule differences from the null hypothesis become less likely to be attributed to random chance, yielding low p-values. This can lead to findings that are *statistically* significant but have *negligible* real-world impact.

- **Example**: Suppose an economist is testing if a policy intervention changes employment rates by 0.1%. With a small sample size, the test might not detect this difference. But with a massive dataset, the same 0.1% difference might yield a p-value <0.05, even though a 0.1% change may not be economically meaningful.

7.1.6 p-Value Hacking

p-Hacking refers to the process of manipulating data analysis until a statistically significant result (p-value <0.05) is achieved. This can include:

- Running multiple tests on the same dataset and only reporting those that yield significance.
- Stopping data collection once a significant p-value is reached.
- Trying various model specifications (e.g., adding or removing control variables) until one finds a significant effect.
- Selectively reporting outcomes (publication bias).

With large datasets, the "search space" for potential analyses grows exponentially. If researchers test many hypotheses or sift through a wide range of variables and subgroups, they can almost always find a "significant" result by chance alone.

- **Multiple Comparison Problem**: When multiple tests are conducted, the chance of finding at least one "significant" result purely by coincidence increases. For instance, with 20 independent tests at $\alpha = 0.05$, there is a 64% chance ($1 - 0.95^{20}$) of incorrectly rejecting at least one null hypothesis.

7.1.7 Practical vs. Statistical Significance

In economics and finance, it is crucial to distinguish between results that are *statistically significant* and those that are *economically meaningful*. Economic or financial significance asks: *Does this effect have tangible importance to policymakers, businesses, or investors?*

- A result might show that a new trading algorithm yields returns that are statistically different from zero, but if that difference is 0.0001% on average, it might not be profitable after accounting for transaction fees, taxes, or other frictions—hence lacking *economic significance*.

7.1.8 Mitigating the Misuse of p-Values

7.1.8.1 Pre-Registration and Replication

- **Pre-Registration**: Researchers specify hypotheses and analytical methods before seeing the data, reducing the temptation to p-hack.
- **Replication**: Independent replication studies help confirm whether a result is robust or merely a fluke.

7.1.8.2 Using Alternatives to (or Supplements for) p-Values

- **Bayesian Methods**: Provide posterior probabilities that incorporate prior information, often giving a more nuanced understanding of uncertainty.
- **Effect Size & Confidence Intervals**: Shift the focus from "Is it significant?" to "How large is the effect, and what is its plausible range?"
- Equivalence Testing: Sometimes, the goal is to show the effect is *not* larger than a certain threshold. Equivalence tests can be used to conclude "no clinically (or economically) significant difference."

7.1.8.3 Adjusting for Multiple Comparisons

- **Bonferroni Correction**: Requires using a more stringent significance threshold when multiple tests are performed (e.g., α/m for m tests).
- False Discovery Rate Control: Allows a more flexible approach, controlling the expected proportion of false positives among significant findings.

7.1.8.4 Emphasizing Relevance Over Statistical "Stars"

Encourage journals, reviewers, and academic circles to stress the *magnitude of effects* and *robustness checks* over whether the result crosses a conventional p-value threshold (like 0.05).

There are three commonly used methods for hypothesis testing:

1. Likelihood Ratio Test: Compares the likelihood under the null and alternative models. Often used for nested models.

2. Wald Test: Assesses whether an estimated parameter is significantly different from a hypothesized value. Requires only one maximization (under the full model).
3. Lagrange Multiplier (Score) Test: Evaluates the slope of the likelihood function at the null hypothesis value. Performs well in small to moderate samples.

7.1.9 Wald Test

The Wald test assesses whether estimated parameters are significantly different from hypothesized values, based on the asymptotic distribution of the estimator.

The general form of the Wald statistic is:

$$W = (\hat{\theta} - \theta_0)'[cov(\hat{\theta})]^{-1}(\hat{\theta} - \theta_0)$$
$$W \sim \chi^2_q$$

where:

- $cov(\hat{\theta})$ is given by the inverse Fisher Information matrix evaluated at $\hat{\theta}$,
- q is the rank of $cov(\hat{\theta})$, which corresponds to the number of non-redundant parameters in θ.

The Wald statistic can also be expressed in different ways:

1. Quadratic form of the test statistic:

$$t_W = \frac{(\hat{\theta} - \theta_0)^2}{I(\theta_0)^{-1}} \sim \chi^2_{(v)}$$

where v is the degree of freedom.

2. Standardized Wald test statistic:

$$s_W = \frac{\hat{\theta} - \theta_0}{\sqrt{I(\hat{\theta})^{-1}}} \sim Z$$

This represents how far the sample estimate is from the hypothesized population parameter.

Significance Level and Confidence Level

- The significance level (α) is the probability threshold at which we reject the null hypothesis.
- The confidence level ($1 - \alpha$) determines the range within which the population parameter is expected to fall with a given probability.

To standardize the estimator and null value, we define the test statistic for the OLS estimator:

$$T = \frac{\sqrt{n}(\hat{\beta}_j - \beta_{j0})}{\sqrt{n}SE(\hat{\beta}_j)} \sim^a N(0, 1)$$

Equivalently:

$$T = \frac{(\hat{\beta}_j - \beta_{j0})}{SE(\hat{\beta}_j)} \sim^a N(0, 1)$$

where:

- T is the test statistic (a function of the data and null hypothesis),
- t is the observed realization of T.

7.1.9.1 Evaluating the Test Statistic

There are three equivalent methods for evaluating hypothesis tests:

1. Critical Value Method

   For a given significance level α, determine the critical value (c):

- One-sided test: $H_0 : \beta_j \geq \beta_{j0}$

$$P(T < c|H_0) = \alpha$$

  Reject H_0 if $t < c$.

- One-sided test: $H_0 : \beta_j \leq \beta_{j0}$

$$P(T > c|H_0) = \alpha$$

  Reject H_0 if $t > c$.

- Two-sided test: $H_0 : \beta_j \neq \beta_{j0}$

$$P(|T| > c|H_0) = \alpha$$

  Reject H_0 if $|t| > c$.

2. p-value Method

   The p-value is the probability of observing a test statistic as extreme as the one obtained, given that the null hypothesis is true.

- One-sided test: $H_0 : \beta_j \geq \beta_{j0}$

$$\text{p-value} = P(T < t|H_0)$$

- One-sided test: $H_0 : \beta_j \leq \beta_{j0}$

$$\text{p-value} = P(T > t|H_0)$$

- Two-sided test: $H_0 : \beta_j \neq \beta_{j0}$

$$\text{p-value} = P(|T| > |t||H_0)$$

Reject H_0 if p-value $< \alpha$.

3. Confidence Interval Method

Using the critical value associated with a given significance level, construct a confidence interval:

$$CI(\hat{\beta}_j)_\alpha = \left[\hat{\beta}_j - c \times SE(\hat{\beta}_j), \hat{\beta}_j + c \times SE(\hat{\beta}_j)\right]$$

Reject H_0 if the hypothesized value falls outside the confidence interval.

- We are not testing whether the true population value is close to the estimate. Instead, we are testing:
  Given a fixed true population value of the parameter, how likely is it that we observed this estimate?
- This can be interpreted as:
  We believe with $(1 - \alpha) \times 100\%$ probability that the confidence interval captures the true parameter value.

Finite Sample Properties
Under stronger assumptions (A1–A6), we can consider finite sample properties:

$$T = \frac{\hat{\beta}_j - \beta_{j0}}{SE(\hat{\beta}_j)} \sim T(n-k)$$

where:

- The derivation of this distribution depends strongly on:
  - A4 (Homoskedasticity)
  - A5 (Data Generation via Random Sampling)
- The T-statistic follows a Student's t-distribution because:

  - The numerator is normally distributed.

- The denominator follows a χ^2 distribution.
- Critical values and p-values will be computed using the Student's t-distribution instead of the standard normal distribution.
- As $n \to \infty$, the $T(n-k)$ distribution converges to a standard normal distribution.

Rule of Thumb

- If $n - k > 120$:
  - The t-distribution critical values and p-values closely approximate those from the standard normal distribution.
- If $n - k < 120$:
  - If (A1–A6) hold, the t-test is an exact finite-sample test.
  - If (A1–A3a, A5) hold, the t-distribution is asymptotically normal.

    Using the t-distribution for critical values is a valid asymptotic test. The discrepancy in critical values disappears as $n \to \infty$.

7.1.9.2 Multiple Hypothesis Testing

We often need to test multiple parameters simultaneously:

- Example 1: $H_0 : \beta_1 = 0$ and $\beta_2 = 0$
- Example 2: $H_0 : \beta_1 = 1$ and $\beta_2 = 0$

Performing separate hypothesis tests on individual parameters does not answer the question of joint significance.

We need a test that accounts for joint distributions rather than evaluating two marginal distributions separately.

Consider the multiple regression model:

$$y = \beta_0 + x_1\beta_1 + x_2\beta_2 + x_3\beta_3 + \epsilon$$

The null hypothesis $H_0 : \beta_1 = 0$ and $\beta_2 = 0$ can be rewritten in matrix form as:

$$H_0 : \mathbf{R}\beta - \mathbf{q} = 0$$

where:

- $\mathbf{R}$ is an $m \times k$ matrix, where:
  - m = number of restrictions.
  - k = number of parameters.
- $\mathbf{q}$ is a $k \times 1$ vector that contains the null hypothesis values.

For the example $H_0 : \beta_1 = 0$ and $\beta_2 = 0$, we define:

$$\mathbf{R} = \begin{bmatrix} 0 & 1 & 0 & 0 \\ 0 & 0 & 1 & 0 \end{bmatrix}, \quad \mathbf{q} = \begin{bmatrix} 0 \\ 0 \end{bmatrix}$$

For the OLS estimator under multiple hypotheses, we use the F-statistic:

$$F = \frac{(\mathbf{R}\hat{\beta} - q)'\hat{\Sigma}^{-1}(\mathbf{R}\hat{\beta} - q)}{m} \sim^a F(m, n-k)$$

where:

- $\hat{\Sigma}^{-1}$ is the estimator for the asymptotic variance–covariance matrix.
- m is the number of restrictions.
- $n - k$ is the residual degrees of freedom.

Assumptions for Variance Estimation

- If A4 (Homoskedasticity) holds:
  - Both the homoskedastic and heteroskedastic variance estimators are valid.
- If A4 does not hold:
  - Only the heteroskedastic variance estimator remains valid.

Relationship Between F and t-Tests

- When $m = 1$ (only one restriction), the F-statistic is simply the squared t-statistic:

$$F = t^2$$

- Since the F-distribution is strictly positive, it is one-sided by definition.

7.1.9.3 Linear Combination Testing

When testing multiple parameters simultaneously, we often assess linear combinations of parameters rather than testing them individually.

For example, consider the following hypotheses:

$$H_0 : \beta_1 - \beta_2 = 0$$
$$H_0 : \beta_1 - \beta_2 > 0$$
$$H_0 : \beta_1 - 2\beta_2 = 0$$

Each of these represents a single restriction on a function of the parameters.

The null hypothesis:

$$H_0 : \beta_1 - \beta_2 = 0$$

can be rewritten in matrix form as:

$$H_0 : \mathbf{R}\beta - \mathbf{q} = 0$$

where:

$$\mathbf{R} = \begin{bmatrix} 0 \ 1 \ -1 \ 0 \ 0 \end{bmatrix}, \quad \mathbf{q} = \begin{bmatrix} 0 \end{bmatrix}$$

Interpretation

- **R** is a $1 \times k$ matrix that selects the relevant parameters for the hypothesis.
- **q** is a $k \times 1$ vector containing the hypothesized values of the linear combination.
- This formulation allows us to use a generalized Wald test to assess whether the constraint holds.

  The Wald test statistic for a linear hypothesis:

$$W = \frac{(\mathbf{R}\hat{\beta} - \mathbf{q})' \left(\mathbf{R}\hat{\Sigma}\mathbf{R}'\right)^{-1} (\mathbf{R}\hat{\beta} - \mathbf{q})}{s^2 q} \sim F_{q,n-k}$$

where:

- $\hat{\beta}$ is the vector of estimated coefficients.
- $\hat{\Sigma}$ is the variance–covariance matrix of $\hat{\beta}$.
- s^2 is the estimated error variance.
- q is the number of restrictions.
- The test follows an F-distribution with degrees of freedom $(q, n - k)$.

```
library(car)
#> Loading required package: carData

# Fit a multiple regression model
mod.duncan <- lm(prestige ~ income + education, data=Duncan)

# Test whether income and education coefficients are equal
linearHypothesis(mod.duncan, "1*income - 1*education = 0")
#>
#> Linear hypothesis test:
#> income - education = 0
#>
#> Model 1: restricted model
#> Model 2: prestige ~ income + education
```

```
#>
#>   Res.Df    RSS Df Sum of Sq      F Pr(>F)
#> 1     43 7518.9
#> 2     42 7506.7  1    12.195 0.0682 0.7952
```

- This tests whether $\beta_1 = \beta_2$ (i.e., whether income and education have the same effect on prestige).
- If the p-value is low, we reject the null hypothesis and conclude that income and education contribute differently to prestige.

7.1.9.4 Estimating the Difference Between Two Coefficients

In some cases, we may be interested in comparing two regression coefficients directly rather than evaluating them separately. For example, we might want to test:

$$H_0 : \beta_1 = \beta_2$$

which is equivalent to testing whether their difference is zero:

$$H_0 : \beta_1 - \beta_2 = 0$$

Alternatively, we can directly estimate the difference between two regression coefficients.

```
difftest_lm <- function(x1, x2, model) {
    # Compute coefficient difference
    diffest <-
        summary(model)$coef[x1, "Estimate"] -
        ↪  summary(model)$coef[x2, "Estimate"]

    # Compute variance of the difference
    vardiff <- (summary(model)$coef[x1, "Std. Error"] ^ 2 +
                    summary(model)$coef[x2, "Std. Error"] ^ 2)
                    ↪  - (2 * vcov(model)[x1, x2])

    # Compute standard error of the difference
    diffse <- sqrt(vardiff)

    # Compute t-statistic
    tdiff <- diffest / diffse

    # Compute p-value (two-sided test)
```

```
    ptdiff <- 2 * (1 - pt(abs(tdiff), model$df.residual))

    # Compute confidence interval
    upr <- diffest + qt(0.975, df = model$df.residual) *
↪   diffse
    lwr <- diffest - qt(0.975, df = model$df.residual) *
↪   diffse

    # Return results as a named list
    return(
        list(
            estimate = round(diffest, 2),
            t_stat = round(tdiff, 2),
            p_value = round(ptdiff, 4),
            lower_CI = round(lwr, 2),
            upper_CI = round(upr, 2),
            df = model$df.residual
        )
    )
}
```

We demonstrate this function using the Duncan dataset from the {car} package:

```
library(car)

# Load Duncan dataset
data(Duncan)

# Fit a linear regression model
mod.duncan <- lm(prestige ~ income + education, data = Duncan)

# Compare the effects of income and education
difftest_lm("income", "education", mod.duncan)
#> $estimate
#> [1] 0.05
#>
#> $t_stat
#> [1] 0.26
#>
#> $p_value
#> [1] 0.7952
#>
#> $lower_CI
#> [1] -0.36
```

```
#>
#> $upper_CI
#> [1] 0.46
#>
#> $df
#> [1] 42
```

7.1.9.5 Nonlinear Hypothesis Testing

In many applications, we may need to test nonlinear restrictions on parameters. These can be expressed as a set of q nonlinear functions:

$$\mathbf{h}(\theta) = \{h_1(\theta), \ldots, h_q(\theta)\}'$$

where each $h_j(\theta)$ is a nonlinear function of the parameter vector θ.

To approximate nonlinear restrictions, we use the Jacobian matrix, denoted as $\mathbf{H}(\theta)$, which contains the first-order partial derivatives of $\mathbf{h}(\theta)$ with respect to the parameters:

$$\mathbf{H}_{q \times p}(\theta) = \begin{bmatrix} \frac{\partial h_1(\theta)}{\partial \theta_1} & \cdots & \frac{\partial h_1(\theta)}{\partial \theta_p} \\ \vdots & \ddots & \vdots \\ \frac{\partial h_q(\theta)}{\partial \theta_1} & \cdots & \frac{\partial h_q(\theta)}{\partial \theta_p} \end{bmatrix}$$

where:

- q is the number of nonlinear restrictions,
- p is the number of estimated parameters.

The Jacobian matrix linearizes the nonlinear restrictions and allows for an approximation of the hypothesis test using a Wald statistic.

We test the null hypothesis:

$$H_0 : \mathbf{h}(\theta) = 0$$

against the two-sided alternative using the Wald statistic:

$$W = \frac{\mathbf{h}(\hat{\theta})' \left\{ \mathbf{H}(\hat{\theta}) \left[\mathbf{F}(\hat{\theta})' \mathbf{F}(\hat{\theta}) \right]^{-1} \mathbf{H}(\hat{\theta})' \right\}^{-1} \mathbf{h}(\hat{\theta})}{s^2 q} \sim F_{q,n-p}$$

where:

- $\hat{\theta}$ is the estimated parameter vector,
- $\mathbf{H}(\hat{\theta})$ is the Jacobian matrix evaluated at $\hat{\theta}$,
- $\mathbf{F}(\hat{\theta})$ is the [Fisher Information Matrix],
- s^2 is the estimated error variance,
- q is the number of restrictions,
- n is the sample size,
- p is the number of parameters.

The test statistic follows an F-distribution with degrees of freedom $(q, n - p)$.

```
library(car)
library(nlWaldTest)

# Load example data
data(Duncan)

# Fit a multiple regression model
mod.duncan <- lm(prestige ~ income + education, data = Duncan)

# Define a nonlinear hypothesis: income squared equals
↪   education
nl_hypothesis <- "b[2]^2 - b[3] = 0"

# Conduct the nonlinear Wald test
nlWaldtest(mod.duncan, texts = nl_hypothesis)
#>
#>  Wald Chi-square test of a restriction on model
#>  parameters
#>
#> data:  mod.duncan
#> Chisq = 0.69385, df = 1, p-value = 0.4049
```

- If the Wald statistic is large, we reject H_0 and conclude that the nonlinear restriction does not hold.
- The p-value provides the probability of observing such an extreme test statistic under the null hypothesis.
- The F-distribution accounts for the fact that multiple nonlinear restrictions are being tested.

7.1.10 Likelihood Ratio Test

The Likelihood Ratio Test (LRT) is a general method for comparing two nested models:

- The reduced model under the null hypothesis (H_0), which imposes constraints on parameters.
- The full model, which allows more flexibility under the alternative hypothesis (H_a).

The test evaluates how much more likely the data are under the full model compared to the restricted model.

The likelihood ratio test statistic is given by:

$$t_{LR} = 2[l(\hat{\theta}) - l(\theta_0)] \sim \chi_v^2$$

where:

- $l(\hat{\theta})$ is the log-likelihood evaluated at the estimated parameter $\hat{\theta}$ (from the full model),
- $l(\theta_0)$ is the log-likelihood evaluated at the hypothesized parameter θ_0 (from the reduced model),
- v is the degrees of freedom (the difference in the number of parameters between the full and reduced models).

This test compares the height of the log-likelihood of the sample estimate versus the hypothesized population parameter.

This test also considers the ratio of two maximized likelihoods:

$$L_r = \text{maximized likelihood under } H_0 \text{ (reduced model)}$$

$$L_f = \text{maximized likelihood under } H_0 \cup H_a \text{ (full model)}$$

Then, the likelihood ratio is defined as:

$$\Lambda = \frac{L_r}{L_f}$$

where:

- Λ cannot exceed 1, because L_f (the likelihood of the full model) is always at least as large as L_r.

The likelihood ratio test statistic is then:

$$-2\ln(\Lambda) = -2\ln\left(\frac{L_r}{L_f}\right) = -2(l_r - l_f)$$

$$\lim_{n\to\infty}(-2\ln(\Lambda)) \sim \chi_v^2$$

where:

- v is the difference in the number of parameters between the full and reduced models.

  If the likelihood ratio is small (i.e., L_r is much smaller than L_f), then:

- The test statistic exceeds the critical value from the χ_v^2 distribution.
- We reject the reduced model and accept the full model at the $\alpha \times 100\%$ significance level.

```
library(lmtest)
#> Loading required package: zoo
#>
#> Attaching package: 'zoo'
#> The following objects are masked from 'package:base':
#>
#>     as.Date, as.Date.numeric

# Load example dataset
data(mtcars)

# Fit a full model with two predictors
full_model <- lm(mpg ~ hp + wt, data = mtcars)

# Fit a reduced model with only one predictor
reduced_model <- lm(mpg ~ hp, data = mtcars)

# Perform the likelihood ratio test
lrtest(reduced_model, full_model)
#> Likelihood ratio test
#>
#> Model 1: mpg ~ hp
#> Model 2: mpg ~ hp + wt
#>   #Df  LogLik Df  Chisq Pr(>Chisq)
#> 1   3 -87.619
#> 2   4 -74.326  1 26.586   2.52e-07 ***
#> ---
#> Signif. codes:
#> 0 '***' 0.001 '**' 0.01 '*' 0.05 '.' 0.1 ' ' 1
```

- If the p-value is small, the reduced model is significantly worse, and we reject H_0.
- A large test statistic indicates that removing a predictor leads to a substantial drop in model fit.

7.1.11 Lagrange Multiplier (Score) Test

The Lagrange Multiplier (LM) Test, also known as the Score Test, evaluates whether a restricted model (under H_0) significantly underperforms compared to an unrestricted model (under H_a) without estimating the full model.

Unlike the Likelihood Ratio Test, which requires estimating both models, the LM test only requires estimation under the restricted model (H_0).

The LM test statistic is based on the first derivative (score function) of the log-likelihood function, evaluated at the parameter estimate under the null hypothesis (θ_0):

$$t_S = \frac{S(\theta_0)^2}{I(\theta_0)} \sim \chi_v^2$$

where:

- $S(\theta_0) = \left.\frac{\partial l(\theta)}{\partial \theta}\right|_{\theta=\theta_0}$ is the score function, i.e., the first derivative of the log-likelihood function evaluated at θ_0.
- $I(\theta_0)$ is the [Fisher Information Matrix], which quantifies the curvature (second derivative) of the log-likelihood.
- v is the degrees of freedom, equal to the number of constraints imposed by H_0.

This test compares:

- The slope of the log-likelihood function at θ_0 (which should be flat under H_0).
- The curvature of the log-likelihood function (captured by $I(\theta_0)$).

Interpretation of the LM Test

- If t_S is large, the slope of the log-likelihood function at θ_0 is steep, indicating that the model fit improves significantly when moving away from θ_0.
- If t_S is small, the log-likelihood function remains nearly flat at θ_0, meaning that the additional parameters in the unrestricted model do not substantially improve the fit.

If the score function $S(\theta_0)$ is significantly different from zero, then we reject H_0 because it suggests that the likelihood function is increasing, implying a better model fit when moving away from θ_0.

```
# Load necessary libraries
library(lmtest)  # For the Lagrange Multiplier test
library(car)     # For example data

# Load example data
data(Prestige)

# Fit a linear regression model
model <- lm(prestige ~ income + education, data = Prestige)
```

```r
# Perform the Lagrange Multiplier test for heteroscedasticity
# Using the Breusch-Pagan test (a type of LM test)
lm_test <- bptest(model)

# Print the results
print(lm_test)
#>
#>  studentized Breusch-Pagan test
#>
#> data:  model
#> BP = 4.1838, df = 2, p-value = 0.1235
```

- `bptest`: This function from the `lmtest` package performs the Breusch–Pagan test, which is a Lagrange Multiplier test for heteroscedasticity.
- Null Hypothesis: The null hypothesis is that the variance of the residuals is constant (homoscedasticity).
- Alternative Hypothesis: The alternative hypothesis is that the variance of the residuals is not constant (heteroscedasticity).

7.1.12 Comparing Hypothesis Tests

A visual comparison of hypothesis tests is shown below:

```r
# Load required libraries
library(ggplot2)

# Generate data for a normal likelihood function
theta <- seq(-3, 3, length.out = 200)  # Theta values

# Likelihood function with theta_hat = 1
likelihood <-
    dnorm(theta, mean = 1, sd = 1)
df <- data.frame(theta, likelihood)

# Define key points
theta_0   <- 0  # Null hypothesis value
theta_hat <- 1  # Estimated parameter (full model)
likelihood_0 <-
    dnorm(theta_0, mean = 1, sd = 1)  # Likelihood at theta_0
likelihood_hat <-
    dnorm(theta_hat, mean = 1, sd = 1)  # Likelihood at
    ↪ theta_hat
```

```
# Plot likelihood function
ggplot(df, aes(x = theta, y = likelihood)) +
    geom_line(color = "blue", linewidth = 1.2) +  # Likelihood
    ↪  curve

    # Vertical lines for theta_0 and theta_hat
    geom_vline(
        xintercept = theta_0,
        linetype = "dashed",
        color = "black",
        linewidth = 1
    ) +
    geom_vline(
        xintercept = theta_hat,
        linetype = "dashed",
        color = "red",
        linewidth = 1
    ) +

    # Labels for theta_0 and theta_hat
    annotate(
        "text",
        x = theta_0 - 0.1,
        y = -0.02,
        label = expression(theta[0]),
        color = "black",
        size = 5,
        fontface = "bold"
    ) +
    annotate(
        "text",
        x = theta_hat + 0.1,
        y = -0.02,
        label = expression(hat(theta)),
        color = "red",
        size = 5,
        fontface = "bold"
    ) +

    # LRT: Compare heights of likelihood at theta_0 and
    ↪  theta_hat
    annotate(
        "segment",
        x = theta_0,
```

```
        xend = theta_0,
        y = likelihood_0,
        yend = likelihood_hat,
        color = "purple",
        linewidth = 1.2,
        arrow = arrow(length = unit(0.15, "inches"))
    ) +
    annotate(
        "text",
        x = -2,
        y = (likelihood_0 + likelihood_hat) / 2 + 0.02,
        label = "LRT: Height",
        color = "purple",
        hjust = 0,
        fontface = "bold",
        size = 5
    ) +
    # Add horizontal lines at both ends of LRT height
    ↪  comparison
    annotate(
        "segment",
        x = -2.5,
        xend = 2.5,
        y = likelihood_0,
        yend = likelihood_0,
        color = "purple",
        linetype = "dotted",
        linewidth = 1
    ) +
    annotate(
        "segment",
        x = -2.5,
        xend = 2.5,
        y = likelihood_hat,
        yend = likelihood_hat,
        color = "purple",
        linetype = "dotted",
        linewidth = 1
    ) +

    # Wald Test: Distance between theta_0 and theta_hat
    annotate(
        "segment",
        x = theta_0,
```

```
        xend = theta_hat,
        y = 0.05,
        yend = 0.05,
        color = "green",
        linewidth = 1.2,
        arrow = arrow(length = unit(0.15, "inches"))
    ) +
    annotate(
        "text",
        x = (theta_0 + theta_hat) / 2,
        y = 0.07,
        label = "Wald: Distance",
        color = "green",
        hjust = 0.5,
        fontface = "bold",
        size = 5
    ) +

    # LM Test: Slope at theta_0
    annotate(
        "segment",
        x = theta_0 - 0.2,
        xend = theta_0 + 0.2,
        y = dnorm(theta_0 - 0.2, mean = 1, sd = 1),
        yend = dnorm(theta_0 + 0.2, mean = 1, sd = 1),
        color = "orange",
        linewidth = 1.2,
        arrow = arrow(length = unit(0.15, "inches"))
    ) +
    annotate(
        "text",
        x = -1.5,
        y = dnorm(-1, mean = 1, sd = 1) + .2,
        label = "LM: Slope",
        color = "orange",
        hjust = 0,
        fontface = "bold",
        size = 5
    ) +

    # Titles and themes
    theme_minimal() +
    labs(title = "Comparison of Hypothesis Tests",
         x = expression(theta),
```

```
        y = "Likelihood") +
    theme(
        plot.title = element_text(size = 16, face = "bold",
        ↪ hjust = 0.5),
        axis.title = element_text(size = 14),
        axis.text = element_text(size = 12)
    )
#> Warning in is.na(x): is.na() applied to non-(list or
#> vector) of type 'expression'
#> Warning in is.na(x): is.na() applied to non-(list or
#> vector) of type 'expression'
```

Each test approaches hypothesis evaluation differently (Fig. 7.1):

1. Likelihood Ratio Test: Compares the **heights** of the log-likelihood at $\hat{\theta}$ (full model) vs. θ_0 (restricted model).
2. Wald Test: Measures the **distance** between $\hat{\theta}$ and θ_0.
3. Lagrange Multiplier Test: Examines the **slope** of the log-likelihood at θ_0 to check if movement toward $\hat{\theta}$ significantly improves fit.

The Likelihood Ratio Test and Lagrange Multiplier Test perform well in small to moderate samples, while the Wald Test is computationally simpler as it only requires one model estimation (Table 7.1).

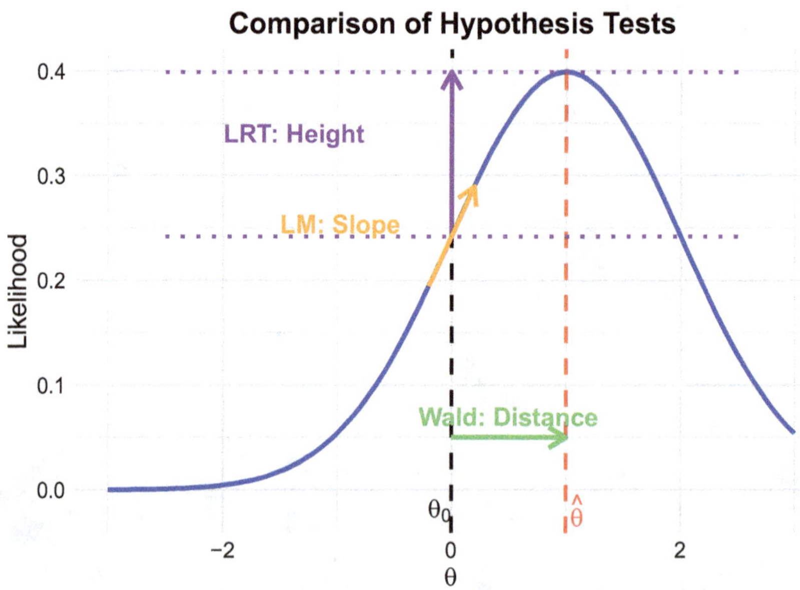

Fig. 7.1 Comparison of hypothesis tests

Table 7.1 Comparison of likelihood ratio, Wald, and Lagrange multiplier tests for hypothesis testing in econometric models

Test	Key idea	Computation	Best use case
Likelihood Ratio Test	Compares log-likelihoods of full vs. restricted models	Estimates both models	When both models can be estimated
Wald Test	Checks if parameters significantly differ from H_0	Estimates only the full model	When the full model is available
Lagrange Multiplier Test	Tests if the score function suggests moving away from H_0	Estimates only the restricted model	When the full model is difficult to estimate

7.2 Two One-Sided Tests Equivalence Testing

The Two One-Sided Test (TOST) procedure is a method used in equivalence testing to determine whether a population effect size falls within a range of practical equivalence.

Unlike traditional null hypothesis significance testing (NHST), which focuses on detecting differences, TOST tests for similarity by checking whether an effect is small enough to be practically insignificant.

7.2.1 When to Use TOST?

1. Bioequivalence Testing
   - Example: Determining whether a generic drug is equivalent to a brand-name drug in terms of effectiveness.
2. Non-Inferiority Testing
   - Example: Assessing whether a new teaching method is not worse than a traditional method by a meaningful margin.
3. Equivalence in Business & Finance
   - Example: Comparing the performance of two financial models to determine if they produce practically the same results.
4. Psychological & Behavioral Research
   - Example: Determining whether a new intervention is equally effective as an existing one.

In traditional hypothesis testing, we assess:

$$H_0 : \theta = \theta_0 \quad vs. \quad H_a : \theta \neq \theta_0$$

where θ is a population parameter (e.g., mean difference, regression coefficient, or effect size).

However, in equivalence testing, we are interested in whether θ falls within a predefined equivalence margin $(- \Delta, \Delta)$.

This leads to the TOST procedure, where we conduct two one-sided tests:

1st One-Sided Test:

$$H_0 : \theta \leq -\Delta \quad vs. \quad H_a : \theta > -\Delta$$

2nd One-Sided Test:

$$H_0 : \theta \geq \Delta \quad vs. \quad H_a : \theta < \Delta$$

If both null hypotheses are rejected, then we conclude equivalence (i.e., θ is within the equivalence range).

7.2.2 Interpretation of the TOST Procedure

- If the p-value for both one-sided tests is less than α, then we conclude that the effect size falls within the equivalence bounds.
- If one or both p-values are greater than α, we fail to reject the null hypothesis and cannot claim equivalence.
- The TOST procedure provides stronger evidence of similarity than traditional NHST, which only assesses whether an effect is *statistically different from zero* rather than *practically insignificant.*

7.2.3 Relationship to Confidence Intervals

Another way to interpret TOST is through confidence intervals (CIs):

- If the entire $(1 - 2\alpha) \times 100\%$ confidence interval lies within $[-\Delta, \Delta]$, we conclude equivalence.
- If the confidence interval extends beyond the equivalence range, we fail to establish equivalence.

This relationship ensures that TOST is consistent with CI-based inference.

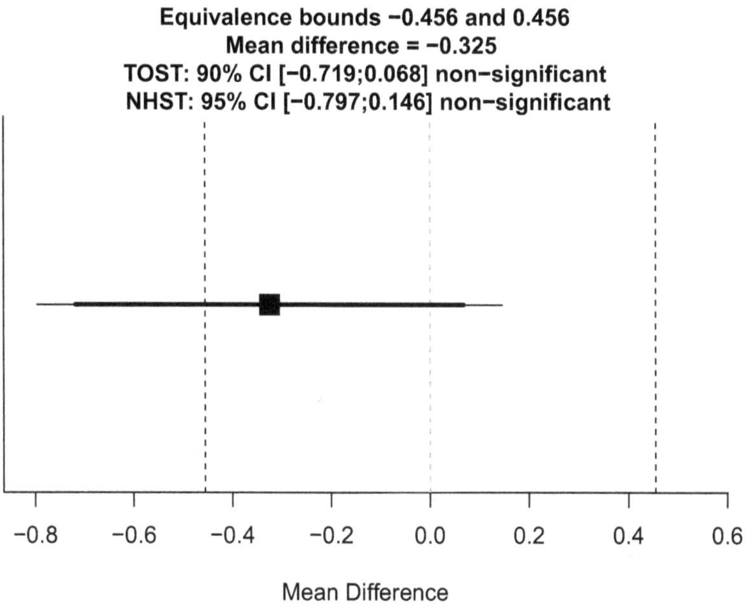

Fig. 7.2 Mean difference analysis

7.2.4 Example 1: Testing the Equivalence of Two Means

Suppose we have two groups and want to test whether their mean difference is practically insignificant within a range of [−0.5, 0.5] (Fig. 7.2).

```
library(TOSTER)

# Simulated data: Two groups with similar means
set.seed(123)
group1 <- rnorm(30, mean = 5, sd = 1)
group2 <- rnorm(30, mean = 5.1, sd = 1)
```

```
# Perform TOST equivalence test
TOSTtwo(
    m1 = mean(group1),
    sd1 = sd(group1),
    n1 = length(group1),
    m2 = mean(group2),
    sd2 = sd(group2),
    n2 = length(group2),
    low_eqbound = -0.5,
```

```
    high_eqbound = 0.5,
    alpha = 0.05
)
#> Warning: `TOSTtwo()` was deprecated in TOSTER 0.4.0.
#> i Please use `tsum_TOST()` instead.
#> This warning is displayed once every 8 hours.
#> Call `lifecycle::last_lifecycle_warnings()` to see where
#> this warning was generated.
```

```
#> TOST results:
#> t-value lower bound: 0.553   p-value lower bound: 0.291
#> t-value upper bound: -3.32   p-value upper bound: 0.0008
#> degrees of freedom : 56.56
#>
#> Equivalence bounds (Cohen's d):
#> low eqbound: -0.5
#> high eqbound: 0.5
#>
#> Equivalence bounds (raw scores):
#> low eqbound: -0.4555
#> high eqbound: 0.4555
#>
#> TOST confidence interval:
#> lower bound 90% CI: -0.719
#> upper bound 90% CI:  0.068
#>
#> NHST confidence interval:
#> lower bound 95% CI: -0.797
#> upper bound 95% CI:  0.146
#>
#> Equivalence Test Result:
#> The equivalence test was non-significant, t(56.56) = 0.553,
↪  p = 0.291, given equivalence bounds of -0.456 and 0.456 (on
↪  a raw scale) and an alpha of 0.05.
#>
#>
#> Null Hypothesis Test Result:
#> The null hypothesis test was non-significant, t(56.56) =
↪  -1.384, p = 0.172, given an alpha of 0.05.
#>
#> NHST: don't reject null significance hypothesis that the
↪  effect is equal to 0
#> TOST: don't reject null equivalence hypothesis
```

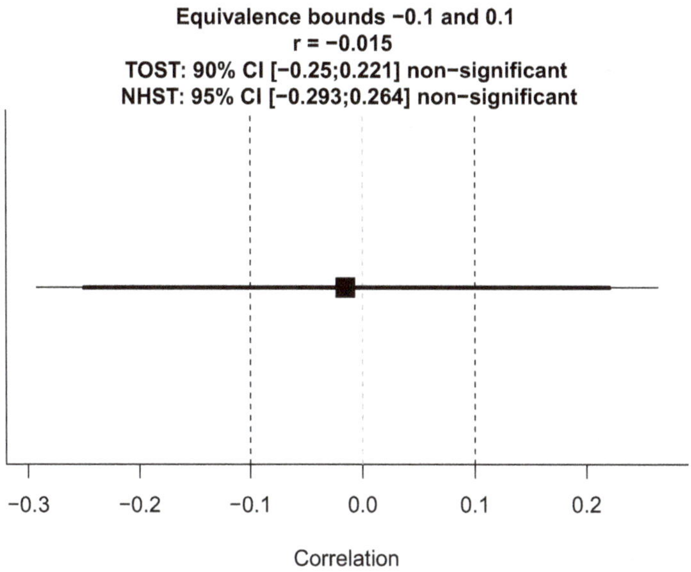

Fig. 7.3 Equivalence bounds

- If both p-values are less than 0.05, we conclude that the groups are equivalent within the given range.
- The confidence interval helps visualize whether the effect size falls entirely within $[-0.5, 0.5]$.

7.2.4.1 Example 2: TOST for Correlation Equivalence

We can also use TOST to test whether a correlation coefficient is effectively zero (Fig. 7.3).

```
# Simulated correlation data
set.seed(123)
x <- rnorm(50)
y <- x * 0.02 + rnorm(50, sd = 1)  # Very weak correlation
```

```
# TOST for correlation
TOSTr(
    n = length(x),
    r = cor(x, y),
    low_eqbound_r = -0.1,
    high_eqbound_r = 0.1,
```

```
    alpha = 0.05
)
#> Warning: `TOSTr()` was deprecated in TOSTER 0.4.0.
#> i Please use `corsum_test()` instead.
#> This warning is displayed once every 8 hours.
#> Call `lifecycle::last_lifecycle_warnings()` to see where
#> this warning was generated.
```

```
#> TOST results:
#> p-value lower bound: 0.280
#> p-value upper bound: 0.214
#>
#> Equivalence bounds (r):
#> low eqbound: -0.1
#> high eqbound: 0.1
#>
#> TOST confidence interval:
#> lower bound 90% CI: -0.25
#> upper bound 90% CI:  0.221
#>
#> NHST confidence interval:
#> lower bound 95% CI: -0.293
#> upper bound 95% CI:  0.264
#>
#> Equivalence Test Result:
#> The equivalence test was non-significant, p = 0.280, given
↪  equivalence bounds of -0.100 and 0.100 and an alpha of
↪  0.05.
#>
#>
#> Null Hypothesis Test Result:
#> The null hypothesis test was non-significant, p = 0.915,
↪  given an alpha of 0.05.
#>
#> NHST: don't reject null significance hypothesis that the
↪  correlation is equal to 0
#> TOST: don't reject null equivalence hypothesis
```

- This tests whether the correlation is within $[-0.1, 0.1]$, meaning "practically zero."
- If both p-values are significant, we conclude that the correlation is effectively negligible.

Table 7.2 Comparison between traditional NHST and TOST equivalence testing

Feature	Traditional NHST	TOST Equivalence Testing
Null hypothesis	H_0: No effect ($\theta = 0$)	H_0: Effect is outside equivalence bounds
Alternative hypothesis	H_a: There is an effect ($\theta \neq 0$)	H_a: Effect is within equivalence bounds
Goal	Detect difference	Establish similarity
p-value Interpretation	Small p means evidence for an effect	Small p means evidence for equivalence

7.2.5 *Advantages of TOST Equivalence Testing*

1. Avoids Misinterpretation of Non-Significance (Table 7.2)
   - Traditional NHST failing to reject H_0 does not imply equivalence.
   - TOST explicitly tests for equivalence, preventing misinterpretation.
2. Aligned with Confidence Intervals
   - TOST conclusions align with confidence interval-based reasoning.
3. Applicable to Various Statistical Tests
   - Can be used for means, correlations, regression coefficients, and more.
4. Commonly Used in Regulatory & Clinical Studies
   - Required for bioequivalence trials by organizations like the FDA (Schuirmann, 1987).

7.2.6 *When* Not *to Use TOST*

- If your research question is about detecting a difference rather than establishing equivalence.
- If the equivalence bounds are too wide to be meaningful in practice.
- If the sample size is too small, making it difficult to detect equivalence reliably.

7.3 False Discovery Rate

When conducting multiple hypothesis tests simultaneously, we increase the probability of false positives (Type I errors). Traditional correction methods like Bonferroni correction are too conservative, reducing statistical power.

The False Discovery Rate (FDR), introduced by Benjamini and Hochberg (1995), is a more flexible and powerful approach that controls the proportion of false discoveries (incorrect rejections of the null hypothesis) while maintaining a reasonable chance of detecting true effects (Table 7.3).

Suppose we perform m independent hypothesis tests, each with a significance level α. The probability of making at least one Type I error (false positive) is:

$$P(\text{at least one false positive}) = 1 - (1 - \alpha)^m$$

For example, with $\alpha = 0.05$ and $m = 20$ tests:

$$P(\text{at least one false positive}) = 1 - (0.95)^{20} \approx 0.64$$

Thus, if we do not adjust for multiple testing, we are highly likely to reject at least one true null hypothesis just by chance.

Why FDR?

- FWER control (Bonferroni, Holm) is too strict, reducing true discoveries.
- FDR control allows a small fraction of false positives while keeping most discoveries valid.

Let:

- m = total number of hypotheses tested
- V = number of false discoveries (Type I errors)
- R = total number of rejected null hypotheses

Then, the False Discovery Rate is:

$$\text{FDR} = E\left[\frac{V}{\max(R, 1)}\right]$$

Table 7.3 Family-wise error rate vs. false discovery rate

Approach	Controls	Method	Pros	Cons
Bonferroni correction	FWER (Probability of ≥ 1 false positive)	Adjusts α: α/m	Very conservative, reduces false positives	Low power, increases false negatives
False discovery rate	Expected proportion of false discoveries	Adjusts p-values dynamically	Higher power, fewer false negatives	Some false positives allowed

- If no null hypotheses are rejected ($R = 0$), we define FDR = 0.
- Unlike FWER, which controls the probability of any false positives, FDR controls the expected proportion of false positives.

7.3.1 *Benjamini–Hochberg Procedure*

The Benjamini–Hochberg (BH) procedure is the most widely used FDR-controlling method (Benjamini & Hochberg, 1995). It works as follows:

Step-by-Step Algorithm

1. Perform m hypothesis tests and obtain p-values: $p_1, p_2, \ldots, p_m$.
2. Rank the p-values in ascending order: $p_{(1)} \leq p_{(2)} \leq \ldots \leq p_{(m)}$.
3. Calculate the Benjamini–Hochberg critical value for each test:

$$p_{(i)} \leq \frac{i}{m}\alpha$$

4. Find the largest i where $p_{(i)} \leq \frac{i}{m}\alpha$.
5. Reject all hypotheses with $p \leq p_{(i)}$.

Interpretation

- This ensures that the expected proportion of false discoveries is controlled at level α.
- Unlike Bonferroni, it does not require independence of tests, making it more powerful.

7.3.1.1 Example 1: FDR Correction on Simulated Data

```
set.seed(123)

# Generate 20 random p-values
p_values <-
    runif(20, 0, 0.1)  # Simulating p-values from multiple
    ↪  tests

# Apply FDR correction (Benjamini-Hochberg)
adjusted_p <- p.adjust(p_values, method = "BH")

# Compare raw and adjusted p-values
data.frame(Raw_p = round(p_values, 3),
           Adjusted_p = round(adjusted_p, 3)) |> head()
```

```
#>   Raw_p Adjusted_p
#> 1 0.029      0.095
#> 2 0.079      0.096
#> 3 0.041      0.095
#> 4 0.088      0.096
#> 5 0.094      0.096
#> 6 0.005      0.046
```

- Adjusted p-values control the expected proportion of false discoveries.
- If an adjusted p-value is below α, we reject the null hypothesis.

7.3.1.2 Example 2: FDR Correction in Gene Expression Analysis

FDR is widely used in genomics, where thousands of genes are tested for differential expression.

```
library(multtest)
#> Loading required package: BiocGenerics
#>
#> Attaching package: 'BiocGenerics'
#> The following objects are masked from 'package:stats':
#>
#>     IQR, mad, sd, var, xtabs
#> The following objects are masked from 'package:base':
#>
#>     anyDuplicated, aperm, append, as.data.frame,
#>     basename, cbind, colnames, dirname, do.call,
#>     duplicated, eval, evalq, Filter, Find, get, grep,
#>     grepl, intersect, is.unsorted, lapply, Map,
#>     mapply, match, mget, order, paste, pmax,
#>     pmax.int, pmin, pmin.int, Position, rank, rbind,
#>     Reduce, rownames, sapply, saveRDS, setdiff,
#>     table, tapply, union, unique, unsplit, which.max,
#>     which.min
#> Loading required package: Biobase
#> Welcome to Bioconductor
#>
#>     Vignettes contain introductory material; view
#>     with 'browseVignettes()'. To cite Bioconductor,
#>     see 'citation("Biobase")', and for packages
#>     'citation("pkgname")'.

# Simulated gene expression study with 1000 genes
```

```r
set.seed(42)
p_values <- runif(100, 0, 0.1)

# Apply different multiple testing corrections

# Bonferroni (very strict)
p_bonf <- p.adjust(p_values, method = "bonferroni")

# Holm's method
p_holm <- p.adjust(p_values, method = "holm")

# Benjamini-Hochberg (FDR)
p_fdr <- p.adjust(p_values, method = "BH")

# Compare significance rates
sum(p_bonf < 0.05)  # Strictest correction
#> [1] 3
sum(p_holm < 0.05)  # Moderately strict
#> [1] 3
sum(p_fdr < 0.05)   # Most discoveries, controlled FDR
#> [1] 5
```

- Bonferroni results in few discoveries (low power).
- Benjamini–Hochberg allows more discoveries, while controlling the proportion of false positives.

Use FDR when:

- You perform many hypothesis tests (e.g., genomics, finance, A/B testing).
- You want to balance false positives and false negatives.
- Bonferroni is too strict, leading to low power.

Do not use FDR if:

- Strict control of any false positives is required (e.g., drug approval studies).
- There are only a few tests, where Bonferroni is appropriate.

7.3.2 Benjamini–Yekutieli Procedure

The Benjamini–Yekutieli (BY) method modifies the Benjamini–Hochberg (BH) procedure to account for correlated test statistics (Benjamini & Yekutieli, 2001).

The key adjustment is a larger critical value for significance, making it more conservative than BH.

Similar to BH, the BY procedure ranks m p-values in ascending order:

$$p_{(1)} \leq p_{(2)} \leq \cdots \leq p_{(m)}$$

Instead of using $\alpha \frac{i}{m}$ as in BH, BY introduces a **correction factor**:

$$p_{(i)} \leq \frac{i}{mC(m)}\alpha$$

where:

$$C(m) = \sum_{j=1}^{m} \frac{1}{j} \approx \ln(m) + 0.577$$

- $C(m)$ is the harmonic series correction (ensures control under dependence).
- This makes the BY threshold larger than BH, reducing false positives.
- Recommended when tests are positively correlated (e.g., in fMRI, finance).

We can apply BY correction using the built-in `p.adjust()` function.

```
set.seed(123)

# Simulate 50 random p-values
p_values <- runif(50, 0, 0.1)

# Apply BY correction
p_by <- p.adjust(p_values, method = "BY")

# Compare raw and adjusted p-values
data.frame(Raw_p = round(p_values, 3), Adjusted_BY =
↪   round(p_by, 3))
#>    Raw_p Adjusted_BY
#> 1  0.029       0.430
#> 2  0.079       0.442
#> 3  0.041       0.430
#> 4  0.088       0.442
#> 5  0.094       0.442
#> 6  0.005       0.342
#> 7  0.053       0.442
#> 8  0.089       0.442
#> 9  0.055       0.442
#> 10 0.046       0.430
#> 11 0.096       0.442
#> 12 0.045       0.430
```

```
#> 13 0.068        0.442
#> 14 0.057        0.442
#> 15 0.010        0.429
#> 16 0.090        0.442
#> 17 0.025        0.430
#> 18 0.004        0.342
#> 19 0.033        0.430
#> 20 0.095        0.442
#> 21 0.089        0.442
#> 22 0.069        0.442
#> 23 0.064        0.442
#> 24 0.099        0.447
#> 25 0.066        0.442
#> 26 0.071        0.442
#> 27 0.054        0.442
#> 28 0.059        0.442
#> 29 0.029        0.430
#> 30 0.015        0.429
#> 31 0.096        0.442
#> 32 0.090        0.442
#> 33 0.069        0.442
#> 34 0.080        0.442
#> 35 0.002        0.342
#> 36 0.048        0.430
#> 37 0.076        0.442
#> 38 0.022        0.430
#> 39 0.032        0.430
#> 40 0.023        0.430
#> 41 0.014        0.429
#> 42 0.041        0.430
#> 43 0.041        0.430
#> 44 0.037        0.430
#> 45 0.015        0.429
#> 46 0.014        0.429
#> 47 0.023        0.430
#> 48 0.047        0.430
#> 49 0.027        0.430
#> 50 0.086        0.442
```

Comparison: BH vs. BY

```
p_bh <- p.adjust(p_values, method = "BH")  # BH correction

# Visualize differences
```

```
data.frame(Raw_p = round(p_values, 3),
           BH_Adjusted = round(p_bh, 3),
           BY_Adjusted = round(p_by, 3)) |> head()
#>   Raw_p BH_Adjusted BY_Adjusted
#> 1 0.029       0.096       0.430
#> 2 0.079       0.098       0.442
#> 3 0.041       0.096       0.430
#> 4 0.088       0.098       0.442
#> 5 0.094       0.098       0.442
#> 6 0.005       0.076       0.342
```

- Benjamini–Yekutieli is more conservative than Benjamini-Hochberg.
- If BH identifies significant results but BY does not, the tests are likely correlated.

7.3.3 Storey's q-value Approach

Storey's q-value method directly estimates False Discovery Rate rather than adjusting individual p-values (Benjamini & Yekutieli, 2001).

Unlike BH/BY, it does not assume a fixed threshold (α) but estimates the FDR dynamically from the data.

Define:

- m_0: Number of **true null hypotheses**.
- π_0: Estimated **proportion of true nulls** in the dataset.

Storey's q-value adjusts p-values based on:

$$q(p) = \frac{\pi_0 \cdot m \cdot p}{\sum_{i=1}^{m} 1_{p_i \leq p}}$$

where:

- π_0 is estimated from the distribution of large p-values.
- Unlike BH/BY, Storey's method dynamically estimates the null proportion.

```
# remotes::install_github("StoreyLab/qvalue")
library(qvalue)

# Simulated data: 1000 hypothesis tests
set.seed(123)
p_values <- runif(1000, 0, 0.1)

# Compute q-values
qvals <- qvalue_truncp(p = p_values)$qvalues
```

```
# Summary of q-values
summary(qvals)
#>    Min. 1st Qu.  Median    Mean 3rd Qu.    Max.
#>  0.3126  0.9689  0.9771  0.9684  0.9847  1.0000
```

Comparison: BH vs. BY vs. Storey

```
# Apply multiple corrections
p_bh <- p.adjust(p_values, method = "BH")    #
↪  Benjamini-Hochberg
p_by <- p.adjust(p_values, method = "BY")    #
↪  Benjamini-Yekutieli
q_vals <- qvalue_truncp(p_values)$qvalues    # Storey's q-value

# Compare significance rates
data.frame(
  Raw_p = round(p_values[1:10], 3),
  BH = round(p_bh[1:10], 3),
  BY = round(p_by[1:10], 3),
  q_value = round(q_vals[1:10], 3)
)
#>     Raw_p    BH    BY q_value
#> 1  0.029 0.097 0.725   0.969
#> 2  0.079 0.098 0.737   0.985
#> 3  0.041 0.097 0.725   0.969
#> 4  0.088 0.099 0.740   0.989
#> 5  0.094 0.100 0.746   0.998
#> 6  0.005 0.094 0.700   0.936
#> 7  0.053 0.098 0.737   0.985
#> 8  0.089 0.099 0.740   0.989
#> 9  0.055 0.098 0.737   0.985
#> 10 0.046 0.098 0.731   0.977
```

7.3.4 Summary of False Discovery Rate Methods

FDR control methods balance Type I and Type II errors, making them more powerful than conservative Family-Wise Error Rate (FWER) methods like Bonferroni (Table 7.4).

Table 7.4 Summary of false discovery rate methods

Method	Type	Strength	Best for
Benjamini–Hochberg	Adjusted p-values	Most powerful	Independent tests (e.g., surveys, psychology)
Benjamini–Yekutieli	Adjusted p-values	More conservative	Correlated tests (e.g., fMRI, finance)
Storey's q-value	Direct FDR estimation	Most flexible	Large-scale studies (e.g., genomics, proteomics)
Bonferroni	Family-Wise Error Rate (FWER)	Very conservative	Small number of tests, strict control
Holm's method	FWER (less strict than Bonferroni)	Moderate	Moderately strict correction

7.4 Comparison of Testing Frameworks

See Table 7.5.

Table 7.5 Comparison of testing frameworks

Framework	Key concept	Strengths	Weaknesses
NHST	Tests if an effect is statistically significant (p-value based)	Simple, widely used	Over-reliance on p-values, arbitrary thresholds (e.g., 0.05)
TOST	Tests whether two means are sufficiently close (equivalence)	Useful in equivalence testing	Requires pre-specified equivalence margin
Bayesian testing	Uses posterior probabilities and Bayes Factor	Incorporates prior knowledge, intuitive	Requires prior distribution, computationally expensive
Decision-theoretic	Minimizes expected loss based on cost functions	Practical for decision-making	Needs subjective cost assignment
False Discovery Rate	Controls proportion of false positives in multiple tests	Useful in high-dimensional data	Can still lead to false discoveries

Chapter 8
Marginal Effects

Marginal effects play a fundamental role in **interpreting regression models**, particularly when analyzing the impact of explanatory variables on an outcome variable. These effects provide a precise measure of **how a small change in an independent variable influences the dependent variable**.

The concept of a marginal effect is closely linked to **derivatives** in calculus. In simple linear models, marginal effects correspond directly to the estimated regression coefficients. However, in nonlinear models, computing marginal effects requires careful consideration, often involving either analytical differentiation or numerical approximation.

8.1 Definition of Marginal Effects

Mathematically, the **marginal effect** of an independent variable X on the expected value of a dependent variable Y is given by:

$$\frac{\partial E[Y|X]}{\partial X}$$

which represents the instantaneous rate of change of $E[Y|X]$ with respect to X.

For a **linear regression model**:

$$Y = \beta_0 + \beta_1 X_1 + \beta_2 X_2 + \cdots + \beta_k X_k + \varepsilon$$

the marginal effect of X_j is simply β_j. However, in more complex cases, such as nonlinear models, interaction effects, or transformations, marginal effects are not directly given by the regression coefficients and must be computed explicitly.

© The Author(s), under exclusive license to Springer Nature Switzerland AG 2025

M. Nguyen, *Advanced Modeling and Data Challenges*,
https://doi.org/10.1007/978-3-032-01719-2_8

8.1.1 Analytical Derivation of Marginal Effects

In models where $E[Y|X]$ is a differentiable function of X, marginal effects are computed using calculus. The derivative of a function $f(x)$ is given by:

$$f'(x) \equiv \lim_{h \to 0} \frac{f(x+h) - f(x)}{h}$$

Example: Quadratic Function Consider the function:

$$f(x) = x^2.$$

The marginal effect is derived as follows:

$$\begin{aligned} f'(x) &= \lim_{h \to 0} \frac{(x+h)^2 - x^2}{h} \\ &= \frac{x^2 + 2xh + h^2 - x^2}{h} \\ &= \frac{2xh + h^2}{h} \\ &= 2x + h. \end{aligned}$$

As $h \to 0$, the marginal effect simplifies to:

$$f'(x) = 2x.$$

Thus, for small changes in x, the effect on $f(x)$ depends on x itself.

8.1.2 Numerical Approximation of Marginal Effects

In practice, analytical differentiation may be infeasible, particularly when dealing with **complex functions, large datasets, or models without closed-form derivatives**. In such cases, numerical differentiation provides an alternative.

8.1.2.1 One-Sided Numerical Approximation

A simple way to approximate the derivative is the forward difference formula:

$$
\begin{aligned}
f'(x) &= \lim_{h \to 0} \frac{(x+h)^2 - x^2}{h} \\
&\approx \frac{f(x+h) - f(x)}{h}
\end{aligned}
$$

where h is a small step size.

8.1.2.2 Two-Sided Numerical Approximation

A more accurate method is the central difference formula:

$$
f_2'(x) \approx \frac{f(x+h) - f(x-h)}{2h}.
$$

This approach reduces numerical error and is generally preferred in computational implementations.

8.1.2.3 Choosing an Appropriate *h*

The choice of h is critical (Gould et al., 2010, chapter 1):

- **Too small**: Can lead to numerical instability due to floating-point precision limitations.
- **Too large**: Reduces the accuracy of the approximation.

A common heuristic is to set $h = 10^{-5}$ or a small fraction of the standard deviation of X.

Numerical derivatives are often preferred in empirical applications, especially when working with **complex models or machine learning algorithms** (Tables 8.1 and 8.2).

Table 8.1 Comparison of analytical and numerical methods

Method	Advantages	Disadvantages
Analytical	Provides exact expressions	Requires differentiability, not always feasible
Numerical	Works for any function, easy to implement	Requires careful choice of step size h

Table 8.2 Comparison of analytical and numerical approaches for marginal effects and standard errors

	Analytical Derivation	Numerical Approximation
Marginal effects	Uses calculus (rules of differentiation)	Uses finite differences to approximate derivatives
Standard errors	Derived using variance rules	Estimated via the delta method using the variance-covariance matrix

8.2 Marginal Effects in Different Contexts

1. **Linear Regression Models**

   For a simple linear regression:

$$E[Y|X] = \beta_0 + \beta_1 X,$$

the marginal effect is **constant** and equal to β_1. This makes interpretation straightforward.

2. **Logit and Probit Models**

   In logistic regression, the expected value of Y is modeled as:

$$E[Y|X] = P(Y = 1|X) = \frac{1}{1 + e^{-\beta_0 - \beta_1 X}}.$$

   The marginal effect is given by:

$$\frac{\partial E[Y|X]}{\partial X} = \beta_1 P(Y = 1|X)(1 - P(Y = 1|X)).$$

   Unlike linear models, the effect **varies with** X, requiring evaluation at specific values (e.g., means or percentiles).

3. **Interaction Effects and Nonlinear Terms**

   When models include interactions (e.g., $X_1 X_2$) or transformations (e.g., $\log(X)$), marginal effects become more complex. For example, in:

$$E[Y|X] = \beta_0 + \beta_1 X + \beta_2 X^2,$$

the marginal effect of X is:

$$\frac{\partial E[Y|X]}{\partial X} = \beta_1 + 2\beta_2 X.$$

   This means the marginal effect **depends on the value of** X.

8.3 Marginal Effects Interpretation

The interpretation of marginal effects differs depending on whether X is continuous or discrete (Table 8.3):

Table 8.3 Interpretation of marginal effects by variable type

Variable type	Definition of marginal effect
Continuous	The derivative $\frac{\partial E[Y\|X]}{\partial X}$ represents an **infinitesimal** change in X.
Discrete	The change in $E[Y\|X]$ when X increases by one unit (also called an **incremental effect**).

For example, in a **binary variable case** (e.g., a dummy variable for gender), the marginal effect is:

$$E[Y|X = 1] - E[Y|X = 0].$$

which quantifies the expected change in Y when switching from $X = 0$ to $X = 1$.

8.4 Delta Method

The Delta Method is a statistical technique for approximating the **mean and variance** of a function of random variables. It is particularly useful in regression analysis when estimating the standard errors of **nonlinear functions** of estimated coefficients, such as:

- **Marginal effects in nonlinear models** (e.g., logistic regression)
- **Elasticities and risk measures** (e.g., in finance)
- **Transformation of regression coefficients** (e.g., log transformations)

This method is based on a first-order [Taylor Series] approximation, which allows us to estimate the variance of a transformed parameter without requiring explicit distributional assumptions.

Let $G(\beta)$ be a function of the estimated parameters β, where β follows an asymptotically normal distribution:

$$\beta \sim N(\hat{\beta}, \mathrm{Var}(\hat{\beta})).$$

Using a **first-order Taylor expansion**, we approximate $G(\beta)$ around its expectation:

$$G(\beta) \approx G(\hat{\beta}) + \nabla G(\beta)(\beta - \hat{\beta}),$$

where $\nabla G(\beta)$ is the gradient (also known as the **Jacobian**) of $G(\beta)$, i.e., the vector of partial derivatives:

$$\nabla G(\beta) = \left(\frac{\partial G}{\partial \beta_1}, \frac{\partial G}{\partial \beta_2}, \ldots, \frac{\partial G}{\partial \beta_k} \right).$$

The variance of $G(\beta)$ is then approximated as:

$$\mathrm{Var}(G(\beta)) \approx \nabla G(\beta)\, \mathrm{Cov}(\beta)\, \nabla G(\beta)'.$$

where:

- $\nabla G(\beta)$ is the gradient vector of $G(\beta)$.
- $\mathrm{Cov}(\beta)$ is the variance–covariance matrix of $\hat{\beta}$.
- The expression $\nabla G(\beta)'$ denotes the transpose of the gradient.

Key Properties of the Delta Method

- **Semi-parametric approach**: It does not require full knowledge of the distribution of $G(\beta)$.
- **Widely applicable**: Useful for computing standard errors in regression models.
- **Alternative approaches**:
  - **Analytical derivation**: Directly deriving a probability function for the margin.
  - **Simulation/Bootstrapping**: Using Monte Carlo methods to approximate standard errors.

8.5 Comparison: Delta Method vs. Alternative Approaches

See Table 8.4.

Table 8.4 Comparison of methods for estimating variance of marginal effects

Method	Description	Pros	Cons
Delta Method	Uses Taylor expansion to approximate variance	Computationally efficient	Accuracy depends on linearity assumption
Analytical Derivation	Directly derives probability function	Exact solution (if feasible)	Can be mathematically complex
Simulation/bootstrapping	Uses repeated sampling from estimated distribution	No assumptions on functional form	Computationally expensive

When to Use the Delta Method

- When you need a **quick approximation** for standard errors.
- When the function $G(\beta)$ is **smooth and differentiable**.
- When working with **large sample sizes**, where asymptotic normality holds.

For deeper exploration, refer to these excellent resources:

- **Advanced**: modmarg package documentation—covers implementation of the Delta Method in R.
- **Intermediate**: UCLA Statistical Consulting—a practical FAQ on the Delta Method.

8.5.1 Example: Applying the Delta Method in a Logistic Regression

To illustrate, let's apply the **Delta Method** to compute the standard error of a nonlinear transformation of regression coefficients.

In logistic regression, the estimated coefficient $\hat{\beta}$ represents the **log-odds** change for a one-unit increase in X. However, we often want the **odds ratio**, which is:

$$G(\beta) = e^{\beta}.$$

By the **Delta Method**, the variance of e^{β} is:

$$\text{Var}(e^{\beta}) \approx e^{2\beta} \cdot \text{Var}(\beta).$$

```
# Load necessary packages
library(ggplot2)
library(margins)
library(sandwich)
library(lmtest)
#> Loading required package: zoo
#>
#> Attaching package: 'zoo'
#> The following objects are masked from 'package:base':
#>
#>     as.Date, as.Date.numeric

# Simulate data
set.seed(123)
n <- 100
X <- rnorm(n)  # Simulate independent variable
```

```
# Generate binary outcome using logistic model
Y <-
    rbinom(n, 1, plogis(0.5 + 0.8 * X))

# Logistic regression
logit_model <- glm(Y ~ X, family = binomial(link = "logit"))

# Extract coefficient and variance
beta_hat <- coef(logit_model)["X"]   # Estimated coefficient
var_beta_hat <- vcov(logit_model)["X", "X"]  # Variance of
↪  beta_hat

# Apply Delta Method
odds_ratio <- exp(beta_hat)  # Transform beta to odds ratio
se_odds_ratio <-
    sqrt(odds_ratio ^ 2 * var_beta_hat)  # Delta Method SE

# Compute 95% Confidence Interval
lower_CI <- exp(beta_hat - 1.96 * sqrt(var_beta_hat))
upper_CI <- exp(beta_hat + 1.96 * sqrt(var_beta_hat))

# Display results
results <- data.frame(
    Term = "X",
    Odds_Ratio = odds_ratio,
    SE = se_odds_ratio,
    Lower_CI = lower_CI,
    Upper_CI = upper_CI
)

print(results)
#>   Term Odds_Ratio        SE Lower_CI Upper_CI
#> X    X   2.655431 0.7677799 1.506669 4.680069

# ---- VISUALIZATION 1: Distribution of Simulated Odds Ratios
↪  ----
set.seed(123)
# Simulate beta estimates
simulated_betas <-
    rnorm(1000, mean = beta_hat, sd = sqrt(var_beta_hat))
simulated_odds_ratios <-
    exp(simulated_betas)  # Apply transformation
```

```
ggplot(data.frame(Odds_Ratio = simulated_odds_ratios),
       aes(x = Odds_Ratio)) +
    geom_histogram(
        color = "black",
        fill = "skyblue",
        bins = 50,
        alpha = 0.7
    ) +
    geom_vline(
        xintercept = odds_ratio,
        color = "red",
        linetype = "dashed",
        linewidth = 1.2
    ) +
    geom_vline(
        xintercept = lower_CI,
        color = "blue",
        linetype = "dotted",
        linewidth = 1.2
    ) +
    geom_vline(
        xintercept = upper_CI,
        color = "blue",
        linetype = "dotted",
        linewidth = 1.2
    ) +
    labs(title = "Distribution of Simulated Odds Ratios",
         x = "Odds Ratio",
         y = "Frequency") +
    theme_minimal()
```

- **Odds Ratio Computation**
  - The odds ratio is the exponentiated coefficient $e^{\hat{\beta}}$, which represents the multiplicative change in the odds of $Y = 1$ for a one-unit increase in X.
  - If $\hat{\beta} = 0.8$, then $e^{0.8} \approx 2.23$, meaning a one-unit increase in X increases the odds of $Y = 1$ by **123%**.
- **Standard Error via Delta Method**
  - Since β follows a normal distribution, its transformation e^{β} is **not normally distributed** but rather follows a log-normal shape.
  - The Delta Method approximates the **standard error** of e^{β} using:

$$SE(e^{\beta}) = \sqrt{e^{2\beta} \cdot \text{Var}(\beta)}$$

- **Confidence Intervals**
  - The confidence interval is obtained by:

$$[e^{\beta-1.96\cdot SE}, e^{\beta+1.96\cdot SE}]$$

  - This helps interpret the uncertainty around the odds ratio estimate.
- **Visualization** (Fig. 8.1)
  - The **histogram of simulated odds ratios** shows how the transformation affects variance:

    The **red dashed line** represents the estimated odds ratio.
    The **blue dotted lines** show the **confidence interval bounds**.
    The **right-skewed distribution** reflects the nonlinear transformation, meaning higher uncertainty for larger values.

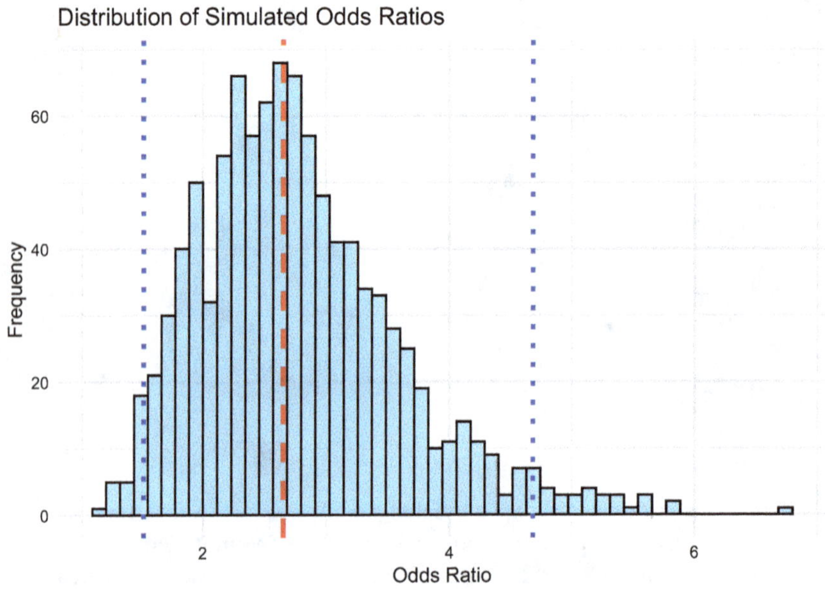

Fig. 8.1 Distribution of simulated odds ratios

8.6 Types of Marginal Effect

8.6.1 Average Marginal Effect

The Average Marginal Effect (AME) measures the **expected change in the predicted probability** when an independent variable **increases by a small amount** while holding all other variables constant. Unlike marginal effects at the mean (MEM), AMEs average the marginal effects **across all observations**, providing a more representative measure.

Applications of AMEs

- **Marketing**: How much does increasing ad spend change the probability of a customer purchase?
- **Finance**: How does an interest rate change impact the probability of loan approval?
- **Econometrics**: What is the effect of education on the probability of employment?

Since nonlinear models like **logit** and **probit** do not have constant marginal effects, AMEs require numerical differentiation. There are two common approaches:

1. One-Sided Numerical Derivative: Uses a small forward step h to estimate the derivative.
2. Two-Sided Numerical Derivative: Takes both a forward step and a backward step to improve accuracy.

8.6.1.1 One-Sided Numerical Derivative

To estimate $\frac{\partial p(\mathbf{X},\beta)}{\partial X}$ numerically:

Algorithm

1. Estimate the model using logistic (or probit) regression.
2. For each observation i:

    1. Compute predicted probability using the observed data:

    $$\hat{Y}_{i0} = p(\mathbf{X}_i, \hat{\beta}).$$

    2. Increase X by a small step h, where:

        - If X is continuous, choose:

        $$h = (|\bar{X}| + 0.001) \times 0.001.$$

        - If X is discrete, set $h = 1$.

3. Compute the new predicted probability:

$$\hat{Y}_{i1} = p(\mathbf{X}_i + h, \hat{\beta}).$$

4. Compute the numerical derivative:

$$\frac{\hat{Y}_{i1} - \hat{Y}_{i0}}{h}.$$

3. Average across all observations:

$$E\left[\frac{\hat{Y}_{i1} - \hat{Y}_{i0}}{h}\right] \approx \frac{\partial p(Y|\mathbf{X}, \beta)}{\partial X}.$$

```
# Load necessary packages
library(margins)
library(sandwich)
library(lmtest)

# Simulate data
set.seed(123)
n <- 100
X <- rnorm(n)
Y <- rbinom(n, 1, plogis(0.5 + 0.8 * X))  # Logistic function

# Logistic regression
logit_model <- glm(Y ~ X, family = binomial(link = "logit"))

# Define step size h for continuous variable
X_mean <- mean(X)
h <- (abs(X_mean) + 0.001) * 0.001

# Compute predicted probabilities at original X
pred_Y0 <- predict(logit_model, type = "response")

# Compute predicted probabilities at X + h
X_new <- X + h
data_new <- data.frame(X = X_new)
pred_Y1 <-
    predict(logit_model, newdata = data_new, type = "response")

# Compute marginal effects
marginal_effects <- (pred_Y1 - pred_Y0) / h
```

```r
# Compute Average Marginal Effect (AME)
AME_one_sided <- mean(marginal_effects)

# Display results
data.frame(Method = "One-Sided AME", Estimate = AME_one_sided)
#>          Method  Estimate
#> 1 One-Sided AME 0.1921614
```

- The AME is the average effect of X on the probability of $Y = 1$.
- Since logistic regression is nonlinear, the effect varies across observations.
- This method assumes that a small h provides a good approximation.

8.6.1.2 Two-Sided Numerical Derivative

To improve accuracy, we use the **two-sided derivative**:

Algorithm

1. Estimate the model using logistic (or probit) regression.
2. For each observation i:
    1. Compute the original predicted probability:
    $$\hat{Y}_{i0} = p(\mathbf{X}_i, \hat{\beta}).$$
    2. Compute the new predicted probabilities:
        - Increase X by h:
        $$\hat{Y}_{i1} = p(\mathbf{X}_i + h, \hat{\beta}).$$
        - Decrease X by h:
        $$\hat{Y}_{i2} = p(\mathbf{X}_i - h, \hat{\beta}).$$
    3. Compute the numerical derivative:
    $$\frac{\hat{Y}_{i1} - \hat{Y}_{i2}}{2h}.$$
3. Average across all observations:
$$E\left[\frac{\hat{Y}_{i1} - \hat{Y}_{i2}}{2h}\right] \approx \frac{\partial p(Y|\mathbf{X}, \beta)}{\partial X}.$$

```
# Compute predicted probabilities at X - h
X_new_minus <- X - h
data_new_minus <- data.frame(X = X_new_minus)
pred_Y2 <-
    predict(logit_model, newdata = data_new_minus, type =
     ↪  "response")

# Compute two-sided marginal effects
marginal_effects_2sided <- (pred_Y1 - pred_Y2) / (2 * h)

# Compute Average Marginal Effect (AME) - Two-Sided
AME_two_sided <- mean(marginal_effects_2sided)

# Display results
data.frame(Method = "Two-Sided AME", Estimate = AME_two_sided)
#>          Method  Estimate
#> 1 Two-Sided AME 0.1921633
```

- **One-sided** AME is computationally simpler but can introduce bias.
- **Two-sided** AME reduces bias but requires two function evaluations per observation (Table 8.5).

8.6.2 Marginal Effects at the Mean

Marginal effects in nonlinear models are not constant, as they depend on the values of independent variables. One way to summarize them is by computing Marginal Effects at the Mean (MEM), which estimates marginal effects **at the average values of the independent variables** (Table 8.6).

MEM is commonly used in:

- **Econometrics**: Evaluating the effect of education on wages at the average level of experience.
- **Finance**: Assessing the impact of credit scores on loan approval probability for a typical applicant.
- **Marketing**: Estimating the effect of price on purchase probability for an average customer.

Table 8.5 Comparison of one-sided vs. two-sided AME

Method	Accuracy	Computational Cost	Bias
One-sided	Lower	Faster	Higher
Two-sided	Higher	Slightly slower	Lower

Table 8.6 MEM vs. AME

Method	Description	Pros	Cons
MEM	Marginal effect computed at the mean values of predictors.	Easy to interpret, based on a single reference point.	Not representative if data is skewed or nonlinear.
AME	Average of marginal effects across all observations.	More generalizable, considers full distribution of data.	Computationally more expensive.

Unlike the Average Marginal Effect, which averages marginal effects over all observations, MEM computes the effect **at a single point**—the mean of the explanatory variables.

Let $p(\mathbf{X}, \beta)$ be the predicted probability in a nonlinear model (e.g., logistic regression). The **MEM** is computed as:

$$\frac{\partial p(\bar{\mathbf{X}}, \beta)}{\partial X}$$

where $\bar{\mathbf{X}}$ is the **vector of mean values** of all explanatory variables.

For a logistic regression model:

$$E[Y|X] = \frac{1}{1 + e^{-(\beta_0 + \beta_1 X)}}$$

the MEM for a continuous variable X is:

$$\left.\frac{\partial E[Y|X]}{\partial X}\right|_{X=\bar{X}} = \beta_1 \cdot p(\bar{X}) \cdot (1 - p(\bar{X})).$$

Step 1: Estimate the Model

```
# Load necessary packages
library(margins)
library(sandwich)
library(lmtest)

# Simulate data
set.seed(123)
n <- 100
X <- rnorm(n)
Y <- rbinom(n, 1, plogis(0.5 + 0.8 * X))  # Logistic function

# Logistic regression
logit_model <- glm(Y ~ X, family = binomial(link = "logit"))
```

Step 2: Compute Marginal Effects at the Mean

```
# Compute mean of X
X_mean <- mean(X)

# Compute predicted probability at mean X
p_mean <-
    predict(logit_model,
            newdata = data.frame(X = X_mean),
            type = "response")

# Compute MEM for X
MEM <- coef(logit_model)["X"] * p_mean * (1 - p_mean)

# Display result
data.frame(Method = "MEM", Estimate = MEM)
#>   Method  Estimate
#> X    MEM 0.2146628
```

Interpretation

- The MEM tells us the effect of X on the probability of $Y = 1$ at the mean value of X.
- It provides a simple interpretation but may not capture the variability in marginal effects across different values of X.

A third approach is **Marginal Effects at Representative Values (MER)**, where we calculate marginal effects at **specific percentiles** (e.g., median, quartiles). Below is a comparison (Table 8.7):

When to Use Each Method

- Use **MEM** when you need a quick, interpretable summary at an "average" individual.
- Use **AME** when marginal effects **vary widely across individuals**.
- Use **MER** when you need to understand effects at **specific values of interest**.

Table 8.7 Comparison of marginal effect estimation methods: MEM, AME, and MER

Method	Description	Pros	Cons
MEM	Marginal effect at mean values of predictors.	Simple to compute, useful if data is symmetric.	Not informative if mean is not representative.
AME	Average of marginal effects over all observations.	More generalizable.	More computationally expensive.
MER	Marginal effect at specific values (e.g., median, percentiles).	Captures effects at different levels of X.	Requires choosing relevant reference values.

8.6.3 *Marginal Effects at the Average*

Marginal effects summarize how an independent variable influences the probability of an outcome in **nonlinear models** (e.g., logistic regression). We have already discussed:

- Marginal Effects at the Mean (MEM): Marginal effects computed at the mean values of the independent variables.
- Average Marginal Effects (AME): The mean of marginal effects computed at each observation.

A third approach is Marginal Effects at the Average(MAE), where we **first average the independent variables** across all observations and then compute the marginal effect at that single averaged observation (Table 8.8).

Let $p(\mathbf{X}, \beta)$ be the probability function of a model (e.g., a logistic regression). The Marginal Effect at the Average is computed as:

$$\frac{\partial p(\bar{\mathbf{X}}, \beta)}{\partial X}$$

where $\bar{\mathbf{X}}$ is the vector of **averaged independent variables across all observations**.

Key Differences Between AME and MAE

- AME answers a **general** question: "How does X affect Y across the entire dataset?"
- MAE answers a more **specific** question: "How does X affect Y for a typical (average) person in our dataset?"

MAE is particularly relevant when we want a single, interpretable effect for a representative individual.

Table 8.8 Comparison: MAE vs. MEM vs. AME

Method	Definition	Pros	Cons
MEM	Compute marginal effects at the mean values of X.	Simple and interpretable.	Mean values may not represent actual observations.
AME	Compute marginal effects for each observation, then take the average.	More robust, accounts for variability.	Computationally more expensive.
MAE	Compute probability at averaged X values, then compute the marginal effect.	Accounts for interactions better than MEM.	Less commonly used, can be misleading if X values are skewed.

Use Cases for MAE

1. **Policy & Business Decision-Making**: If policymakers or business leaders want to know the effect of a tax increase on a "typical" consumer, MAE gives an effect for a single representative individual.
2. **Marketing Campaigns**: If a marketing team wants to know how much increasing ad spend affects the purchase probability of an "average" customer, MAE provides this insight.
3. **Simplified Reporting**: AMEs vary across individuals, which can make reporting complex. MAE condenses everything into one easy-to-interpret number.

Intuition Behind MAE

- Instead of computing individual marginal effects (as in AME), MAE computes the **marginal effect for a single averaged observation**.
- This method is **somewhat similar to MEM**, but instead of taking the mean of each independent variable separately, it **first computes a single averaged observation and then derives the marginal effect at that observation**.

Step 1: Estimate the Model

```
# Load necessary packages
library(margins)

# Simulate data
set.seed(123)
n <- 100
X1 <- rnorm(n)  # Continuous variable
X2 <- rbinom(n, 1, 0.5)  # Binary variable
# Logistic function
Y <-
    rbinom(n, 1, plogis(0.5 + 0.8 * X1 - 0.5 * X2))

# Logistic regression
logit_model <- glm(Y ~ X1 + X2, family = binomial(link =
↪  "logit"))
```

Step 2: Compute MAE

```
# Compute the average of independent variables
X_mean <- data.frame(X1 = mean(X1), X2 = mean(X2))

# Compute predicted probability at averaged X
p_mean <- predict(logit_model, newdata = X_mean, type =
↪  "response")
```

```
# Compute MAE for X1
MAE_X1 <- coef(logit_model)["X1"] * p_mean * (1 - p_mean)

# Compute MAE for X2
MAE_X2 <- coef(logit_model)["X2"] * p_mean * (1 - p_mean)

# Display results
data.frame(
    Method = "MAE",
    Variable = c("X1", "X2"),
    Estimate = c(MAE_X1, MAE_X2)
)
#>    Method Variable    Estimate
#> X1    MAE       X1  0.20280618
#> X2    MAE       X2 -0.06286593
```

- The MAE for X_1 represents the change in probability when increasing X_1 at the average values of X_1 and X_2.
- The MAE for X_2 (a binary variable) represents the probability change when switching from $X_2 = 0$ to $X_2 = 1$, holding all other variables at their average (Table 8.9).

When to Use MAE

- When you want a single number summary that reflects a realistic scenario.
- When there are interaction effects, and you want to account for the joint impact of predictors.
- However, if predictor distributions are skewed, AME is usually preferred.

Table 8.9 Comparison of MEM, AME, and MAE in marginal effect computation

Method	Computes marginal effect...	Accounts for variability?	Best for...
MEM	At the mean of each independent variable.	No	Quick interpretation at a reference point. When individual means are meaningful (e.g., symmetric data).
AME	At each observation, then averages.	Yes	Generalizable results.
MAE	At a single averaged observation.	No	Simple summary when interactions exist. When we need a single interpretable summary that accounts for interactions

8.7 Packages for Marginal Effects

Several R packages compute **marginal effects** for regression models, each with different features and functionalities. The primary packages include:

1. **`marginaleffects`**—A modern, flexible, and efficient package.
2. **`margins`**—A widely used package for replicating Stata's `margins` command.
3. **`mfx`**—A package tailored for Generalized Linear Models (`glm`).

These tools help analyze how small changes in explanatory variables impact the dependent variable.

8.7.1 `marginaleffects` Package (Recommended)

The **`marginaleffects`** package is the successor to `margins` and `emtrends`, offering a faster, more efficient, and more flexible approach to estimating marginal effects.

Why Use `marginaleffects`?

- Supports interaction effects and complex models
- Computes marginal effects, marginal means, and counterfactuals
- Integrates well with ggplot2 for visualization
- Works with many model types (linear, logistic, Poisson, etc.)

Limitation

- No built-in function for **multiple comparisons correction**, but you can use `p.adjust()` for adjustment.

Key Definitions

- **Marginal Effects**: The partial derivative (slope) of the outcome with respect to each variable.
  - `margins` package defines marginal effects as "partial derivatives of the regression equation with respect to each variable in the model for each unit in the data."
- **Marginal Means**: The expected outcome averaged over a grid of predictor values.

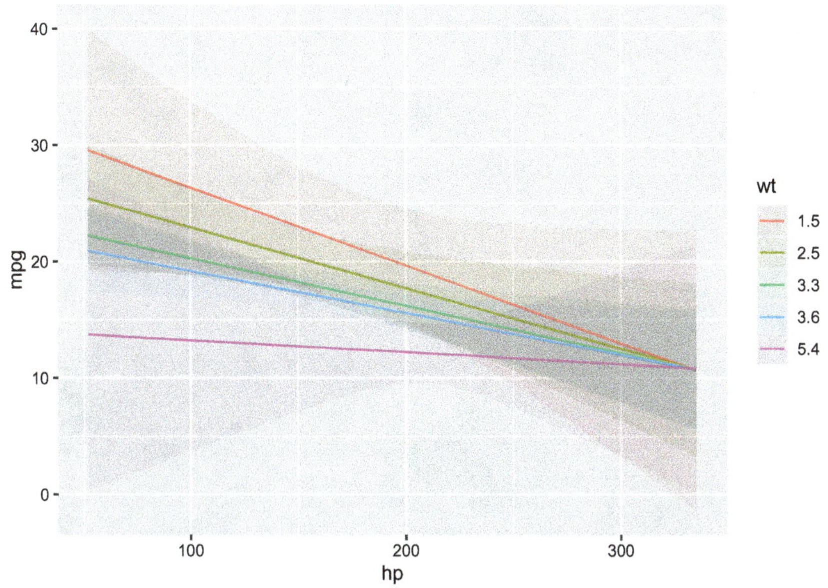

Fig. 8.2 Line chart coefficients

Computing Predictions and Marginal Effects (Fig. 8.2)

```
library(marginaleffects)
library(tidyverse)
#> -- Attaching core tidyverse packages ---- tidyverse 2.0.0
↪  --
#> v dplyr     1.1.4     v readr     2.1.5
#> v forcats   1.0.0     v stringr   1.5.1
#> v lubridate 1.9.4     v tibble    3.2.1
#> v purrr     1.0.4     v tidyr     1.3.1
#> -- Conflicts ---------------------- tidyverse_conflicts()
↪  --
#> x dplyr::filter() masks stats::filter()
#> x dplyr::lag()    masks stats::lag()
#> i Use the conflicted package
↪  (<http://conflicted.r-lib.org/>) to force all conflicts to
↪  become errors
data(mtcars)


# Fit a regression model with interaction terms
mod <- lm(mpg ~ hp * wt * am, data = mtcars)
```

```r
# Get predicted values
predictions(mod) %>% head()
#>
#>  Estimate Std. Error    z Pr(>|z|)     S 2.5 % 97.5 %
#>      22.5      0.884 25.4   <0.001 471.7  20.8   24.2
#>      20.8      1.194 17.4   <0.001 223.3  18.5   23.1
#>      25.3      0.709 35.7   <0.001 922.7  23.9   26.7
#>      20.3      0.704 28.8   <0.001 601.5  18.9   21.6
#>      17.0      0.712 23.9   <0.001 416.2  15.6   18.4
#>      19.7      0.875 22.5   <0.001 368.8  17.9   21.4
#>
#> Type:  response

# Create a reference grid for prediction
newdata <- datagrid(am = 0,
                    wt = c(2, 4),
                    model = mod)
```

```r
# Plot predictions for 'hp' and 'wt'
marginaleffects::plot_predictions(mod,
                                  newdata = newdata,
                                  condition = c("hp", "wt"))
```

Computing Marginal Effects

```r
# Compute Average Marginal Effects (AME)
mfx <- marginaleffects::slopes(mod, variables = c("hp", "wt"))
head(mfx)
#>
#>  Estimate Std. Error     z Pr(>|z|)   S   2.5 %    97.5 %
#>   -0.0369     0.0185 -2.00  0.04598 4.4 -0.0732 -0.000659
#>   -0.0287     0.0156 -1.84  0.06630 3.9 -0.0593  0.001931
#>   -0.0466     0.0226 -2.06  0.03945 4.7 -0.0909 -0.002249
#>   -0.0423     0.0133 -3.18  0.00146 9.4 -0.0683 -0.016232
#>   -0.0390     0.0134 -2.91  0.00363 8.1 -0.0653 -0.012730
#>   -0.0387     0.0135 -2.87  0.00409 7.9 -0.0652 -0.012289
#>
#> Term: hp
#> Type:  response
#> Comparison: dY/dX

# Compute Group-Average Marginal Effects
head(marginaleffects::slopes(mod, by = "hp", variables =
↪  "am"))
```

```
#>
#>  hp Estimate Std. Error      z Pr(>|z|)   S 2.5 % 97.5 %
#>  52     3.98       5.20  0.764    0.445 1.2 -6.22  14.18
#>  62    -2.77       2.51 -1.107    0.268 1.9 -7.68   2.14
#>  65     3.00       4.13  0.725    0.468 1.1 -5.10  11.10
#>  66     2.03       3.48  0.582    0.561 0.8 -4.80   8.85
#>  91     1.86       2.76  0.674    0.500 1.0 -3.54   7.26
#>  93     1.20       2.35  0.511    0.609 0.7 -3.40   5.80
#>
#> Term: am
#> Type:  response
#> Comparison: 1 - 0

# Marginal Effects at Representative Values (MER)
marginaleffects::slopes(mod, newdata = datagrid(am = 0, wt =
↪  c(2, 4)))
#>
#>  Term Contrast am wt Estimate Std. Error      z Pr(>|z|)
#>    am    1 - 0  0  2   2.5465     2.7860  0.914   0.3607
#>    am    1 - 0  0  4  -2.9661     3.0381 -0.976   0.3289
#>    hp    dY/dX  0  2  -0.0598     0.0283 -2.115   0.0344
#>    hp    dY/dX  0  4  -0.0309     0.0187 -1.654   0.0981
#>    wt    dY/dX  0  2  -2.6762     1.4199 -1.885   0.0595
#>    wt    dY/dX  0  4  -2.6762     1.4206 -1.884   0.0596
#>    S   2.5 %   97.5 %
#>  1.5 -2.9139  8.00694
#>  1.6 -8.9207  2.98852
#>  4.9 -0.1153 -0.00439
#>  3.4 -0.0676  0.00571
#>  4.1 -5.4591  0.10676
#>  4.1 -5.4605  0.10816
#>
#> Type:  response

# Marginal Effects at the Mean (MEM)
marginaleffects::slopes(mod, newdata = "mean")
#>
#>  Term Contrast Estimate Std. Error      z Pr(>|z|)   S
#>    am    1 - 0  -0.8086     1.5238 -0.531  0.59568 0.7
#>    hp    dY/dX  -0.0422     0.0133 -3.181  0.00147 9.4
#>    wt    dY/dX  -2.6762     1.4193 -1.886  0.05935 4.1
#>    2.5 %  97.5 %
#>  -3.7952  2.1781
#>  -0.0683 -0.0162
```

```
#>  -5.4579  0.1055
#>
#> Type:  response
```

Counterfactual Comparisons

```
# Counterfactual comparison: Effect of changing 'am' from 0 to
↪ 1
comparisons(mod, variables = list(am = 0:1))
#>
#>  Estimate Std. Error      z Pr(>|z|)   S 2.5 % 97.5 %
#>     0.325       1.68  0.193    0.847 0.2 -2.97   3.62
#>    -0.544       1.57 -0.347    0.729 0.5 -3.62   2.53
#>     1.201       2.35  0.511    0.609 0.7 -3.40   5.80
#>    -1.703       1.87 -0.912    0.362 1.5 -5.36   1.96
#>    -0.615       1.68 -0.366    0.715 0.5 -3.91   2.68
#> --- 22 rows omitted. See ?print.marginaleffects ---
#>     4.081       3.94  1.037    0.300 1.7 -3.63  11.79
#>     2.106       2.29  0.920    0.358 1.5 -2.38   6.59
#>     0.895       1.64  0.544    0.586 0.8 -2.33   4.12
#>     4.027       3.24  1.243    0.214 2.2 -2.32  10.38
#>    -0.237       1.59 -0.149    0.881 0.2 -3.35   2.87
#> Term: am
#> Type:  response
#> Comparison: 1 - 0
```

8.7.2 `margins` *Package*

The `margins` package is a popular choice, designed to replicate Stata's `margins` command in R. It provides marginal effects for each variable in a model, including interaction terms. `margins` define:

1. **Average Partial Effects:** The contribution of each variable to the outcome scale, conditional on the other variables involved in the link function transformation of the linear predictor.
2. **Average Marginal Effects:** The marginal contribution of each variable on the scale of the linear predictor (Fig. 8.3).

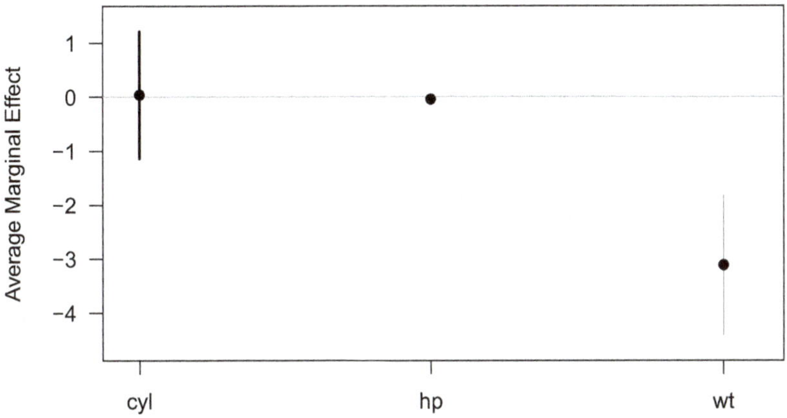

Fig. 8.3 Average marginal effect

Key Features

- Computes Average Partial Effects (APE).
- Supports Marginal Effects at the Mean (MEM).
- Provides visualizations of marginal effects.

Limitation

- Slower than `marginaleffects`, especially with large datasets.

```
library(margins)

# Fit a linear model with interactions
mod <- lm(mpg ~ cyl * hp + wt, data = mtcars)

# Compute marginal effects
summary(margins(mod))
#>  factor     AME     SE       z      p   lower   upper
#>     cyl  0.0381 0.5999  0.0636 0.9493 -1.1376  1.2139
#>      hp -0.0463 0.0145 -3.1909 0.0014 -0.0748 -0.0179
#>      wt -3.1198 0.6613 -4.7175 0.0000 -4.4160 -1.8236

# Equivalent function for summary
margins_summary(mod)
#>  factor     AME     SE       z      p   lower   upper
#>     cyl  0.0381 0.5999  0.0636 0.9493 -1.1376  1.2139
#>      hp -0.0463 0.0145 -3.1909 0.0014 -0.0748 -0.0179
#>      wt -3.1198 0.6613 -4.7175 0.0000 -4.4160 -1.8236
```

```
# Plot marginal effects
plot(margins(mod))
```

Marginal Effects at Representative Values

```
# Compute marginal effects when 'hp' = 150
margins(mod, at = list(hp = 150)) %>% summary()
#>  factor       hp     AME     SE       z      p   lower
#>     cyl 150.0000  0.1009 0.6128  0.1647 0.8692 -1.1001
#>      hp 150.0000 -0.0463 0.0145 -3.1909 0.0014 -0.0748
#>      wt 150.0000 -3.1198 0.6613 -4.7175 0.0000 -4.4160
#>    upper
#>   1.3019
#>  -0.0179
#>  -1.8236
```

8.7.3 `mfx` *Package*

The **`mfx`** package is specialized for GLMs, computing marginal effects for probit, logit, Poisson, and other count models.

Limitation

- Computes only marginal effects for each variable, not the average marginal effect.

Supported Models in `mfx` (Table 8.10)

Table 8.10 Marginal effects functions by model and outcome type

Model	Outcome type	Function
Probit	Binary	`probitmfx()`
Logit	Binary	`logitmfx()`
Poisson	Count	`poissonmfx()`
Negative binomial	Count	`negbinmfx()`
Beta	Rate	`betamfx()`

Example: Poisson Regression

```
library(mfx)
#> Loading required package: MASS
#>
#> Attaching package: 'MASS'
#> The following object is masked from 'package:dplyr':
#>
#>     select
#> Loading required package: betareg
data("mtcars")

# Fit a Poisson model and compute marginal effects
poissonmfx(formula = vs ~ mpg * cyl * disp, data = mtcars)
#> Warning: glm.fit: fitted rates numerically 0 occurred
#> Call:
#> poissonmfx(formula = vs ~ mpg * cyl * disp, data = mtcars)
#>
#> Marginal Effects:
#>                     dF/dx   Std. Err.       z  P>|z|
#> mpg            1.4722e-03  8.7531e-03  0.1682 0.8664
#> cyl            6.6420e-03  3.9263e-02  0.1692 0.8657
#> disp           1.5899e-04  9.4555e-04  0.1681 0.8665
#> mpg:cyl       -3.4698e-04  2.0564e-03 -0.1687 0.8660
#> mpg:disp      -7.6794e-06  4.5545e-05 -0.1686 0.8661
#> cyl:disp      -3.3837e-05  1.9919e-04 -0.1699 0.8651
#> mpg:cyl:disp  1.6812e-06  9.8919e-06  0.1700 0.8650
```

For more details, see the `mfx` vignette.

8.7.4 Comparison of Packages

See Table 8.11.

Table 8.11 Comparison of R packages for estimating marginal effects

Feature	`marginaleffects` (Recommended)	`margins`	`mfx`
Computes AME	Yes	Yes	No
Computes MEM	Yes	Yes	No
Computes MER	Yes	Yes	No
Supports GLM models	Yes	Yes	Yes (but limited to `glm`)
Works with interactions	Yes	Yes	No
Fast and efficient	Yes	No (slower)	Yes (for GLMs)
Supports counterfactuals	Yes	No	No

Chapter 9
Moderation

Moderation analysis is essential for understanding interaction effects, when the relationship between two variables depends on the value of a third. This chapter introduces the concept through real-world examples. After outlining common types of moderation (binary, continuous, hierarchical), the chapter walks through the key terminology, including moderators, focal predictors, and conditional effects. It covers the classic moderation model and introduces interaction terms in regression. Later sections delve into two-way and three-way interactions, providing detailed guidance on specification, estimation, and interpretation. Graphical methods for exploring interaction effects are emphasized, using interaction plots and marginal effects visualization. The chapter ensures that readers are able not only to model interaction effects correctly but also to communicate them clearly to non-technical stakeholders.

Moderation analysis examines how the relationship between an independent variable (X) and a dependent variable (Y) changes depending on a third variable, the *moderator* (M). In regression terms, moderation is represented as an interaction effect.

9.1 Types of Moderation Analyses

There are two primary approaches to analyzing moderation:

1. Spotlight Analysis
    - Also known as **Simple Slopes Analysis**.
    - Examines the effect of X on Y at specific values of M (e.g., mean, $\pm$ 1 SD, percentiles).
    - Typically used for categorical or discretized moderators.

© The Author(s), under exclusive license to Springer Nature Switzerland AG 2025

M. Nguyen, *Advanced Modeling and Data Challenges*,
https://doi.org/10.1007/978-3-032-01719-2_9

2. Floodlight Analysis
   - Extends spotlight analysis to examine moderation across the entire range of M.
   - Based on Johnson–Neyman Intervals, identifying values of M where the effect of X on Y is statistically significant.
   - Useful when the moderator is continuous and no specific cutoffs are predefined.

9.2 Key Terminology

- **Main Effect**: The effect of an independent variable without considering interactions.
- **Interaction Effect**: The combined effect of X and M on Y.
- **Simple Slope**: The slope of X on Y at a specific value of M (used when M is continuous).
- **Simple Effect**: The effect of X on Y at a particular level of M when X is categorical.

9.3 Moderation Model

A typical moderation model is represented as:

$$Y = \beta_0 + \beta_1 X + \beta_2 M + \beta_3 X \times M + \varepsilon$$

where:

- β_0: Intercept
- β_1: Main effect of X
- β_2: Main effect of M
- β_3: Interaction effect of X and M

  If β_3 is significant, it suggests that the effect of X on Y depends on M.

9.4 Types of Interactions

1. **Continuous by Continuous**: Both X and M are continuous (e.g., age moderating the effect of income on spending).
2. **Continuous by Categorical**: X is continuous, and M is categorical (e.g., gender moderating the effect of education on salary).

3. **Categorical by Categorical**: Both X and M are categorical (e.g., the effect of a training program on performance, moderated by job role).

9.5 Three-Way Interactions

For models with a second moderator (W), we examine:

$$
\begin{aligned}
Y = \beta_0 + \beta_1 X + \beta_2 M + \beta_3 W + \beta_4 X \times M \\
+ \beta_5 X \times W + \beta_6 M \times W + \beta_7 X \times M \times W + \varepsilon
\end{aligned}
$$

- To interpret three-way interactions, the **slope difference test** can be used (Dawson & Richter, 2006).

9.6 Additional Resources

- **Bayesian ANOVA models**: `BANOVAL` package allows floodlight analysis.
- **Structural Equation Modeling**: `cSEM` package includes `doFloodlightAnalysis`.

  For more details, refer to Spiller et al. (2013).

9.7 Application

9.7.1 *`emmeans` Package*

The `emmeans` package (Estimated Marginal Means) is a powerful tool for post-hoc analysis of linear models, enabling researchers to explore interaction effects through simple slopes and estimated marginal means. To install and load the package:

```
install.packages("emmeans")
```

The dataset used in this section is sourced from the UCLA Statistical Consulting Group, where:

- `gender` (male, female) and `prog` (exercise program: jogging, swimming, reading) are categorical variables.
- `loss` represents weight loss, and `hours` and `effort` are continuous predictors.

```r
library(tidyverse)
#> -- Attaching core tidyverse packages ---- tidyverse 2.0.0 --
#> v dplyr     1.1.4     v readr     2.1.5
#> v forcats   1.0.0     v stringr   1.5.1
#> v ggplot2   3.5.2     v tibble    3.2.1
#> v lubridate 1.9.4     v tidyr     1.3.1
#> v purrr     1.0.4
#> -- Conflicts ---------------------- tidyverse_conflicts() --
#> x dplyr::filter() masks stats::filter()
#> x dplyr::lag()    masks stats::lag()
#> i Use the conflicted package (<http://conflicted.r-lib.org/>) to force
↪   all conflicts to become errors

dat <- readRDS("data/exercise.rds") %>%
    mutate(prog = factor(prog, labels = c("jog", "swim", "read"))) %>%
    mutate(gender = factor(gender, labels = c("male", "female")))
```

9.7.1.1 Continuous by Continuous Interaction

We begin with an interaction model between two continuous variables: `hours` (exercise duration) and `effort` (self-reported effort level).

```r
contcont <- lm(loss ~ hours * effort, data = dat)
summary(contcont)
#>
#> Call:
#> lm(formula = loss ~ hours * effort, data = dat)
#>
#> Residuals:
#>    Min     1Q Median     3Q    Max
#> -29.52 -10.60  -1.78  11.13  34.51
#>
#> Coefficients:
#>              Estimate Std. Error t value Pr(>|t|)
#> (Intercept)   7.79864   11.60362   0.672   0.5017
#> hours        -9.37568    5.66392  -1.655   0.0982 .
#> effort       -0.08028    0.38465  -0.209   0.8347
#> hours:effort  0.39335    0.18750   2.098   0.0362 *
#> ---
#> Signif. codes:
#> 0 '***' 0.001 '**' 0.01 '*' 0.05 '.' 0.1 ' ' 1
#>
#> Residual standard error: 13.56 on 896 degrees of freedom
#> Multiple R-squared:  0.07818,    Adjusted R-squared:  0.07509
#> F-statistic: 25.33 on 3 and 896 DF,  p-value: 9.826e-16
```

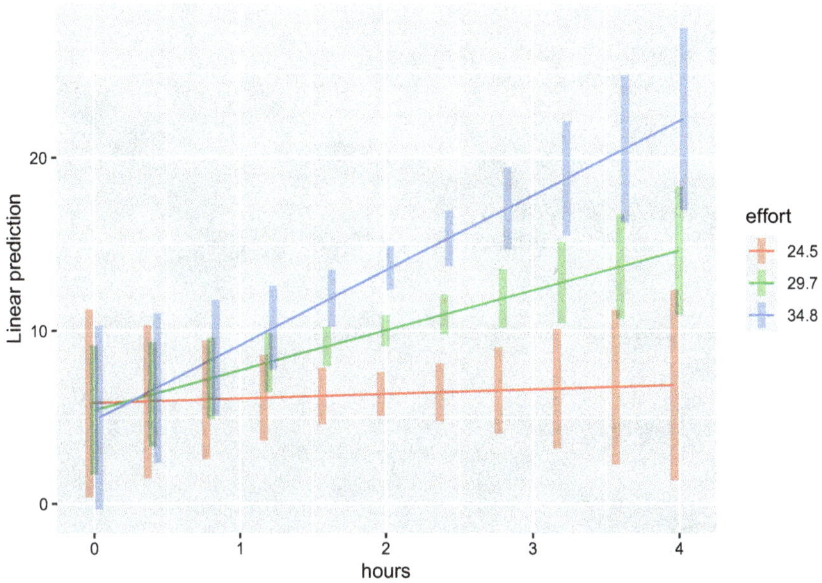

Fig. 9.1 Linear prediction plot

9.7.1.1.1 Simple Slopes Analysis (Spotlight Analysis)

Following Aiken and West (2005), the spotlight analysis examines the effect of `hours` on `loss` at three levels of `effort` (Fig. 9.1):

- Mean of `effort` plus one standard deviation
- Mean of `effort`
- Mean of `effort` minus one standard deviation

```
library(emmeans)
#> Welcome to emmeans.
#> Caution: You lose important information if you filter this package's
↪  results.
#> See '? untidy'
effar <- round(mean(dat$effort) + sd(dat$effort), 1)
effr  <- round(mean(dat$effort), 1)
effbr <- round(mean(dat$effort) - sd(dat$effort), 1)



# Define values for estimation
mylist <- list(effort = c(effbr, effr, effar))

# Compute simple slopes
emtrends(contcont, ~ effort, var = "hours", at = mylist)
#>  effort hours.trend    SE  df lower.CL upper.CL
```

```
#>     24.5          0.261 1.350 896    -2.392      2.91
#>     29.7          2.307 0.915 896     0.511      4.10
#>     34.8          4.313 1.310 896     1.745      6.88
#>
#> Confidence level used: 0.95

# Visualization of the interaction
mylist <- list(hours = seq(0, 4, by = 0.4),
               effort = c(effbr, effr, effar))
```

```
emmip(contcont, effort ~ hours, at = mylist, CIs = TRUE)
```

```
# Test statistical differences in slopes
emtrends(
    contcont,
    pairwise ~ effort,
    var = "hours",
    at = mylist,
    adjust = "none"
)
#> $emtrends
#>  effort hours.trend    SE  df lower.CL upper.CL
#>     24.5          0.261 1.350 896    -2.392      2.91
#>     29.7          2.307 0.915 896     0.511      4.10
#>     34.8          4.313 1.310 896     1.745      6.88
#>
#> Results are averaged over the levels of: hours
#> Confidence level used: 0.95
#>
#> $contrasts
#>  contrast                  estimate    SE  df t.ratio p.value
#>  effort24.5 - effort29.7     -2.05 0.975 896  -2.098  0.0362
#>  effort24.5 - effort34.8     -4.05 1.930 896  -2.098  0.0362
#>  effort29.7 - effort34.8     -2.01 0.956 896  -2.098  0.0362
#>
#> Results are averaged over the levels of: hours
```

The three p-values obtained above correspond to the interaction term in the regression model.

For a professional figure, we refine the visualization using `ggplot2` (Fig. 9.2):

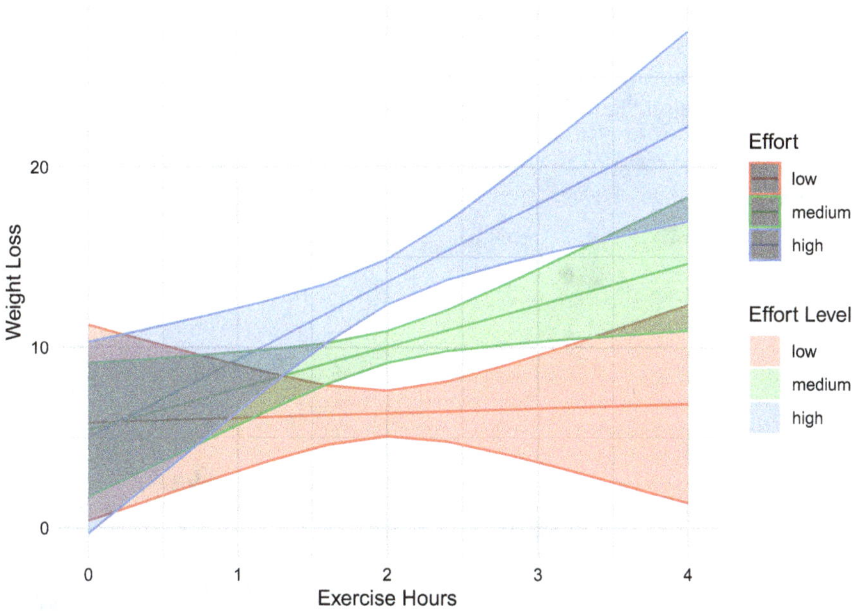

Fig. 9.2 Weight loss vs exercise hours

```
library(ggplot2)

# Prepare data for plotting
mylist <- list(hours = seq(0, 4, by = 0.4),
               effort = c(effbr, effr, effar))
contcontdat <-
    emmip(contcont,
          effort ~ hours,
          at = mylist,
          CIs = TRUE,
          plotit = FALSE)

# Convert effort levels to factors
contcontdat$feffort <- factor(contcontdat$effort)
levels(contcontdat$feffort) <- c("low", "medium", "high")
```

```
ggplot(data = contcontdat,
       aes(x = hours, y = yvar, color = feffort)) +
    geom_line() +
    geom_ribbon(aes(ymax = UCL, ymin = LCL, fill = feffort),
                alpha = 0.4) + labs(x = "Exercise Hours",
                                    y = "Weight Loss",
                                    color = "Effort",
                                    fill = "Effort Level") +
    theme_minimal()
```

9.7.1.2 Continuous by Categorical Interaction

Next, we examine an interaction where hours (continuous) interacts with gender (categorical). We set "Female" as the reference category:

```
dat$gender <- relevel(dat$gender, ref = "female")
contcat <- lm(loss ~ hours * gender, data = dat)
summary(contcat)
#>
#> Call:
#> lm(formula = loss ~ hours * gender, data = dat)
#>
#> Residuals:
#>     Min      1Q  Median      3Q     Max
#> -27.118 -11.350  -1.963  10.001  42.376
#>
#> Coefficients:
#>                  Estimate Std. Error t value Pr(>|t|)
#> (Intercept)         3.335      2.731   1.221    0.222
#> hours               3.315      1.332   2.489    0.013 *
#> gendermale          3.571      3.915   0.912    0.362
#> hours:gendermale   -1.724      1.898  -0.908    0.364
#> ---
#> Signif. codes:
#> 0 '***' 0.001 '**' 0.01 '*' 0.05 '.' 0.1 ' ' 1
#>
#> Residual standard error: 14.06 on 896 degrees of freedom
#> Multiple R-squared:  0.008433,	Adjusted R-squared:  0.005113
#> F-statistic:  2.54 on 3 and 896 DF,  p-value: 0.05523
```

Simple Slopes by Gender

```
# Compute simple slopes for each gender
emtrends(contcat, ~ gender, var = "hours")
#>  gender hours.trend   SE  df lower.CL upper.CL
#>  female        3.32 1.33 896    0.702     5.93
#>  male          1.59 1.35 896   -1.063     4.25
#>
#> Confidence level used: 0.95

# Test slope differences
emtrends(contcat, pairwise ~ gender, var = "hours")
#> $emtrends
#>  gender hours.trend   SE  df lower.CL upper.CL
#>  female        3.32 1.33 896    0.702     5.93
#>  male          1.59 1.35 896   -1.063     4.25
#>
#> Confidence level used: 0.95
#>
#> $contrasts
#>  contrast      estimate  SE  df t.ratio p.value
#>  female - male     1.72 1.9 896   0.908  0.3639
```

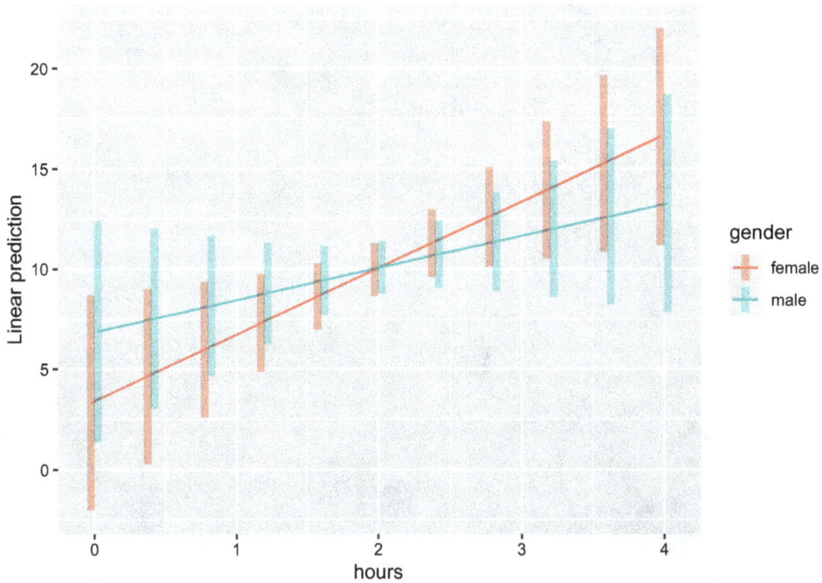

Fig. 9.3 Linear prediction vs hours

Since this test is equivalent to the interaction term in the regression model, a significant result confirms a moderating effect of `gender` (Fig. 9.3).

```
mylist <- list(hours = seq(0, 4, by = 0.4),
               gender = c("female", "male"))
```

```
emmip(contcat, gender ~ hours, at = mylist, CIs = TRUE)
```

9.7.1.3 Categorical by Categorical Interaction

Now, we examine the interaction between two categorical variables: `gender` (male, female) and `prog` (exercise program). We set "Read" as the reference category for `prog` and "Female" for `gender`:

```
dat$prog   <- relevel(dat$prog, ref = "read")
dat$gender <- relevel(dat$gender, ref = "female")

catcat <- lm(loss ~ gender * prog, data = dat)
summary(catcat)
#>
#> Call:
#> lm(formula = loss ~ gender * prog, data = dat)
```

```
#>
#> Residuals:
#>      Min       1Q   Median       3Q      Max
#> -19.1723  -4.1894  -0.0994   3.7506  27.6939
#>
#> Coefficients:
#>                      Estimate Std. Error t value Pr(>|t|)
#> (Intercept)           -3.6201     0.5322  -6.802 1.89e-11
#> gendermale            -0.3355     0.7527  -0.446    0.656
#> progjog                7.9088     0.7527  10.507  < 2e-16
#> progswim              32.7378     0.7527  43.494  < 2e-16
#> gendermale:progjog     7.8188     1.0645   7.345 4.63e-13
#> gendermale:progswim   -6.2599     1.0645  -5.881 5.77e-09
#>
#> (Intercept)          ***
#> gendermale
#> progjog              ***
#> progswim             ***
#> gendermale:progjog   ***
#> gendermale:progswim  ***
#> ---
#> Signif. codes:
#> 0 '***' 0.001 '**' 0.01 '*' 0.05 '.' 0.1 ' ' 1
#>
#> Residual standard error: 6.519 on 894 degrees of freedom
#> Multiple R-squared:  0.7875, Adjusted R-squared:  0.7863
#> F-statistic: 662.5 on 5 and 894 DF,  p-value: < 2.2e-16
```

Simple Effects and Contrast Analysis (Fig. 9.4)

```
# Estimated marginal means for all combinations of gender and program
emcatcat <- emmeans(catcat, ~ gender * prog)

# Compare effects of gender within each program
contrast(emcatcat, "revpairwise", by = "prog", adjust = "bonferroni")
#> prog = read:
#>  contrast      estimate    SE  df t.ratio p.value
#>  male - female   -0.335 0.753 894  -0.446  0.6559
#>
#> prog = jog:
#>  contrast      estimate    SE  df t.ratio p.value
#>  male - female    7.483 0.753 894   9.942  <.0001
#>
#> prog = swim:
#>  contrast      estimate    SE  df t.ratio p.value
#>  male - female   -6.595 0.753 894  -8.762  <.0001
```

```
emmip(catcat, prog ~ gender, CIs = TRUE)
```

For a more intuitive presentation, we use a bar graph with error bars (Fig. 9.5)

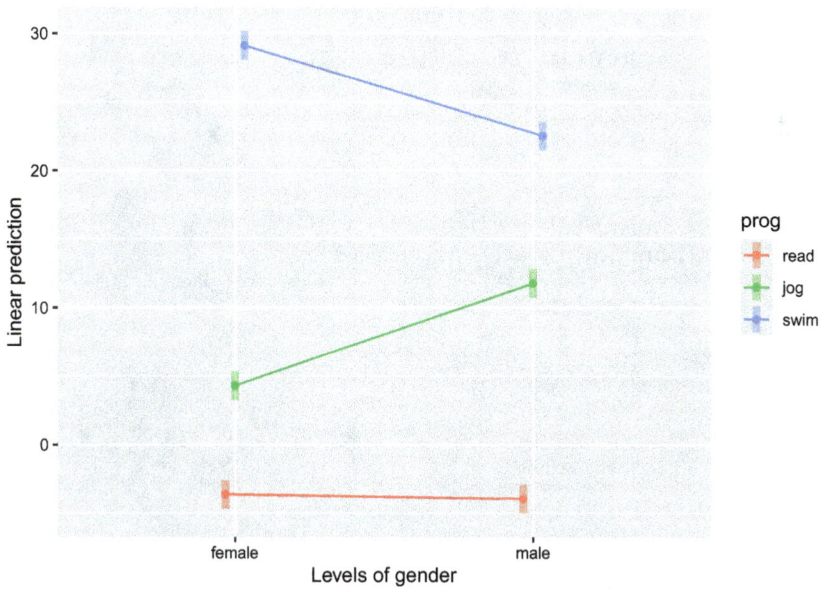

Fig. 9.4 Linear prediction vs gender

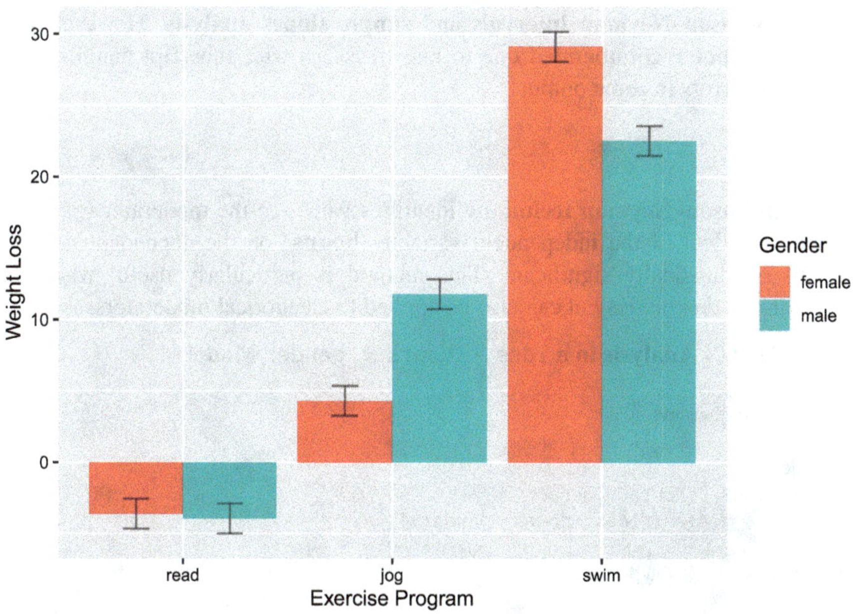

Fig. 9.5 Weigth loss vs exercise program

```
# Prepare data
catcatdat <- emmip(catcat,
                   gender ~ prog,
                   CIs = TRUE,
                   plotit = FALSE)
```

```
ggplot(data = catcatdat,
       aes(x = prog, y = yvar, fill = gender)) +
    geom_bar(stat = "identity", position = "dodge") + geom_errorbar(
        position = position_dodge(.9),
        width = .25,
        aes(ymax = UCL, ymin = LCL),
        alpha = 0.3
    ) + labs(x = "Exercise Program",
             y = "Weight Loss",
             fill = "Gender")
```

9.7.2 probemod Package

The probemod package is designed for moderation analysis, particularly focusing on **Johnson–Neyman intervals** and **simple slopes analysis**. However, **this package is not recommended** due to known issues with subscript handling and formatting errors in some outputs.

```
install.packages("probemod", dependencies = T)
```

The **Johnson–Neyman technique** identifies values of the moderator (gender) where the effect of the independent variable (hours) on the dependent variable (loss) is statistically significant. This method is particularly useful when the moderator is continuous but can also be applied to categorical moderators.

Example: J-N Analysis in a `loss ~ hours * gender` Model

```
library(probemod)

myModel <-
    lm(loss ~ hours * gender, data = dat %>%
           select(loss, hours, gender))

jnresults <- jn(myModel,
                dv = 'loss',
                iv = 'hours',
                mod = 'gender')
```

The `jn()` function computes Johnson–Neyman intervals, highlighting the values of `gender` at which the relationship between `hours` and `loss` is statistically significant.

The **Pick-a-Point** method tests the simple effect of `hours` at specific values of `gender`, akin to spotlight analysis.

```
pickapoint(
    myModel,
    dv = 'loss',
    iv = 'hours',
    mod = 'gender',
    alpha = .01
)

plot(jnresults)
```

9.7.3 `interactions` Package

The `interactions` package is a **recommended** tool for visualizing and interpreting interaction effects in regression models. It provides user-friendly functions for **interaction plots**, **simple slopes analysis**, and **Johnson–Neyman intervals**, making it an excellent choice for moderation analysis.

```
install.packages("interactions")
```

9.7.3.1 Continuous by Continuous Interaction

This section covers interactions where at least one of the two variables is continuous.

Example: Interaction Between `Illiteracy` and `Murder`

We use the `state.x77` dataset to explore how **Illiteracy Rate** and **Murder Rate** interact to predict **Income** across U.S. states.

```
states <- as.data.frame(state.x77)
fiti <- lm(Income ~ Illiteracy * Murder + `HS Grad`, data = states)
summary(fiti)
#>
#> Call:
#> lm(formula = Income ~ Illiteracy * Murder + `HS Grad`, data = states)
#>
#> Residuals:
#>      Min       1Q  Median      3Q     Max
#> -916.27 -244.42   28.42  228.14 1221.16
#>
```

```
#> Coefficients:
#>                     Estimate Std. Error t value Pr(>|t|)
#> (Intercept)          1414.46     737.84   1.917  0.06160 .
#> Illiteracy            753.07     385.90   1.951  0.05724 .
#> Murder                130.60      44.67   2.923  0.00540 **
#> `HS Grad`              40.76      10.92   3.733  0.00053 ***
#> Illiteracy:Murder     -97.04      35.86  -2.706  0.00958 **
#> ---
#> Signif. codes:
#> 0 '***' 0.001 '**' 0.01 '*' 0.05 '.' 0.1 ' ' 1
#>
#> Residual standard error: 459.5 on 45 degrees of freedom
#> Multiple R-squared:  0.4864, Adjusted R-squared:  0.4407
#> F-statistic: 10.65 on 4 and 45 DF,  p-value: 3.689e-06
```

For continuous moderators, the standard values chosen for visualization are:

- **Mean + 1 SD**
- **Mean**
- **Mean - 1 SD**

The `interact_plot()` function provides an easy way to visualize these effects (Fig. 9.6).

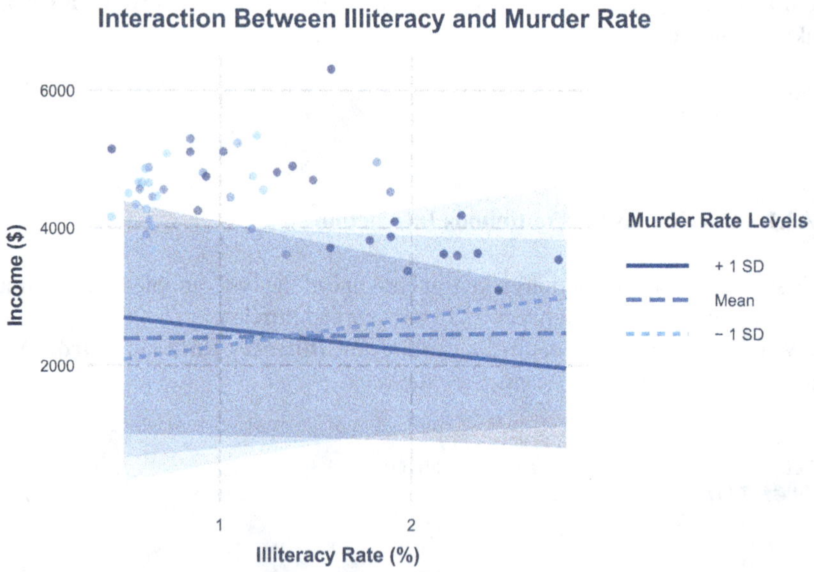

Fig. 9.6 Interaction between illiteracy and murder rate

```
library(interactions)
interact_plot(fiti,
              pred = Illiteracy,
              modx = Murder,

              # Disable automatic mean-centering
              centered = "none",

              # Exclude the mean value of the moderator
              # modx.values = ``plus-minus",

              # Divide the moderator's distribution into three groups
              # modx.values = ``terciles",

              plot.points = TRUE, # Overlay raw data

              # Different shapes for different levels of the moderator
              point.shape = TRUE,

              # Jittering to prevent overplotting
              jitter = 0.1,

              # Custom appearance
              x.label = "Illiteracy Rate (%)",
              y.label = "Income ($)",
              main.title = "Interaction Between Illiteracy and Murder
              ↪ Rate",
              legend.main = "Murder Rate Levels",
              colors = "blue",

              # Confidence bands
              interval = TRUE,
              int.width = 0.9,
              robust = TRUE # Use robust standard errors
              )
```

If the model includes **weights**, they can be incorporated into the visualization (Fig. 9.7).

```
fiti <- lm(Income ~ Illiteracy * Murder,
           data = states,
           weights = Population)
```

```
interact_plot(fiti,
              pred = Illiteracy,
              modx = Murder,
              plot.points = TRUE)
```

A **partial effect plot** shows how the effect of one variable changes across different levels of another variable while controlling for other predictors (Fig. 9.8).

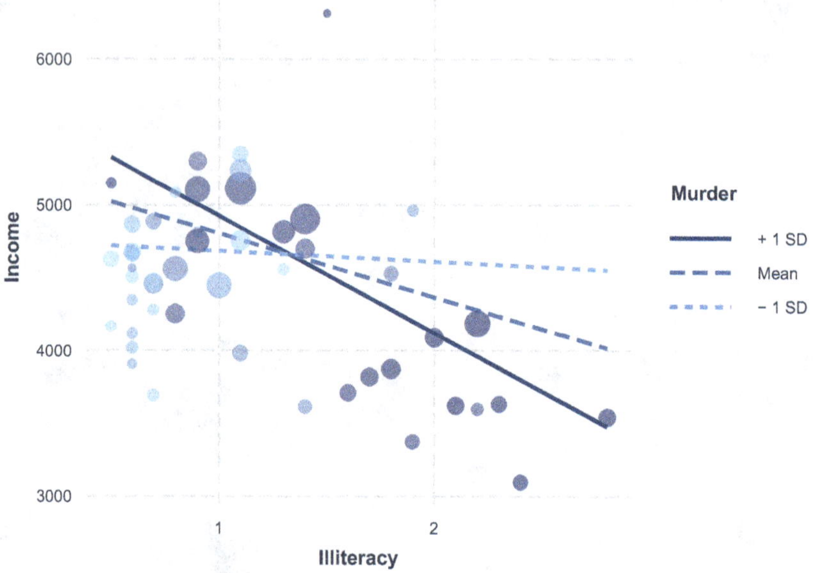

Fig. 9.7 Bubble chart between income and illiteracy

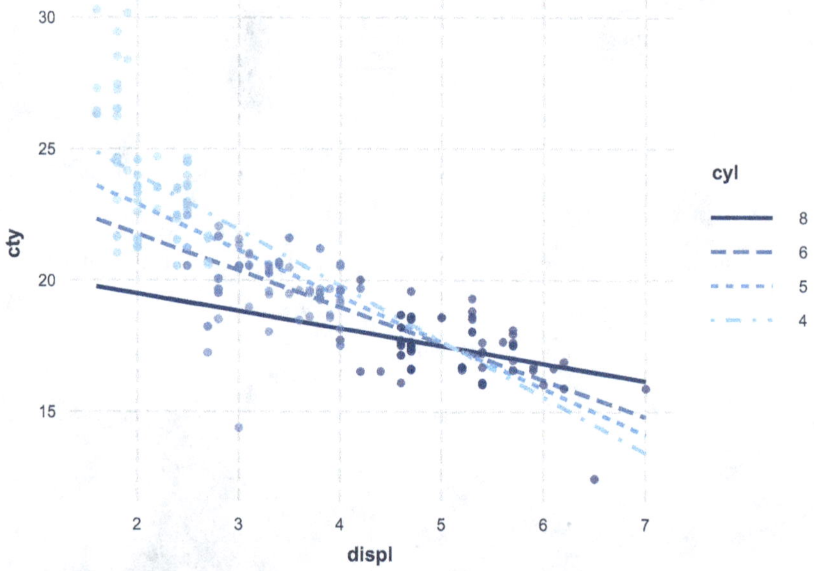

Fig. 9.8 Scatter Plot between cty and displ

```
library(ggplot2)
data(cars)

fitc <- lm(cty ~ year + cyl * displ + class + fl + drv,
           data = mpg)
summary(fitc)
#>
#> Call:
#> lm(formula = cty ~ year + cyl * displ + class + fl + drv, data = mpg)
#>
#> Residuals:
#>     Min      1Q  Median      3Q     Max
#> -5.9772 -0.7164  0.0018  0.7690  5.9314
#>
#> Coefficients:
#>                   Estimate Std. Error t value Pr(>|t|)
#> (Intercept)     -200.97599   47.00954  -4.275 2.86e-05 ***
#> year               0.11813    0.02347   5.033 1.01e-06 ***
#> cyl               -1.85648    0.27745  -6.691 1.86e-10 ***
#> displ             -3.56467    0.65943  -5.406 1.70e-07 ***
#> classcompact      -2.60177    0.92972  -2.798 0.005597 **
#> classmidsize      -2.62996    0.93273  -2.820 0.005253 **
#> classminivan      -4.40817    1.03844  -4.245 3.24e-05 ***
#> classpickup       -4.37322    0.93416  -4.681 5.02e-06 ***
#> classsubcompact   -2.38384    0.92943  -2.565 0.010997 *
#> classsuv          -4.27352    0.86777  -4.925 1.67e-06 ***
#> fld                6.34343    1.69499   3.742 0.000233 ***
#> fle               -4.57060    1.65992  -2.754 0.006396 **
#> flp               -1.91733    1.58649  -1.209 0.228158
#> flr               -0.78873    1.56551  -0.504 0.614901
#> drvf               1.39617    0.39660   3.520 0.000525 ***
#> drvr               0.48740    0.46113   1.057 0.291694
#> cyl:displ          0.36206    0.07934   4.564 8.42e-06 ***
#> ---
#> Signif. codes:
#> 0 '***' 0.001 '**' 0.01 '*' 0.05 '.' 0.1 ' ' 1
#>
#> Residual standard error: 1.526 on 217 degrees of freedom
#> Multiple R-squared:  0.8803, Adjusted R-squared:  0.8715
#> F-statistic: 99.73 on 16 and 217 DF,  p-value: < 2.2e-16
```

```
interact_plot(
    fitc,
    pred = displ,
    modx = cyl,

    # Show observed data as partial residuals
    partial.residuals = TRUE,

    # Specify moderator values manually
    modx.values = c(4, 5, 6, 8)
)
```

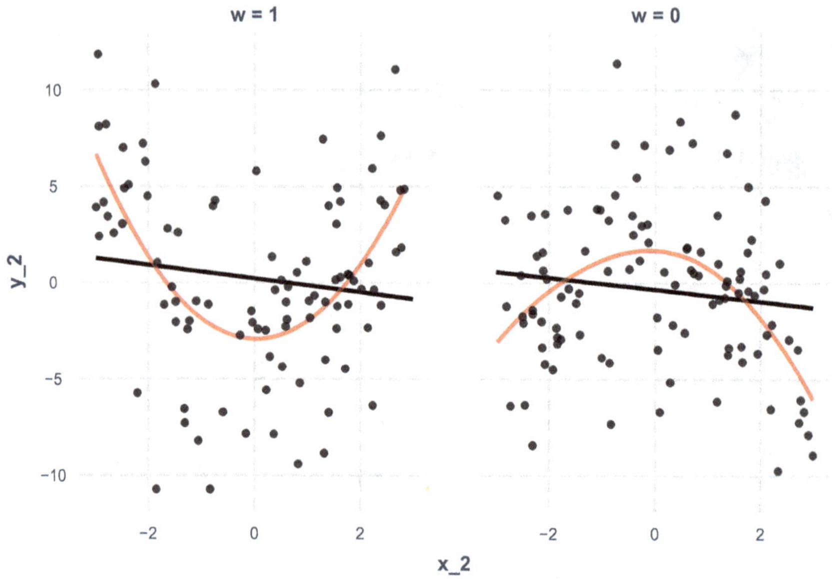

Fig. 9.9 Interaction plot

To check whether an interaction is **truly linear**, we can compare fitted lines based on (Fig. 9.9):

- **The whole sample** (black line)
- **Subsamples** based on the moderator (red line)

```
# Generate synthetic data
x_2 <- runif(n = 200, min = -3, max = 3)
w   <- rbinom(n = 200, size = 1, prob = 0.5)
err <- rnorm(n = 200, mean = 0, sd = 4)
y_2 <- 2.5 - x_2 ^ 2 - 5 * w + 2 * w * (x_2 ^ 2) + err

data_2 <- as.data.frame(cbind(x_2, y_2, w))

# Fit interaction model
model_2 <- lm(y_2 ~ x_2 * w, data = data_2)
summary(model_2)
#>
#> Call:
#> lm(formula = y_2 ~ x_2 * w, data = data_2)
#>
#> Residuals:
#>      Min       1Q   Median       3Q      Max
#> -11.5863  -2.6203   0.1698   2.8023  11.7960
#>
#> Coefficients:
#>             Estimate Std. Error t value Pr(>|t|)
```

```
#> (Intercept) -0.34716    0.43629  -0.796    0.427
#> x_2         -0.30702    0.25620  -1.198    0.232
#> w            0.56013    0.63690   0.879    0.380
#> x_2:w       -0.04348    0.36961  -0.118    0.906
#>
#> Residual standard error: 4.492 on 196 degrees of freedom
#> Multiple R-squared:  0.01929,    Adjusted R-squared:  0.004276
#> F-statistic: 1.285 on 3 and 196 DF,  p-value: 0.2808
```

```
# Linearity check plot
interact_plot(
    model_2,
    pred = x_2,
    modx = w,
    linearity.check = TRUE,
    plot.points = TRUE
)
#> `geom_smooth()` using formula = 'y ~ x'
```

9.7.3.1.1 Simple Slopes Analysis

A **simple slopes analysis** examines the **conditional effect** of an independent variable (X) at specific levels of the moderator (M).

How `sim_slopes()` Works:

- **Continuous moderators**: Analyzes effects at the **mean** and $\pm$ **1SD**.
- **Categorical moderators**: Uses all factor levels.
- **Mean-centers all variables** except the predictor of interest.

Example: Continuous by Continuous Interaction

```
library(interactions)
sim_slopes(fiti,
           pred = Illiteracy,
           modx = Murder,
           johnson_neyman = FALSE)
#> SIMPLE SLOPES ANALYSIS
#>
#> Slope of Illiteracy when Murder =  5.420973 (- 1 SD):
#>
#>     Est.     S.E.   t val.      p
#> -------- -------- -------- ------
#>   -17.43   250.08    -0.07   0.94
#>
#> Slope of Illiteracy when Murder =  8.685043 (Mean):
#>
#>     Est.     S.E.   t val.      p
#> -------- -------- -------- ------
```

```
#>   -399.64   178.86    -2.23   0.03
#> 
#> Slope of Illiteracy when Murder = 11.949113 (+ 1 SD):
#> 
#>      Est.      S.E.    t val.      p
#> --------- -------- -------- ------
#>   -781.85   189.11    -4.13   0.00
```

We can also visualize the simple slopes (Fig. 9.10)

```
# Store results
ss <- sim_slopes(fiti,
                 pred = Illiteracy,
                 modx = Murder,
                 modx.values = c(0, 5, 10))
```

```
plot(ss)
```

For publication-quality results, we convert the simple slopes analysis into a table using `huxtable`.

```
library(huxtable)
#> 
#> Attaching package: 'huxtable'
#> The following object is masked from 'package:dplyr':
#> 
#>     add_rownames
#> The following object is masked from 'package:ggplot2':
#> 
#>     theme_grey

ss <- sim_slopes(fiti,
                 pred = Illiteracy,
                 modx = Murder,
                 modx.values = c(0, 5, 10))

# Convert to a formatted table
print(as_huxtable(ss))
#>             Value of Murder    Slope of Illiteracy
#> 
#>             0.00               617.34 (434.85)
#>             5.00               31.86 (262.63)
#>             10.00              -553.62 (171.42)**
```

9.7.3.1.2 Johnson–Neyman Intervals

The **Johnson–Neyman technique** identifies the range of the moderator (M) where the effect of the predictor (X) on the dependent variable (Y) is statistically

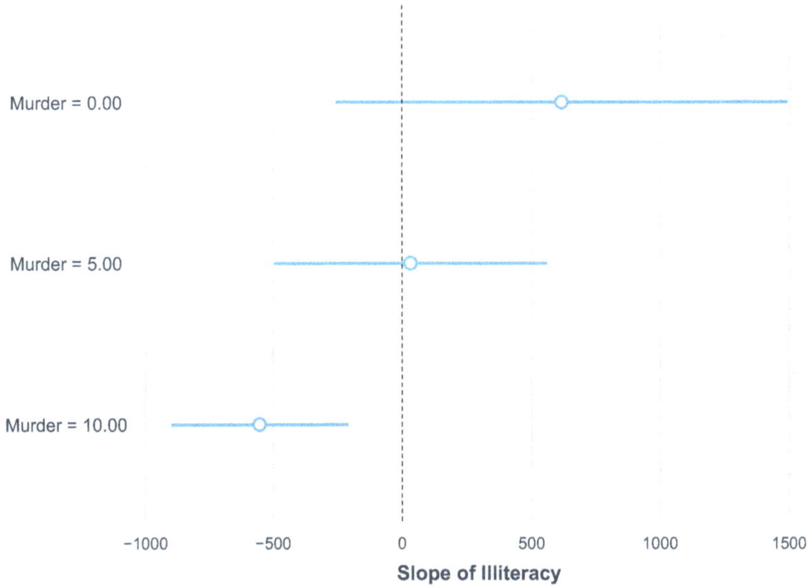

Fig. 9.10 Slope of illiteracy

significant. This approach is useful when the moderator is continuous, allowing us to determine where an effect exists rather than arbitrarily choosing values.

Although the J-N approach has been widely used (Johnson & Neyman, 1936), it has known **inflated Type I error rates** (Bauer & Curran, 2005). A correction method was proposed by Esarey and Sumner (2018) to address these issues.

Since J-N performs multiple comparisons across all values of the moderator, it **inflates Type I error**. To control for this, we use False Discovery Rate correction.

Example: Johnson–Neyman Analysis

```
sim_slopes(
    fiti,
    pred = Illiteracy,
    modx = Murder,
    johnson_neyman = TRUE,
    control.fdr = TRUE,  # Correction for Type I and II errors

    # Include conditional intercepts
    # cond.int = TRUE,

    robust = "HC3",  # Use robust SE

    # Don't mean-center non-focal variables
    # centered = `none",

    jnalpha = 0.05  # Significance level
)
```

```
#> JOHNSON-NEYMAN INTERVAL
#>
#> When Murder is OUTSIDE the interval [-7.87, 8.51], the
#> slope of Illiteracy is p < .05.
#>
#> Note: The range of observed values of Murder is [1.40,
#> 15.10]
#>
#> Interval calculated using false discovery rate adjusted t = 2.35
#>
#> SIMPLE SLOPES ANALYSIS
#>
#> Slope of Illiteracy when Murder =  5.420973 (- 1 SD):
#>
#>      Est.      S.E.    t val.      p
#> -------- --------- --------- ------
#>    -17.43    227.37    -0.08   0.94
#>
#> Slope of Illiteracy when Murder =  8.685043 (Mean):
#>
#>      Est.      S.E.    t val.      p
#> --------- --------- --------- ------
#>    -399.64   158.77    -2.52   0.02
#>
#> Slope of Illiteracy when Murder = 11.949113 (+ 1 SD):
#>
#>      Est.      S.E.    t val.      p
#> --------- --------- --------- ------
#>    -781.85   156.96    -4.98   0.00
```

To Visualize the J-N Intervals (Fig. 9.11)

```
johnson_neyman(fiti,
               pred = Illiteracy,
               modx = Murder,
               control.fdr = TRUE, # Corrects for Type I error
               alpha = .05)
#> JOHNSON-NEYMAN INTERVAL
#>
#> When Murder is OUTSIDE the interval [-22.57, 8.52], the
#> slope of Illiteracy is p < .05.
#>
#> Note: The range of observed values of Murder is [1.40,
#> 15.10]
#>
#> Interval calculated using false discovery rate adjusted t = 2.33
```

- The y-axis represents the conditional slope of the predictor (X).
- The x-axis represents the values of the moderator (M).
- The shaded region represents the range where the effect of X on Y is statistically significant.

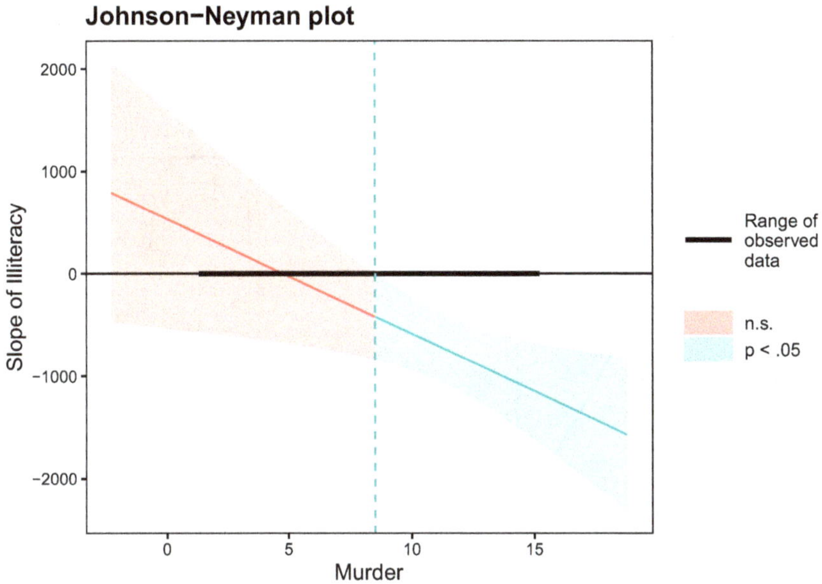

Fig. 9.11 Johnson–Neyman plot

9.7.3.1.3 Three-Way Interactions

In three-way interactions, the effect of X on Y depends on **two moderators** (M_1 and M_2). This allows for a more nuanced understanding of moderation effects.

Example: Three-Way Interaction Visualization (Fig. 9.12)

```
library(jtools)
# Convert 'cyl' to factor
mtcars$cyl <- factor(mtcars$cyl,
                     labels = c("4 cylinder", "6 cylinder", "8
                     ↪  cylinder"))

# Fit the model
fitc3 <- lm(mpg ~ hp * wt * cyl, data = mtcars)
```

```
interact_plot(fitc3,
              pred = hp,
              modx = wt,
              mod2 = cyl) +
    theme_apa(legend.pos = "bottomright")
```

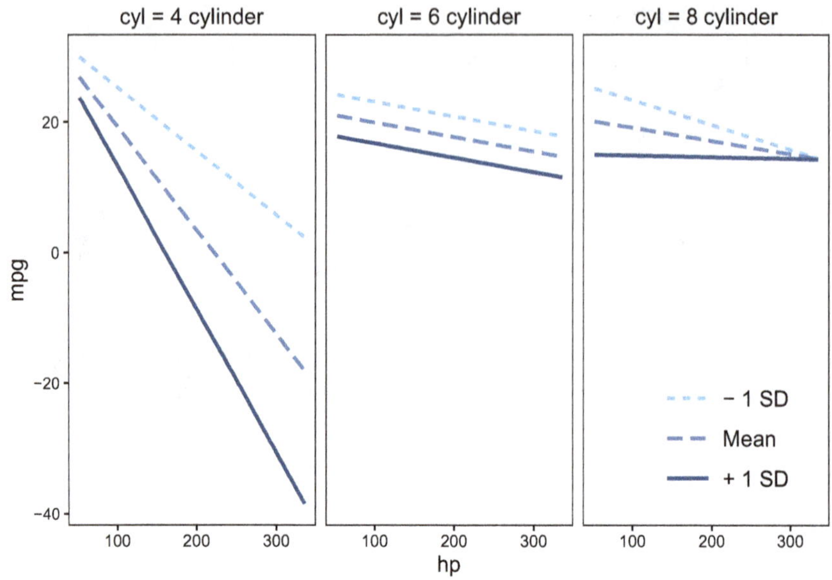

Fig. 9.12 Interaction plot

9.7.3.1.4 Johnson–Neyman for Three-Way Interaction

The Johnson–Neyman technique can also be applied in a **three-way interaction context** (Fig. 9.13)

```
library(survey)
#> Loading required package: grid
#> Loading required package: Matrix
#>
#> Attaching package: 'Matrix'
#> The following objects are masked from 'package:tidyr':
#>
#>     expand, pack, unpack
#> Loading required package: survival
#>
#> Attaching package: 'survey'
#> The following object is masked from 'package:graphics':
#>
#>     dotchart
data(api)

# Define survey design
dstrat <- svydesign(
    id = ~ 1,
    strata = ~ stype,
    weights = ~ pw,
    data = apistrat,
```

```
    fpc = ~ fpc
)

# Fit 3-way interaction model
regmodel3 <-
    survey::svyglm(api00 ~ avg.ed * growth * enroll, design = dstrat)
```

```
# Johnson-Neyman analysis with visualization
sim_slopes(
    regmodel3,
    pred = growth,
    modx = avg.ed,
    mod2 = enroll,
    jnplot = TRUE
)
#> Warning in get_plot_component(plot, "guide-box"): Multiple
#> components found; returning the first one. To return all,
#> use `return_all = TRUE`.
#> ##### While enroll (2nd moderator) =  153.0518 (- 1 SD) ####
#>
#> JOHNSON-NEYMAN INTERVAL
#>
#> When avg.ed is OUTSIDE the interval [2.75, 3.82], the
#> slope of growth is p < .05.
#>
#> Note: The range of observed values of avg.ed is [1.38,
#> 4.44]
#>
#> SIMPLE SLOPES ANALYSIS
#>
#> Slope of growth when avg.ed = 2.085002 (- 1 SD):
#>
#>   Est.   S.E.   t val.      p
#> ------ ------ -------- ------
#>   1.25   0.32     3.86   0.00
#>
#> Slope of growth when avg.ed = 2.787381 (Mean):
#>
#>   Est.   S.E.   t val.      p
#> ------ ------ -------- ------
#>   0.39   0.22     1.75   0.08
#>
#> Slope of growth when avg.ed = 3.489761 (+ 1 SD):
#>
#>    Est.   S.E.   t val.      p
#> ------- ------ -------- ------
#>   -0.48   0.35    -1.37   0.17
#>
#> ###### While enroll (2nd moderator) =  595.2821 (Mean) #####
#>
#> JOHNSON-NEYMAN INTERVAL
#>
```

```
#> When avg.ed is OUTSIDE the interval [2.84, 7.83], the
#> slope of growth is p < .05.
#>
#> Note: The range of observed values of avg.ed is [1.38,
#> 4.44]
#>
#> SIMPLE SLOPES ANALYSIS
#>
#> Slope of growth when avg.ed = 2.085002 (- 1 SD):
#>
#>   Est.   S.E.   t val.      p
#> ------ ------ -------- ------
#>   0.72   0.22     3.29   0.00
#>
#> Slope of growth when avg.ed = 2.787381 (Mean):
#>
#>   Est.   S.E.   t val.      p
#> ------ ------ -------- ------
#>   0.34   0.16     2.16   0.03
#>
#> Slope of growth when avg.ed = 3.489761 (+ 1 SD):
#>
#>    Est.   S.E.   t val.      p
#> ------- ------ -------- ------
#>   -0.04   0.24    -0.16   0.87
#>
#> ##### While enroll (2nd moderator) = 1037.5125 (+ 1 SD) ####
#>
#> JOHNSON-NEYMAN INTERVAL
#>
#> The Johnson--Neyman interval could not be found. Is the
#> p value for your interaction term below the specified
#> alpha?
#>
#> SIMPLE SLOPES ANALYSIS
#>
#> Slope of growth when avg.ed = 2.085002 (- 1 SD):
#>
#>   Est.   S.E.   t val.      p
#> ------ ------ -------- ------
#>   0.18   0.31     0.58   0.56
#>
#> Slope of growth when avg.ed = 2.787381 (Mean):
#>
#>   Est.   S.E.   t val.      p
#> ------ ------ -------- ------
#>   0.29   0.20     1.49   0.14
#>
#> Slope of growth when avg.ed = 3.489761 (+ 1 SD):
#>
#>   Est.   S.E.   t val.      p
#> ------ ------ -------- ------
#>   0.40   0.27     1.49   0.14
```

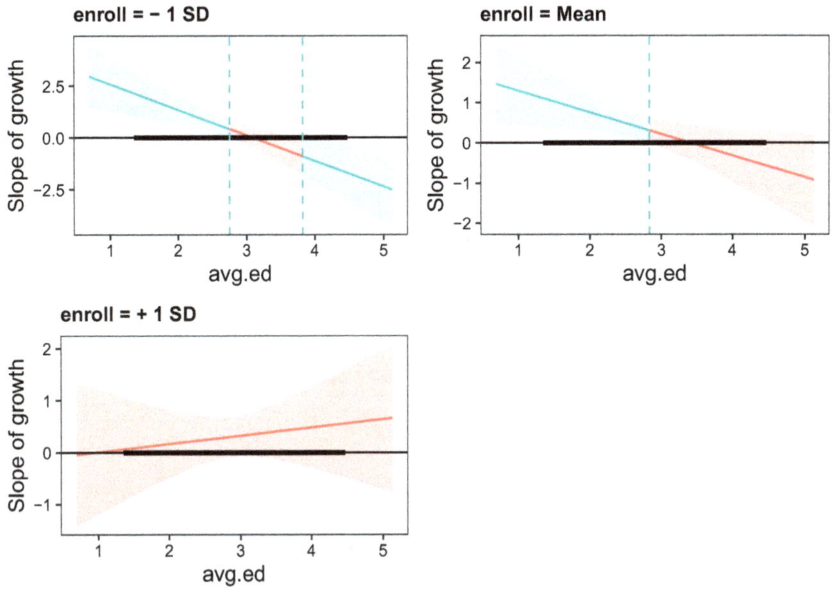

Fig. 9.13 Slope of growth

To present the results in a **publication-ready format**, we generate tables and plots (Fig. 9.14)

```
ss3 <-
    sim_slopes(regmodel3,
               pred = growth,
               modx = avg.ed,
               mod2 = enroll)
```

```
# Plot results
plot(ss3)
```

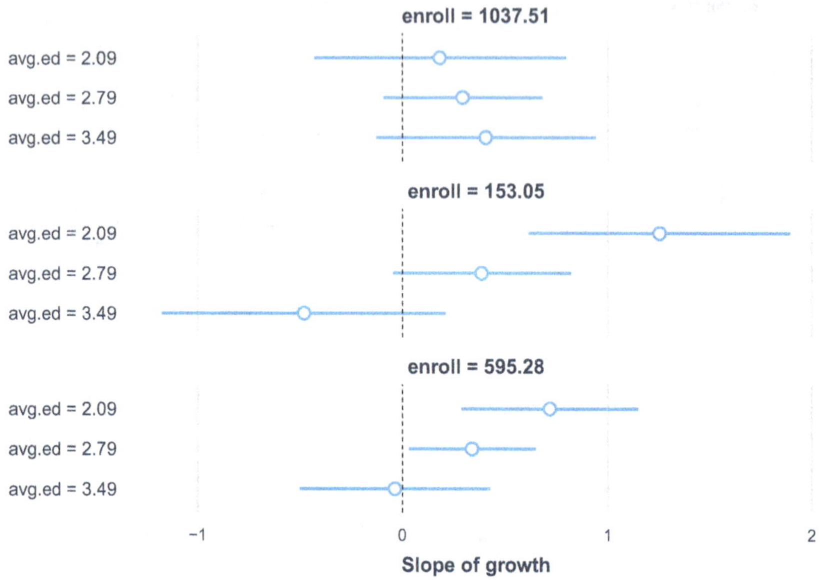

Fig. 9.14 Slope of growth

```
# Convert results into a formatted table
library(huxtable)
print(as_huxtable(ss3))
#>
#>              enroll = 153
#>              Value of avg.ed   Slope of growth
#>
#>              2.09              1.25 (0.32)***
#>              2.79              0.39 (0.22)#
#>              3.49              -0.48 (0.35)
#>
#>              enroll = 595.28
#>              Value of avg.ed   Slope of growth
#>
#>              2.09              0.72 (0.22)**
#>              2.79              0.34 (0.16)*
#>              3.49              -0.04 (0.24)
#>
#>              enroll = 1037.51
#>              Value of avg.ed   Slope of growth
#>
#>              2.09              0.18 (0.31)
#>              2.79              0.29 (0.20)
#>              3.49              0.40 (0.27)
```

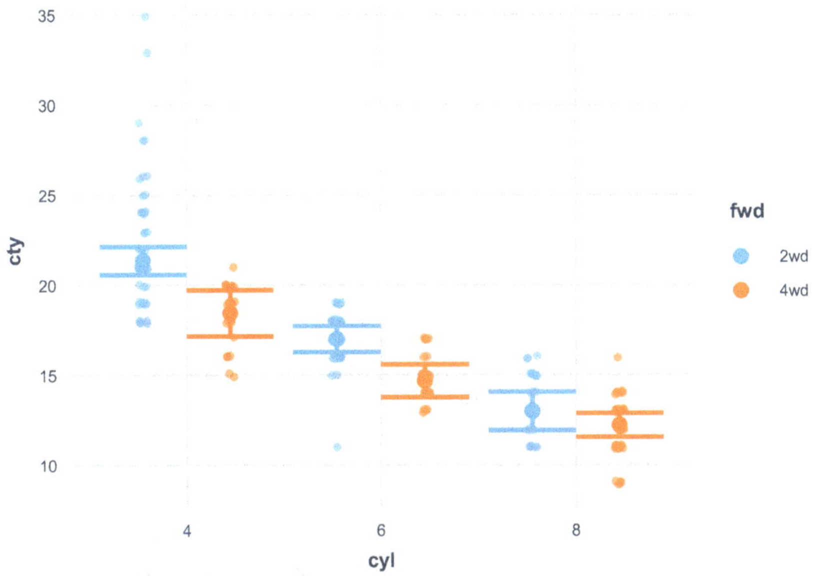

Fig. 9.15 Scatter plot

9.7.3.2 Categorical Interactions

Interactions between categorical predictors can be visualized using **categorical plots**.

Example: Interaction Between cyl, fwd, **and** auto (Fig. 9.15)

```
library(ggplot2)
library(tidyverse)

# Convert variables to factors
mpg2 <- mpg %>%
    mutate(cyl = factor(cyl))

mpg2["auto"] <- "auto"
mpg2$auto[mpg2$trans %in% c("manual(m5)", "manual(m6)")] <- "manual"
mpg2$auto <- factor(mpg2$auto)

mpg2["fwd"] <- "2wd"
mpg2$fwd[mpg2$drv == "4"] <- "4wd"
mpg2$fwd <- factor(mpg2$fwd)

# Drop cars with 5 cylinders (since most have 4, 6, or 8)
mpg2 <- mpg2[mpg2$cyl != "5",]

# Fit the model
fit3 <- lm(cty ~ cyl * fwd * auto, data = mpg2)
```

```
library(jtools) # For summ()
summ(fit3)
```

Observations	230
Dependent variable	cty
Type	OLS linear regression

F(11,218)	61.37
R^2	0.76
$Adj.R^2$	0.74

	Est.	S.E.	t val.	p
(Intercept)	21.37	0.39	54.19	0.00
cyl6	−4.37	0.54	−8.07	0.00
cyl8	−8.37	0.67	−12.51	0.00
fwd4wd	−2.91	0.76	−3.83	0.00
automanual	1.45	0.57	2.56	0.01
cyl6:fwd4wd	0.59	0.96	0.62	0.54
cyl8:fwd4wd	2.13	0.99	2.15	0.03
cyl6:automanual	−0.76	0.90	−0.84	0.40
cyl8:automanual	0.71	1.18	0.60	0.55
fwd4wd:automanual	−1.66	1.07	−1.56	0.12
cyl6:fwd4wd:automanual	1.29	1.52	0.85	0.40
cyl8:fwd4wd:automanual	−1.39	1.76	−0.79	0.43

Standard errors: OLS

```
cat_plot(fit3,
         pred = cyl,
         modx = fwd,
         plot.points = TRUE)
```

Line Plot for Categorical Interaction (Fig. 9.16)

```
cat_plot(
    fit3,
    pred = cyl,
    modx = fwd,
    geom = "line",
    point.shape = TRUE,
    vary.lty = TRUE
)
```

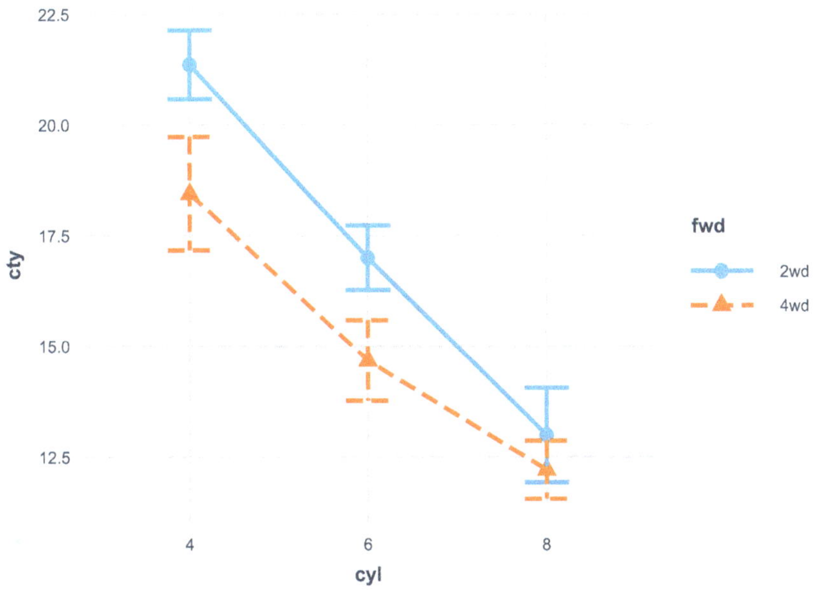

Fig. 9.16 Line chart between cty and cyl

Bar Plot Representation (Fig. 9.17)

```
cat_plot(
    fit3,
    pred = cyl,
    modx = fwd,
    geom = "bar",
    interval = TRUE,
    plot.points = TRUE
)
```

9.7.4 `interactionR` Package

The `interactionR` package is designed for publication-quality reporting of interaction effects, particularly in epidemiology and social sciences. It provides tools for computing interaction measures, confidence intervals, and statistical inference following well-established methodologies.

Key Features

- Publication-Ready Interaction Analysis
- Confidence intervals calculated using:

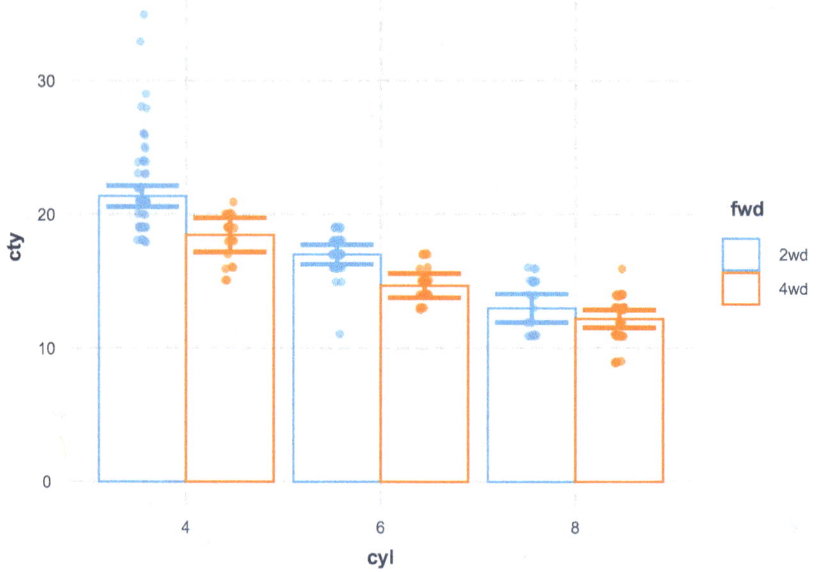

Fig. 9.17 Bar chart between cty and cyl

- Delta method (Hosmer & Lemeshow, 1992)
- Variance recovery ("mover") method (Zou, 2008)
- Bootstrapping (Assmann et al., 1996)

- Standardized reporting guidelines based on (Knol & VanderWeele, 2012).

```
install.packages("interactionR", dependencies = T)
```

9.7.5 `sjPlot` Package

The `sjPlot` package is highly recommended for publication-quality visualizations of interaction effects. It provides enhanced aesthetics and customizable interaction plots suitable for academic journals.

More details: sjPlot interaction visualization

```
install.packages("sjPlot")
```

9.7.6 Summary of Moderation Analysis Packages

See Table 9.1.

Table 9.1 Summary of moderation analysis packages

Package	Purpose	Key features	Recommended?
`emmeans`	Estimated marginal means & simple slopes	Computes simple slopes, spotlight analysis, floodlight analysis (J-N method)	Yes
`probemod`	Johnson–Neyman technique	Tests moderator significance ranges	No (Subscript issues)
`interactions`	Interaction visualization	Produces robust, customizable interaction plots	Yes
`interactionR`	Epidemiological interaction measures	Computes RERI, AP, SI for additive scale interactions	Yes (for public health research)
`sjPlot`	Publication-quality interaction plots	Highly customizable, ideal for academic papers	Highly Recommended

Chapter 10
Mediation

Mediation analysis uncovers the process by which a predictor affects an outcome through one or more intermediate variables. This chapter introduces the traditional approach to mediation, focusing on path analysis, the Baron & Kenny framework, and the Sobel test. Practical limitations of this approach are discussed, especially in the presence of confounding. Then, it presents a modern causal inference framework for mediation, including counterfactual definitions of direct and indirect effects. The use of bootstrapping is demonstrated with business examples. Assumptions such as sequential ignorability are explored in detail. Graphical representations, including path diagrams and causal graphs, are used to aid interpretation. This chapter provides the tools to not only estimate mediation effects but also to assess their robustness, interpret their business relevance, and report them transparently.

10.1 Traditional Approach

The classical mediation analysis follows the approach introduced by Baron and Kenny (1986), though it has limitations, particularly in requiring the first step ($X \rightarrow Y$) to be significant. Despite its shortcomings, this framework provides a useful foundation.

10.1.1 Steps in the Traditional Mediation Model

Mediation is typically assessed through three regression models:

1. **Total Effect**: $X \rightarrow Y$
2. **Path** a: $X \rightarrow M$
3. **Path** b and Direct Effect (c'): $X + M \rightarrow Y$

© The Author(s), under exclusive license to Springer Nature Switzerland AG 2025

M. Nguyen, *Advanced Modeling and Data Challenges*,
https://doi.org/10.1007/978-3-032-01719-2_10

where:

- X = independent (causal) variable
- Y = dependent (outcome) variable
- M = mediating variable

Originally, Baron and Kenny (1986) required the direct path $X \rightarrow Y$ to be significant. However, mediation can still occur even if this direct effect is not significant. For example:

- The effect of X on Y might be **fully absorbed** by M.
- Multiple mediators (M_1, M_2) with **opposing effects** could cancel each other out, leading to a non-significant direct effect.

10.1.2 Graphical Representation of Mediation

10.1.2.1 Unmediated Model (Fig. 10.1)

Here, c represents the **total effect** of X on Y.

10.1.2.2 Mediated Model (Fig. 10.2)

Here:

- c' = **direct effect** (effect of X on Y after accounting for mediation)
- ab = **indirect effect** (mediation pathway)

Thus, we can express:

$$\text{total effect} = \text{direct effect} + \text{indirect effect}$$

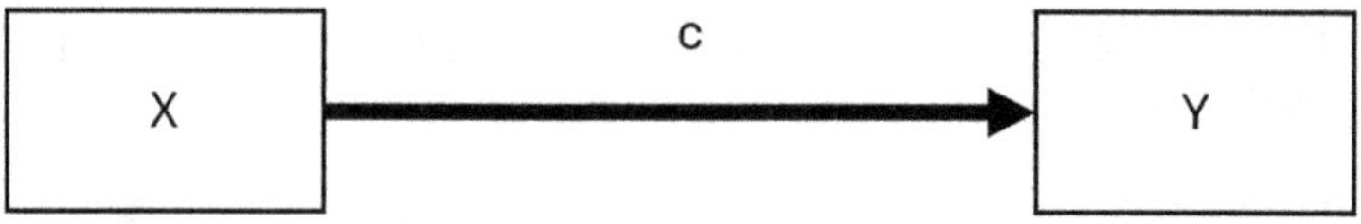

Fig. 10.1 Unmediated model

Fig. 10.2 Mediated model

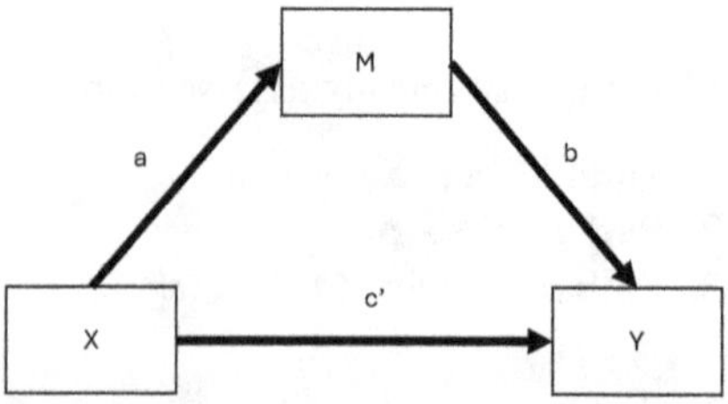

or,

$$c = c' + ab$$

This equation holds under standard linear models but not necessarily in cases such as:

1. Latent variable models
2. Logistic regression (only an approximation)
3. Multilevel models (Bauer et al., 2006)

10.1.3 Measuring Mediation

Several approaches exist for quantifying the **indirect effect** (ab):

1. **Proportional Reduction Approach**:

$$1 - \frac{c'}{c}$$

   Not recommended due to high instability, especially when c is small (MacKinnon et al., 1995).

2. **Product Method**:

$$a \times b$$

   The most common approach.

3. **Difference Method**:

$$c - c'$$

   Conceptually similar to the product method but less precise in small samples.

10.1.4 Assumptions in Linear Mediation Models

For valid mediation analysis, the following assumptions should hold:

1. **No unmeasured confounders** between $X - Y$, $X - M$, and $M - Y$.
2. **No reverse causality**: X should not be influenced by a confounder (C) that also affects $M - Y$.
3. **Measurement reliability**: M should be measured without error (if not, consider errors-in-variables models).

Regression Equations for Mediation Steps
Step 1: Total Effect of X on Y

$$Y = \beta_0 + cX + \epsilon$$

- The significance of c is **not required** for mediation to occur.

  Step 2: Effect of X on M

$$M = \alpha_0 + aX + \epsilon$$

- The coefficient a must be significant for mediation analysis to proceed.

  Step 3: Effect of M on Y (Including X)

$$Y = \gamma_0 + c'X + bM + \epsilon$$

- If c' becomes non-significant after including M, full mediation occurs.
- If c' is reduced but remains significant, partial mediation is present (Table 10.1).

10.1.5 Testing for Mediation

Several statistical tests exist to assess whether the indirect effect (ab) is significant:

1. **Sobel Test** (Sobel, 1982)
   - Based on the standard error of ab.
   - **Limitation**: Assumes normality of ab, which may not hold in small samples.
2. **Joint Significance Test**
   - If both a and b are significant, mediation is likely.
3. **Bootstrapping (Preferred)** (Shrout & Bolger, 2002; Preacher & Hayes, 2004)
   - Estimates the confidence interval for ab.
   - Does not assume normality.
   - Recommended for small-to-moderate sample sizes.

Table 10.1 Interpretation of mediation outcomes

Effect of X on Y	Mediation type
b_4 (from Step 3) is **insignificant**	Full mediation
$b_4 < b_1$ (from Step 1) but still significant	Partial mediation

10.1.6 Additional Considerations

- Proximal mediation (where path a exceeds path b) can lead to multicollinearity and reduced statistical power. In contrast, distal mediation (where path b exceeds path a) tends to maximize power. In fact, slightly distal mediators—where b is somewhat larger than a—often strike an ideal balance for power in mediation analyses (Hoyle, 1999).
- Tests of direct effects (c and c') generally have lower power than tests of the indirect effect (ab). As a result, it is possible for the indirect effect (ab) to be statistically significant even when the direct effect (c) is not. This situation can appear to indicate "complete mediation," yet the lack of a statistically significant direct effect between X and Y (i.e., c') does not definitively rule out other possibilities (Kenny & Judd, 2014).
- Because testing ab essentially combines two tests, it often provides a power advantage over testing c' alone. However, using a non-significant c' as the sole criterion for claiming complete mediation should be done cautiously—if at all—given the importance of adequate sample size and power. Indeed, Hayes and Scharkow (2013) recommend avoiding claims of complete mediation based solely on a non-significant c', particularly when partial mediation may still be present.

10.1.7 Assumptions in Mediation Analysis

Valid mediation analysis requires several key assumptions, which can be categorized into **causal direction**, **interaction effects**, **measurement reliability**, and **confounding control**.

10.1.7.1 Direction

- Causal Order of Variables
- A simple but weak solution is to measure X before M and Y to prevent reverse causality (i.e., M or Y causing X). Similarly, measuring M before Y avoids feedback effects of Y on M.
- However, causal feedback loops between M and Y may still exist.
  - If we assume full mediation ($c' = 0$), models with reciprocal causal effects between M and Y can be estimated using instrumental variables (IV).
  - Smith (1982) suggests treating both M and Y as potential mediators of each other, requiring distinct instrumental variables for each to avoid cross-contamination of causal effects.

10.1.7.2 Interaction Effects in Mediation

- If M interacts with X in predicting Y, then M is both a mediator and a moderator (Baron & Kenny, 1986).
- **The interaction term** $X \times M$ should always be included in the model to account for possible moderation effects.
- For interpreting such interactions in mediation models, see (VanderWeele, 2015), who provides a framework for **moderated mediation** analysis.

10.1.7.3 Reliability

Measurement error in any of the three key variables (X, M, Y) can bias estimates of mediation effects.

1. Measurement Error in the Mediator (M):
   - Biases both b and c'.
   - Potential solution: Model M as a latent variable (reduces bias but may decrease statistical power) (Ledgerwood & Shrout, 2011).
   - Specific effects:
     - b is attenuated (biased toward 0).
     - c' is:

       Overestimated if $ab > 0$.
       Underestimated if $ab < 0$.
2. Measurement Error in the Treatment (X):
   - Biases both a and b.
   - Specific effects:
     - a is attenuated.
     - b is:

       Overestimated if $ac' > 0$.
       Underestimated if $ac' < 0$.
3. Measurement Error in the Outcome (Y):
   - If unstandardized, there is no bias.
   - If standardized, there is attenuation bias (reduced effect sizes due to error variance).

10.1.7.4 Confounding in Mediation Analysis

Omitted variable bias can distort any of the three core relationships ($X \to Y$, $X \to M$, $M \to Y$). Addressing confounding requires either design-based or statistical solutions.

Design-Based Strategies (Preferred if Feasible)

- Randomization of the independent variable (X) reduces confounding bias.
- Randomization of the mediator (M), if possible, further strengthens causal claims.
- Controlling for measured confounders, though this only addresses observable confounding.

Statistical Strategies (When Randomization is Not Possible)

1. Instrumental Variables Approach:
    - Used when a confounder affects both M and Y.
    - Front-door adjustment can be applied if there exists a third variable that fully mediates the effect of M on Y while being independent of the confounder.
2. Weighting Methods (e.g., Inverse Probability Weighting – IPW):
    - Corrects for confounding by reweighting observations to balance confounders across treatment groups.
    - Requires the strong ignorability assumption: All confounders must be measured and correctly specified (Westfall & Yarkoni, 2016).
    - While this assumption cannot be formally tested, sensitivity analyses can help assess robustness.
    - See Heiss for R code on implementing IPW in mediation models.

10.1.8 Indirect Effect Tests

Testing the indirect effect (ab) is crucial in mediation analysis. Several methods exist, each with its advantages and limitations.

10.1.8.1 Sobel Test (Delta Method)

- Developed by Sobel (1982).
- Also known as the delta method.
- Not recommended because it assumes the sampling distribution of ab is normal, which often does not hold (MacKinnon et al., 1995).

  The standard error (SE) of the indirect effect is:

$$SE_{ab} = \sqrt{\hat{b}^2 s_{\hat{a}}^2 + \hat{a}^2 s_{\hat{b}}^2}$$

  The **Z-statistic** for testing whether ab is significantly different from 0 is:

$$z = \frac{\hat{ab}}{\sqrt{\hat{b}^2 s_{\hat{a}}^2 + \hat{a}^2 s_{\hat{b}}^2}}$$

Disadvantages

- Assumes that a and b are independent.
- Assumes that ab follows a normal distribution.
- Poor performance in small samples.
- Lower power and more conservative than bootstrapping.

Special Case: Inconsistent Mediation

- Mediation can occur even when **direct and indirect effects** have **opposite signs**, known as **inconsistent mediation** (MacKinnon et al., 2007).
- This happens when the **mediator acts as a suppressor variable**, leading to counteracting paths.

```
library(bda)
#> Loading required package: boot
#> bda - 18.3.2
library(mediation)
#> Loading required package: MASS
#> Loading required package: Matrix
#> Loading required package: mvtnorm
#> Loading required package: sandwich
#> mediation: Causal Mediation Analysis
#> Version: 4.5.0
data("boundsdata")

# Sobel Test for Mediation
bda::mediation.test(boundsdata$med, boundsdata$ttt,
↪   boundsdata$out) |>
    tibble::rownames_to_column() |>
    causalverse::nice_tab(2)
#> Warning: replacing previous import 'data.table::wday' by
#> 'lubridate::wday' when loading 'causalverse'
#> Warning: replacing previous import 'data.table::second' by
#> 'lubridate::second' when loading 'causalverse'
#> Warning: replacing previous import 'data.table::isoweek' by
#> 'lubridate::isoweek' when loading 'causalverse'
#> Warning: replacing previous import 'data.table::yday' by
#> 'lubridate::yday' when loading 'causalverse'
#> Warning: replacing previous import 'data.table::hour' by
#> 'lubridate::hour' when loading 'causalverse'
#> Warning: replacing previous import 'data.table::year' by
#> 'lubridate::year' when loading 'causalverse'
#> Warning: replacing previous import 'data.table::month' by
#> 'lubridate::month' when loading 'causalverse'
#> Warning: replacing previous import 'data.table::week' by
```

```
#> 'lubridate::week' when loading 'causalverse'
#> Warning: replacing previous import 'data.table::minute' by
#> 'lubridate::minute' when loading 'causalverse'
#> Warning: replacing previous import 'data.table::mday' by
#> 'lubridate::mday' when loading 'causalverse'
#> Warning: replacing previous import 'data.table::quarter' by
#> 'lubridate::quarter' when loading 'causalverse'
#> Warning: replacing previous import 'formattable::digits' by
#> 'xtable::digits' when loading 'causalverse'
#>   rowname Sobel Aroian Goodman
#> 1 z.value  4.05   4.03    4.07
#> 2 p.value  0.00   0.00    0.00
```

10.1.8.2 Joint Significance Test

- Tests if the indirect effect is nonzero by checking whether both a and b are statistically significant.
- Assumes $a \perp b$ (independence of paths).
- Performs similarly to bootstrapping (Hayes & Scharkow, 2013).
- More robust to non-normality but can be sensitive to heteroscedasticity (Fossum & Montoya, 2023).
- Does not provide confidence intervals, making effect size interpretation harder.

10.1.8.3 Bootstrapping (Preferred Method)

- First applied to mediation by Bollen and Stine (1990).
- Uses resampling to empirically estimate the sampling distribution of the indirect effect.
- Does not require normality assumptions or $a \perp b$ independence.
- Works well with small samples.
- Can handle complex models.

Which Bootstrapping Method?

- **Percentile bootstrap** is preferred due to better **Type I error rates** (Tibbe & Montoya, 2022).
- **Bias-corrected bootstrapping** can be too **liberal** (inflates Type I errors) (Fritz et al., 2012).

Special Case: Meta-Analytic Bootstrapping

- Bootstrapping can be applied **without raw data**, using only a, b, $var(a)$, $var(b)$, $cov(a, b)$ from multiple studies.

```r
# Meta-Analytic Bootstrapping for Mediation
library(causalverse)

result <- causalverse::med_ind(
    a = 0.5,
    b = 0.7,
    var_a = 0.04,
    var_b = 0.05,
    cov_ab = 0.01
)
```

```r
result$plot
#> `stat_bin()` using `bins = 30`. Pick better value with
#> `binwidth`.
```

When an **instrumental variable (IV)** is available, the causal effect can be estimated more reliably. Below are visual representations (Fig. 10.3).

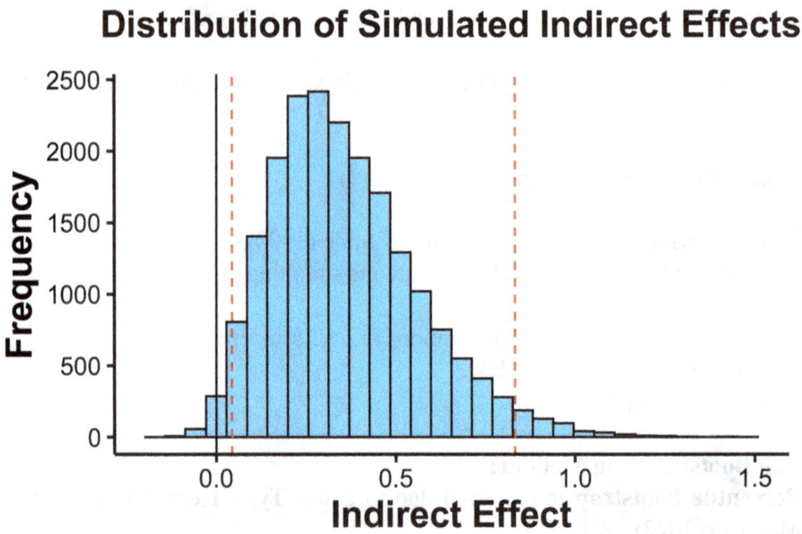

Fig. 10.3 Distribution of simulated indirect effects

```
library(DiagrammeR)

# Simple Treatment-Outcome Model
grViz("
digraph {
  graph []
  node [shape = plaintext]
    X [label = 'Treatment']
    Y [label = 'Outcome']
  edge [minlen = 2]
    X->Y
  { rank = same; X; Y }
}")

# Mediation Model with an Instrument
grViz("
digraph {
  graph []
  node [shape = plaintext]
    X [label ='Treatment', shape = box]
    Y [label ='Outcome', shape = box]
    M [label ='Mediator', shape = box]
    IV [label ='Instrument', shape = box]
  edge [minlen = 2]
    IV->X
    X->M
    M->Y
    X->Y
  { rank = same; X; Y; M }
}")
```

Mediation Analysis with Fixed Effects Models

```
library(mediation)
library(fixest)

data("boundsdata")

# Step 1: Total Effect (c)
out1 <- feols(out ~ ttt, data = boundsdata)

# Step 2: Indirect Effect (a)
out2 <- feols(med ~ ttt, data = boundsdata)
```

```r
# Step 3: Direct & Indirect Effect (c' & b)
out3 <- feols(out ~ med + ttt, data = boundsdata)

# Proportion of Mediation
coef(out2)['ttt'] * coef(out3)['med'] / coef(out1)['ttt'] * 100
#>      ttt
#> 68.63609
```

Bootstrapped Mediation Analysis

```r
library(boot)
set.seed(1)

# Define the bootstrapping function
mediation_fn <- function(data, i) {
    df <- data[i,]

    a_path <- feols(med ~ ttt, data = df)
    a <- coef(a_path)['ttt']

    b_path <- feols(out ~ med + ttt, data = df)
    b <- coef(b_path)['med']

    cp <- coef(b_path)['ttt']

    # Indirect Effect (a * b)
    ind_ef <- a * b
    total_ef <- a * b + cp
    return(c(ind_ef, total_ef))
}

# Perform Bootstrapping
boot_med <- boot(boundsdata, mediation_fn, R = 100, parallel =
↪  "multicore", ncpus = 2)
boot_med
#>
#> ORDINARY NONPARAMETRIC BOOTSTRAP
#>
#>
#> Call:
#> boot(data = boundsdata, statistic = mediation_fn, R = 100,
↪  parallel = "multicore",
#>     ncpus = 2)
#>
```

```
#>
#> Bootstrap Statistics :
#>       original        bias    std. error
#> t1* 0.04112035  0.0006346725 0.009539903
#> t2* 0.05991068 -0.0004462572 0.029556611

# Summary and Confidence Intervals
summary(boot_med) |> causalverse::nice_tab()
#>     R original bootBias bootSE bootMed
#> 1 100     0.04        0   0.01    0.04
#> 2 100     0.06        0   0.03    0.06

# Confidence Intervals (percentile bootstrap preferred)
boot.ci(boot_med, type = c("norm", "perc"))
#> BOOTSTRAP CONFIDENCE INTERVAL CALCULATIONS
#> Based on 100 bootstrap replicates
#>
#> CALL :
#> boot.ci(boot.out = boot_med, type = c("norm", "perc"))
#>
#> Intervals :
#> Level      Normal             Percentile
#> 95%   ( 0.0218,  0.0592 )   ( 0.0249,  0.0623 )
#> Calculations and Intervals on Original Scale
#> Some percentile intervals may be unstable

# Point Estimates (Indirect and Total Effects)
colMeans(boot_med$t)
#> [1] 0.04175502 0.05946442
```

Alternatively, use the `robmed` package for **robust** mediation analysis:

```
library(robmed)
#> Loading required package: ggplot2
#> Loading required package: robustbase
#>
#> Attaching package: 'robustbase'
#> The following object is masked from 'package:boot':
#>
#>     salinity
```

10.1.9 Power Analysis for Mediation

To assess whether the study has sufficient **power** to detect mediation effects, use:

```
library(pwr2ppl)
#>
#> Attaching package: 'pwr2ppl'
#> The following object is masked from 'package:boot':
#>
#>     corr

# Power analysis for the indirect effect (ab path)
medjs(
    rx1m1 = .3,  # Correlation: X → M (path a)
    rx1y  = .1,  # Correlation: X → Y (path c')
    rym1  = .3,  # Correlation: M → Y (path b)
    n     = 100, # Sample size
    alpha = 0.05,
    mvars = 1,   # Number of mediators
    rep   = 1000 # Replications (use 10,000 for accuracy)
)
#> Sample size is 100
#> Power is 0.702
```

For **interactive** power analysis, see Kenny's Mediation Power App.

Summary of Indirect Effect Tests

See Table 10.2.

10.1.10 Multiple Mediation Analysis

In some cases, a single mediator (M) does not fully capture the indirect effect of X on Y. Multiple mediation models extend traditional mediation by including two or more mediators, allowing us to examine how multiple pathways contribute to an outcome.

Several R packages handle multiple mediation models:

Table 10.2 Comparison of mediation testing methods

Test	Pros	Cons
Sobel test	Simple, fast	Assumes normality, low power
Joint significance test	Robust to non-normality	No confidence interval
Bootstrapping (recommended)	No normality assumption, handles small samples	May be liberal if bias-corrected

- **manymome**: A flexible package for multiple mediation modeling.
  - Vignette

```
library(manymome)
```

- **mma**: Used for multiple mediator models.
  - Package PDF
  - Vignette

```
library(mma)
#> Loading required package: gbm
#> Loaded gbm 2.2.2
#> This version of gbm is no longer under development.
↪  Consider transitioning to gbm3,
↪  https://github.com/gbm-developers/gbm3
#> Loading required package: splines
#> Loading required package: survival
#>
#> Attaching package: 'survival'
#> The following object is masked from 'package:robustbase':
#>
#>     heart
#> The following object is masked from 'package:boot':
#>
#>     aml
#> Loading required package: car
#> Loading required package: carData
#>
#> Attaching package: 'car'
#> The following object is masked from 'package:boot':
#>
#>     logit
#> Loading required package: gplots
#>
#> Attaching package: 'gplots'
#> The following object is masked from 'package:stats':
#>
#>     lowess
#>
#> Attaching package: 'mma'
#> The following object is masked from 'package:pwr2ppl':
#>
#>     med
```

10.1.10.1 Multiple Mediators: Structural Equation Modeling Approach

A popular method for estimating multiple mediation models is **Structural Equation Modeling** using **lavaan**.

To test multiple mediation, we first **simulate data** where two mediators (M_1 and M_2) contribute to the outcome (Y).

```
# Load required packages
library(MASS)  # For mvrnorm (generating correlated errors)
library(lavaan)
#> This is lavaan 0.6-19
#> lavaan is FREE software! Please report any bugs.

# Function to generate synthetic data
generate_data <- function(n = 10000, a1 = 0.5, a2 = -0.35,
                          b1 = 0.7, b2 = 0.48,
                          corr = TRUE, correlation_value =
                          ↪  0.7) {
    set.seed(12345)
    X <- rnorm(n)  # Independent variable

    # Generate correlated errors for mediators
    if (corr) {
        Sigma <- matrix(c(1, correlation_value,
↪   correlation_value, 1), nrow = 2)
        errors <- mvrnorm(n, mu = c(0, 0), Sigma = Sigma)
    } else {
        errors <- mvrnorm(n, mu = c(0, 0), Sigma = diag(2))
    }

    M1 <- a1 * X + errors[, 1]
    M2 <- a2 * X + errors[, 2]
    Y  <- b1 * M1 + b2 * M2 + rnorm(n)  # Outcome variable

    return(data.frame(X = X, M1 = M1, M2 = M2, Y = Y))
}
```

We analyze the indirect effects through both mediators (M_1 and M_2).

1. Correctly Modeling Correlated Mediators

```
# Generate data with correlated mediators
Data_corr <- generate_data(n = 10000, corr = TRUE,
↪  correlation_value = 0.7)
```

```
# Define SEM model for multiple mediation
model_corr <- '
  Y ~ b1 * M1 + b2 * M2 + c * X
  M1 ~ a1 * X
  M2 ~ a2 * X
  M1 ~~ M2  # Correlated mediators (modeling correlation
↪  correctly)
'

# Fit SEM model
fit_corr <- sem(model_corr, data = Data_corr)

# Extract parameter estimates
parameterEstimates(fit_corr)[, c("lhs", "rhs", "est", "se",
↪  "pvalue")]
#>     lhs rhs    est    se pvalue
#> 1     Y  M1  0.700 0.014  0.000
#> 2     Y  M2  0.487 0.014  0.000
#> 3     Y   X -0.009 0.015  0.545
#> 4    M1   X  0.519 0.010  0.000
#> 5    M2   X -0.340 0.010  0.000
#> 6    M1  M2  0.677 0.012  0.000
#> 7     Y   Y  0.975 0.014  0.000
#> 8    M1  M1  0.973 0.014  0.000
#> 9    M2  M2  0.982 0.014  0.000
#> 10    X   X  1.000 0.000     NA
```

2. Incorrectly Ignoring Correlation Between Mediators

```
# Define SEM model without modeling mediator correlation
model_uncorr <- '
  Y ~ b1 * M1 + b2 * M2 + c * X
  M1 ~ a1 * X
  M2 ~ a2 * X
'

# Fit incorrect model
fit_uncorr <- sem(model_uncorr, data = Data_corr)

# Compare parameter estimates
parameterEstimates(fit_uncorr)[, c("lhs", "rhs", "est",
↪  "se", "pvalue")]
#>   lhs rhs    est    se pvalue
```

```
#> 1     Y  M1   0.700 0.010  0.000
#> 2     Y  M2   0.487 0.010  0.000
#> 3     Y   X -0.009 0.012  0.443
#> 4    M1   X  0.519 0.010  0.000
#> 5    M2   X -0.340 0.010  0.000
#> 6     Y   Y  0.975 0.014  0.000
#> 7    M1  M1  0.973 0.014  0.000
#> 8    M2  M2  0.982 0.014  0.000
#> 9     X   X  1.000 0.000     NA
```

Comparison of Model Fits
To check whether modeling correlation matters, we compare AIC and RMSEA.

```
# Extract model fit statistics
fit_measures <- function(fit) {
  fitMeasures(fit, c("aic", "bic", "rmsea", "chisq"))
}

# Compare model fits
fit_measures(fit_corr)  # Correct model (correlated mediators)
#>      aic      bic    rmsea    chisq
#> 77932.45 77997.34     0.00     0.00
fit_measures(fit_uncorr)  # Incorrect model (ignores
↪  correlation)
#>       aic       bic     rmsea     chisq
#> 84453.208 84510.891     0.808  6522.762
```

- If AIC and RMSEA are lower in the correlated model, it suggests that accounting for correlated errors improves fit.

  After fitting the model, we assess:

1. **Direct Effect**: The effect of X on Y **after accounting for both mediators** (c').
2. **Indirect Effects**:
   - $a_1 \times b_1$: Effect of $X \to M_1 \to Y$.
   - $a_2 \times b_2$: Effect of $X \to M_2 \to Y$.
3. **Total Effect**: Sum of direct and indirect effects.

```
# Extract indirect and direct effects
parameterEstimates(fit_corr, standardized = TRUE)
#>    lhs op rhs label    est    se       z pvalue ci.lower
#> 1    Y  ~  M1    b1  0.700 0.014  50.489  0.000    0.673
#> 2    Y  ~  M2    b2  0.487 0.014  35.284  0.000    0.460
#> 3    Y  ~   X     c -0.009 0.015  -0.606  0.545   -0.038
#> 4   M1  ~   X    a1  0.519 0.010  52.563  0.000    0.499
#> 5   M2  ~   X    a2 -0.340 0.010 -34.314  0.000   -0.360
#> 6   M1 ~~  M2        0.677 0.012  56.915  0.000    0.654
#> 7    Y ~~   Y        0.975 0.014  70.711  0.000    0.948
#> 8   M1 ~~  M1        0.973 0.014  70.711  0.000    0.946
#> 9   M2 ~~  M2        0.982 0.014  70.711  0.000    0.955
#> 10   X ~~   X        1.000 0.000      NA     NA    1.000
#>    ci.upper std.lv std.all std.nox
#> 1     0.727  0.700   0.528   0.528
#> 2     0.514  0.487   0.345   0.345
#> 3     0.020 -0.009  -0.006  -0.006
#> 4     0.538  0.519   0.465   0.465
#> 5    -0.321 -0.340  -0.325  -0.325
#> 6     0.700  0.677   0.692   0.692
#> 7     1.002  0.975   0.447   0.447
#> 8     1.000  0.973   0.784   0.784
#> 9     1.010  0.982   0.895   0.895
#> 10    1.000  1.000   1.000   1.000
```

If c' is reduced (but still significant), we have partial mediation. If $c' \approx 0$, it suggests full mediation.

```
# Load required packages
library(MASS)  # for mvrnorm
library(lavaan)

# Function to generate synthetic data with correctly
↪  correlated errors for mediators
generate_data <-
  function(n = 10000,
           a1 = 0.5,
           a2 = -0.35,
           b1 = 0.7,
           b2 = 0.48,
           corr = TRUE,
           correlation_value = 0.7) {
    set.seed(12345)
```

```
    X <- rnorm(n)

    # Generate correlated errors using a multivariate normal
     ↪ distribution
    if (corr) {
        Sigma <- matrix(c(1, correlation_value,
↪   correlation_value, 1), nrow = 2)  # Higher covariance
↪   matrix for errors
        errors <- mvrnorm(n, mu = c(0, 0), Sigma = Sigma)  #
↪   Generate correlated errors
    } else {
        errors <- mvrnorm(n, mu = c(0, 0), Sigma = diag(2))  #
↪   Independent errors
    }

    M1 <- a1 * X + errors[, 1]
    M2 <- a2 * X + errors[, 2]
    Y <- b1 * M1 + b2 * M2 + rnorm(n)  # Y depends on M1 and
↪   M2

    data.frame(X = X, M1 = M1, M2 = M2, Y = Y)
}

# Ground truth for comparison
ground_truth <- data.frame(Parameter = c("b1", "b2"),
↪   GroundTruth = c(0.7, 0.48))

# Function to extract relevant estimates, standard errors, and
↪   model fit
extract_estimates_b1_b2 <- function(fit) {
    estimates <- parameterEstimates(fit)
    estimates <- estimates[estimates$lhs == "Y" &
↪   estimates$rhs %in% c("M1", "M2"), c("rhs", "est", "se")]
    estimates$Parameter <- ifelse(estimates$rhs == "M1", "b1",
↪   "b2")
    estimates <- estimates[, c("Parameter", "est", "se")]
    fit_stats <- fitMeasures(fit, c("aic", "bic", "rmsea",
↪   "chisq"))
    return(list(estimates = estimates, fit_stats = fit_stats))
}

# Case 1: Correlated errors for mediators (modeled correctly)
Data_corr <- generate_data(n = 10000, corr = TRUE,
↪   correlation_value = 0.7)
```

```
model_corr <- '
  Y ~ b1 * M1 + b2 * M2 + c * X
  M1 ~ a1 * X
  M2 ~ a2 * X
  M1 ~~ M2  # Correlated mediators (errors)
'
fit_corr <- sem(model = model_corr, data = Data_corr)
results_corr <- extract_estimates_b1_b2(fit_corr)

# Case 2: Uncorrelated errors for mediators (modeled
↪ correctly)
Data_uncorr <- generate_data(n = 10000, corr = FALSE)
model_uncorr <- '
  Y ~ b1 * M1 + b2 * M2 + c * X
  M1 ~ a1 * X
  M2 ~ a2 * X
'
fit_uncorr <- sem(model = model_uncorr, data = Data_uncorr)
results_uncorr <- extract_estimates_b1_b2(fit_uncorr)

# Case 3: Correlated errors, but not modeled as correlated
fit_corr_incorrect <- sem(model = model_uncorr, data =
↪ Data_corr)
results_corr_incorrect <-
↪ extract_estimates_b1_b2(fit_corr_incorrect)

# Case 4: Uncorrelated errors, but modeled as correlated
fit_uncorr_incorrect <- sem(model = model_corr, data =
↪ Data_uncorr)
results_uncorr_incorrect <-
↪ extract_estimates_b1_b2(fit_uncorr_incorrect)

# Combine all estimates for comparison
estimates_combined <- list(
    "Correlated (Correct)" = results_corr$estimates,
    "Uncorrelated (Correct)" = results_uncorr$estimates,
    "Correlated (Incorrect)" =
    ↪ results_corr_incorrect$estimates,
    "Uncorrelated (Incorrect)" =
    ↪ results_uncorr_incorrect$estimates
)

# Combine all into a single table
comparison_table <- do.call(rbind,
↪ lapply(names(estimates_combined), function(case) {
```

```
    df <- estimates_combined[[case]]
    df$Case <- case
    df
}))

# Merge with ground truth for final comparison
comparison_table <- merge(comparison_table, ground_truth, by =
↪  "Parameter")

# Display the comparison table
comparison_table
#>   Parameter       est          se                     Case
#> 1        b1 0.7002984 0.013870433     Correlated (Correct)
#> 2        b1 0.6973612 0.009859426   Uncorrelated (Correct)
#> 3        b1 0.7002984 0.010010367   Correlated (Incorrect)
#> 4        b1 0.6973612 0.009859634 Uncorrelated (Incorrect)
#> 5        b2 0.4871118 0.013805615     Correlated (Correct)
#> 6        b2 0.4868318 0.010009908   Uncorrelated (Correct)
#> 7        b2 0.4871118 0.009963588   Correlated (Incorrect)
#> 8        b2 0.4868318 0.010010119 Uncorrelated (Incorrect)
#>   GroundTruth
#> 1        0.70
#> 2        0.70
#> 3        0.70
#> 4        0.70
#> 5        0.48
#> 6        0.48
#> 7        0.48
#> 8        0.48

# Display model fit statistics for each case
fit_stats_combined <- list(
    "Correlated (Correct)" = results_corr$fit_stats,
    "Uncorrelated (Correct)" = results_uncorr$fit_stats,
    "Correlated (Incorrect)" =
    ↪  results_corr_incorrect$fit_stats,
    "Uncorrelated (Incorrect)" =
    ↪  results_uncorr_incorrect$fit_stats
)

fit_stats_combined
#> $`Correlated (Correct)`
#>      aic      bic    rmsea    chisq
#> 77932.45 77997.34     0.00     0.00
```

```
#>
#> $`Uncorrelated (Correct)`
#>        aic       bic     rmsea     chisq
#> 84664.312 84721.995     0.000     0.421
#>
#> $`Correlated (Incorrect)`
#>        aic       bic     rmsea     chisq
#> 84453.208 84510.891     0.808  6522.762
#>
#> $`Uncorrelated (Incorrect)`
#>       aic      bic   rmsea   chisq
#> 84665.89 84730.78    0.00    0.00
```

10.1.11 Multiple Treatments in Mediation

In some cases, **multiple independent variables** (X_1, X_2) influence the same mediators. This is called **multiple treatments mediation** (Hayes & Preacher, 2014).

For an example in **PROCESS (SPSS/R)**, see:

Process Mediation with Multiple Treatments.

10.2 Causal Inference Approach to Mediation

Traditional mediation models assume that regression-based estimates provide valid causal inference. However, **causal mediation analysis** (CMA) extends beyond traditional models by explicitly defining mediation in terms of **potential outcomes and counterfactuals**.

10.2.1 Example: Traditional Mediation Analysis

We begin with a classic **three-step mediation approach**.

```
# Load data
myData <- read.csv("data/mediationData.csv")

# Step 1 (Total Effect: X → Y) [No longer required]
model.0 <- lm(Y ~ X, data = myData)
```

```
summary(model.0)
#>
#> Call:
#> lm(formula = Y ~ X, data = myData)
#>
#> Residuals:
#>     Min      1Q  Median      3Q     Max
#> -5.0262 -1.2340 -0.3282  1.5583  5.1622
#>
#> Coefficients:
#>             Estimate Std. Error t value Pr(>|t|)
#> (Intercept)   2.8572     0.6932   4.122 7.88e-05 ***
#> X             0.3961     0.1112   3.564 0.000567 ***
#> ---
#> Signif. codes:
#> 0 '***' 0.001 '**' 0.01 '*' 0.05 '.' 0.1 ' ' 1
#>
#> Residual standard error: 1.929 on 98 degrees of freedom
#> Multiple R-squared:  0.1147, Adjusted R-squared:  0.1057
#> F-statistic:  12.7 on 1 and 98 DF,  p-value: 0.0005671

# Step 2 (Effect of X on M)
model.M <- lm(M ~ X, data = myData)
summary(model.M)
#>
#> Call:
#> lm(formula = M ~ X, data = myData)
#>
#> Residuals:
#>     Min      1Q  Median      3Q     Max
#> -4.3046 -0.8656  0.1344  1.1344  4.6954
#>
#> Coefficients:
#>             Estimate Std. Error t value Pr(>|t|)
#> (Intercept)  1.49952    0.58920   2.545   0.0125 *
#> X            0.56102    0.09448   5.938 4.39e-08 ***
#> ---
#> Signif. codes:
#> 0 '***' 0.001 '**' 0.01 '*' 0.05 '.' 0.1 ' ' 1
#>
#> Residual standard error: 1.639 on 98 degrees of freedom
#> Multiple R-squared:  0.2646, Adjusted R-squared:  0.2571
#> F-statistic: 35.26 on 1 and 98 DF,  p-value: 4.391e-08
```

```
# Step 3 (Effect of M on Y, controlling for X)
model.Y <- lm(Y ~ X + M, data = myData)
summary(model.Y)
#>
#> Call:
#> lm(formula = Y ~ X + M, data = myData)
#>
#> Residuals:
#>     Min      1Q  Median      3Q     Max
#> -3.7631 -1.2393  0.0308  1.0832  4.0055
#>
#> Coefficients:
#>             Estimate Std. Error t value Pr(>|t|)
#> (Intercept)   1.9043     0.6055   3.145   0.0022 **
#> X             0.0396     0.1096   0.361   0.7187
#> M             0.6355     0.1005   6.321 7.92e-09 ***
#> ---
#> Signif. codes:
#> 0 '***' 0.001 '**' 0.01 '*' 0.05 '.' 0.1 ' ' 1
#>
#> Residual standard error: 1.631 on 97 degrees of freedom
#> Multiple R-squared:  0.373,	Adjusted R-squared:  0.3601
#> F-statistic: 28.85 on 2 and 97 DF,  p-value: 1.471e-10

# Step 4: Bootstrapping for ACME
library(mediation)
results <-
    mediate(
        model.M,
        model.Y,
        treat = 'X',
        mediator = 'M',
        boot = TRUE,
        sims = 500
    )
summary(results)
#>
#> Causal Mediation Analysis
#>
#> Nonparametric Bootstrap Confidence Intervals with the
↪  Percentile Method
#>
#>                Estimate 95% CI Lower 95% CI Upper p-value
#> ACME             0.3565       0.2119         0.51  <2e-16
```

```
#> ADE                0.0396       -0.1750          0.28   0.760
#> Total Effect       0.3961        0.1743          0.64   0.004
#> Prop. Mediated     0.9000        0.5042          1.94   0.004
#>
#> ACME           ***
#> ADE
#> Total Effect   **
#> Prop. Mediated **
#> ---
#> Signif. codes:
#> 0 '***' 0.001 '**' 0.01 '*' 0.05 '.' 0.1 ' ' 1
#>
#> Sample Size Used: 100
#>
#>
#> Simulations: 500
```

- **Total Effect**: $\hat{c} = 0.3961 \rightarrow$ effect of X on Y **without** controlling for M.
- **Direct Effect (ADE)**: $\hat{c'} = 0.0396 \rightarrow$ effect of X on Y **after** accounting for M.
- **ACME (Average Causal Mediation Effect)**:
  - ACME $= \hat{c} - \hat{c'} = 0.3961 - 0.0396 = 0.3565$
  - Equivalent to **product of paths**: $\hat{a} \times \hat{b} = 0.56102 \times 0.6355 = 0.3565$.

These calculations do not rely on strong causal assumptions. For a causal interpretation, we need a more rigorous framework.

10.2.2 Two Approaches in Causal Mediation Analysis

The `mediation` package (Imai et al., 2010a, 2010b) enables **causal mediation analysis**. It supports two inference types:

1. **Model-Based Inference**
   - Assumptions:
     - Treatment is randomized (or approximated via matching).
     - Sequential Ignorability: No unobserved confounding in:
       1. Treatment $\rightarrow$ Mediator
       2. Treatment $\rightarrow$ Outcome
       3. Mediator $\rightarrow$ Outcome
     - This assumption is hard to justify in observational studies.

2. **Design-Based Inference**
   - Relies on experimental design to isolate the causal mechanism.

Notation
We follow the standard potential outcomes framework:

- $M_i(t)$ = mediator under treatment condition t
- $T_i \in 0, 1$ = treatment assignment
- $Y_i(t, m)$ = outcome under treatment t and mediator value m
- X_i = observed pretreatment covariates

The **treatment effect** for an individual i:

$$\tau_i = Y_i(1, M_i(1)) - Y_i(0, M_i(0)).$$

which decomposes into:

1. **Causal Mediation Effect (ACME)**:

$$\delta_i(t) = Y_i(t, M_i(1)) - Y_i(t, M_i(0))$$

2. **Direct Effect (ADE)**:

$$\zeta_i(t) = Y_i(1, M_i(1)) - Y_i(0, M_i(0))$$

Summing up:

$$\tau_i = \delta_i(t) + \zeta_i(1 - t)$$

Sequential Ignorability Assumption
For CMA to be valid, we assume:

$$\{Y_i(t', m), M_i(t)\} \perp T_i | X_i = x$$
$$Y_i(t', m) \perp M_i(t) | T_i = t, X_i = x$$

- First condition is the standard strong ignorability condition where treatment assignment is random conditional on pretreatment confounders.
- Second condition is stronger where the mediators are also random given the observed treatment and pretreatment confounders. This condition is satisfied only when there is no unobserved pretreatment confounders, and posttreatment confounders, and multiple mediators that are correlated.

Key Challenge
Sequential Ignorability is not testable. Researchers should conduct **sensitivity analysis**.

We now fit a **causal mediation model** using mediation.

```
library(mediation)
set.seed(2014)
data("framing", package = "mediation")

# Step 1: Fit mediator model (M ~ T, X)
med.fit <-
    lm(emo ~ treat + age + educ + gender + income, data =
    ↪  framing)

# Step 2: Fit outcome model (Y ~ M, T, X)
out.fit <-
    glm(
        cong_mesg ~ emo + treat + age + educ + gender +
↪   income,
        data = framing,
        family = binomial("probit")
    )

# Step 3: Causal Mediation Analysis (Quasi-Bayesian)
med.out <-
    mediate(
        med.fit,
        out.fit,
        treat = "treat",
        mediator = "emo",
        robustSE = TRUE,
        sims = 100
    ) # Use sims = 10000 in practice
summary(med.out)
#>
#> Causal Mediation Analysis
#>
#> Quasi-Bayesian Confidence Intervals
#>
#>                          Estimate 95% CI Lower 95% CI Upper
#> ACME (control)             0.0791       0.0351         0.15
#> ACME (treated)             0.0804       0.0367         0.16
#> ADE (control)              0.0206      -0.0976         0.12
#> ADE (treated)              0.0218      -0.1053         0.12
```

```
#> Total Effect                 0.1009       -0.0497          0.23
#> Prop. Mediated (control)     0.6946       -6.3109          3.68
#> Prop. Mediated (treated)     0.7118       -5.7936          3.50
#> ACME (average)               0.0798        0.0359          0.15
#> ADE (average)                0.0212       -0.1014          0.12
#> Prop. Mediated (average)     0.7032       -6.0523          3.59
#>                            p-value
#> ACME (control)              <2e-16 ***
#> ACME (treated)              <2e-16 ***
#> ADE (control)                 0.70
#> ADE (treated)                 0.70
#> Total Effect                  0.14
#> Prop. Mediated (control)      0.14
#> Prop. Mediated (treated)      0.14
#> ACME (average)              <2e-16 ***
#> ADE (average)                 0.70
#> Prop. Mediated (average)      0.14
#> ---
#> Signif. codes:
#> 0 '***' 0.001 '**' 0.01 '*' 0.05 '.' 0.1 ' ' 1
#>
#> Sample Size Used: 265
#>
#>
#> Simulations: 100
```

Alternative: Nonparametric Bootstrap

```
med.out <-
    mediate(
        med.fit,
        out.fit,
        boot = TRUE,
        treat = "treat",
        mediator = "emo",
        sims = 100,
        boot.ci.type = "bca"
    )
#> Running nonparametric bootstrap
summary(med.out)
#>
#> Causal Mediation Analysis
#>
```

```
#> Nonparametric Bootstrap Confidence Intervals with the BCa
↪  Method
#>
#>                           Estimate 95% CI Lower 95% CI Upper
#> ACME (control)              0.0848       0.0424         0.14
#> ACME (treated)              0.0858       0.0410         0.14
#> ADE (control)               0.0117      -0.0726         0.13
#> ADE (treated)               0.0127      -0.0784         0.14
#> Total Effect                0.0975       0.0122         0.25
#> Prop. Mediated (control)    0.8698       1.7460       151.20
#> Prop. Mediated (treated)    0.8804       1.6879       138.91
#> ACME (average)              0.0853       0.0434         0.14
#> ADE (average)               0.0122      -0.0756         0.14
#> Prop. Mediated (average)    0.8751       1.7170       145.05
#>                           p-value
#> ACME (control)             <2e-16 ***
#> ACME (treated)             <2e-16 ***
#> ADE (control)                0.58
#> ADE (treated)                0.58
#> Total Effect                 0.06 .
#> Prop. Mediated (control)     0.06 .
#> Prop. Mediated (treated)     0.06 .
#> ACME (average)             <2e-16 ***
#> ADE (average)                0.58
#> Prop. Mediated (average)     0.06 .
#> ---
#> Signif. codes:
#> 0 '***' 0.001 '**' 0.01 '*' 0.05 '.' 0.1 ' ' 1
#>
#> Sample Size Used: 265
#>
#>
#> Simulations: 100
```

If we suspect **moderation**, we include an **interaction term**.

```
med.fit <-
    lm(emo ~ treat + age + educ + gender + income, data =
    ↪  framing)

out.fit <-
    glm(
        cong_mesg ~ emo * treat + age + educ + gender +
↪   income,
```

```
        data = framing,
        family = binomial("probit")
    )

med.out <-
    mediate(
        med.fit,
        out.fit,
        treat = "treat",
        mediator = "emo",
        robustSE = TRUE,
        sims = 100
    )

summary(med.out)
#>
#> Causal Mediation Analysis
#>
#> Quasi-Bayesian Confidence Intervals
#>
#>                          Estimate 95% CI Lower
#> ACME (control)            0.07417      0.02401
#> ACME (treated)            0.09496      0.02702
#> ADE (control)            -0.01353     -0.11855
#> ADE (treated)             0.00726     -0.11007
#> Total Effect              0.08143     -0.05646
#> Prop. Mediated (control)  0.64510    -14.31243
#> Prop. Mediated (treated)  0.98006    -17.83202
#> ACME (average)            0.08457      0.02738
#> ADE (average)            -0.00314     -0.11457
#> Prop. Mediated (average)  0.81258    -16.07223
#>                         95% CI Upper p-value
#> ACME (control)                  0.14  <2e-16 ***
#> ACME (treated)                  0.16  <2e-16 ***
#> ADE (control)                   0.11    0.76
#> ADE (treated)                   0.11    0.90
#> Total Effect                    0.19    0.26
#> Prop. Mediated (control)        3.13    0.26
#> Prop. Mediated (treated)        4.01    0.26
#> ACME (average)                  0.15  <2e-16 ***
#> ADE (average)                   0.12    1.00
#> Prop. Mediated (average)        3.55    0.26
#> ---
#> Signif. codes:
```

```
#> 0 '***' 0.001 '**' 0.01 '*' 0.05 '.' 0.1 ' ' 1
#>
#> Sample Size Used: 265
#>
#>
#> Simulations: 100
test.TMint(med.out, conf.level = .95)  # Tests for interaction
↪  effect
#>
#>  Test of ACME(1) - ACME(0) = 0
#>
#> data:  estimates from med.out
#> ACME(1) - ACME(0) = 0.020796, p-value = 0.3
#> alternative hypothesis: true ACME(1) - ACME(0) is not equal
↪  to 0
#> 95 percent confidence interval:
#>  -0.01757310  0.07110837
```

Since **sequential ignorability** is untestable, we examine how unmeasured confounding affects ACME estimates (Fig. 10.4).

```
# Load required package
library(mediation)

# Simulate some example data
set.seed(123)
n <- 100
data <- data.frame(
  treat = rbinom(n, 1, 0.5),       # Binary treatment
  med = rnorm(n),                   # Continuous mediator
  outcome = rnorm(n)                 # Continuous outcome
)

# Fit the mediator model (med ~ treat)
med_model <- lm(med ~ treat, data = data)

# Fit the outcome model (outcome ~ treat + med)
outcome_model <- lm(outcome ~ treat + med, data = data)

# Perform mediation analysis
med_out <- mediate(med_model,
                   outcome_model,
                   treat = "treat",
                   mediator = "med",
```

```
                   sims = 100)

# Conduct sensitivity analysis
sens_out <- medsens(med_out, sims = 100)

summary(sens_out)
#>
#> Mediation Sensitivity Analysis for Average Causal Mediation
↪  Effect
#>
#> Sensitivity Region
#>
#>        Rho    ACME 95% CI Lower 95% CI Upper R^2_M*R^2_Y*
#>  [1,] -0.9 -0.6194      -1.3431       0.1043         0.81
#>  [2,] -0.8 -0.3898      -0.8479       0.0682         0.64
#>  [3,] -0.7 -0.2790      -0.6096       0.0516         0.49
#>  [4,] -0.6 -0.2067      -0.4552       0.0418         0.36
#>  [5,] -0.5 -0.1525      -0.3406       0.0355         0.25
#>  [6,] -0.4 -0.1083      -0.2487       0.0321         0.16
#>  [7,] -0.3 -0.0700      -0.1723       0.0323         0.09
#>  [8,] -0.2 -0.0354      -0.1097       0.0389         0.04
#>  [9,] -0.1 -0.0028      -0.0648       0.0591         0.01
#> [10,]  0.0  0.0287      -0.0416       0.0990         0.00
#> [11,]  0.1  0.0603      -0.0333       0.1538         0.01
#> [12,]  0.2  0.0928      -0.0317       0.2173         0.04
#> [13,]  0.3  0.1275      -0.0333       0.2882         0.09
#> [14,]  0.4  0.1657      -0.0369       0.3684         0.16
#> [15,]  0.5  0.2100      -0.0422       0.4621         0.25
#> [16,]  0.6  0.2642      -0.0495       0.5779         0.36
#> [17,]  0.7  0.3364      -0.0601       0.7329         0.49
#> [18,]  0.8  0.4473      -0.0771       0.9717         0.64
#> [19,]  0.9  0.6768      -0.1135       1.4672         0.81
#>       R^2_M~R^2_Y~
#>  [1,]       0.7807
#>  [2,]       0.6168
#>  [3,]       0.4723
#>  [4,]       0.3470
#>  [5,]       0.2409
#>  [6,]       0.1542
#>  [7,]       0.0867
#>  [8,]       0.0386
#>  [9,]       0.0096
#> [10,]       0.0000
#> [11,]       0.0096
```

```
#> [12,]         0.0386
#> [13,]         0.0867
#> [14,]         0.1542
#> [15,]         0.2409
#> [16,]         0.3470
#> [17,]         0.4723
#> [18,]         0.6168
#> [19,]         0.7807
#>
#> Rho at which ACME = 0: -0.1
#> R^2_M*R^2_Y* at which ACME = 0: 0.01
#> R^2_M~R^2_Y~ at which ACME = 0: 0.0096
```

```
plot(sens_out)
```

- If ACME confidence intervals contain 0, the effect is not robust to confounding.

Alternatively, using R^2 interpretation, we need to specify the direction of confounder that affects the mediator and outcome variables in `plot` using `sign.prod = "positive"` (i.e., same direction) or `sign.prod = "negative"` (i.e., opposite direction).

Refer to Table 10.3 for a summary of differences between traditional mediation and causal mediation.

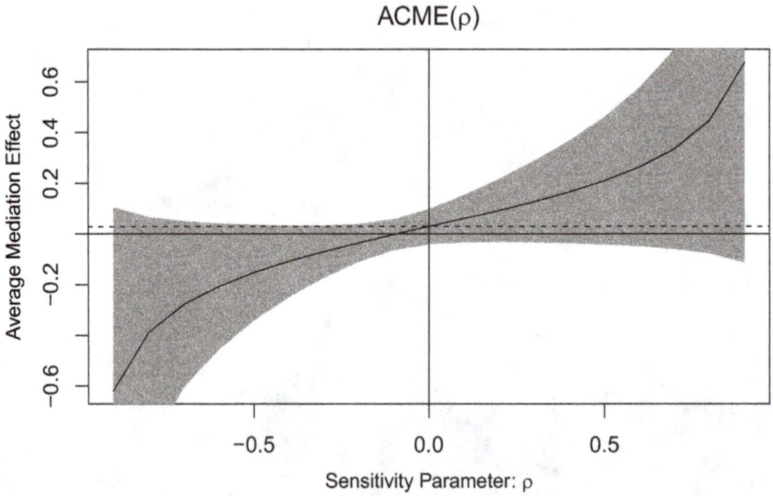

Fig. 10.4 ACME

```
plot(sens.out, sens.par = "R2", r.type = "total", sign.prod =
↪  "positive")
```

Table 10.3 Summary of causal mediation vs. traditional mediation

Aspect	Traditional mediation	Causal mediation
Model assumption	Linear regressions	Potential outcomes framework
Assumptions needed	No omitted confounders	Sequential ignorability
Inference method	Product of coefficients	Counterfactual reasoning
Bootstrapping?	Common	Essential
Sensitivity analysis?	Rarely used	Strongly recommended

Chapter 11
Prediction and Estimation

In modern statistics, econometrics, and machine learning, two primary goals often motivate data analysis:

1. **Prediction**: To build a function $\hat{f}$ that **accurately predicts** an outcome Y from observed features (predictors) X.
2. **Estimation or Causal Inference**: To **uncover and quantify** the relationship (often causal) between X and Y, typically by estimating parameters like β in a model $Y = g(X; \beta)$.

These goals, while superficially similar, rest on distinct philosophical and mathematical foundations. Below, we explore the difference in detail, illustrating key ideas with formal definitions, theorems, proofs (where relevant), and references to seminal works.

11.1 Conceptual Framing

11.1.1 Predictive Modeling

Predictive modeling focuses on building a function $\hat{f} : \mathcal{X} \to \mathcal{Y}$ that maps inputs X to outputs Y. For simplicity, assume:

- $X \in \mathbb{R}^p$ (though in practice X can be images, text, time series, etc.).
- $Y \in \mathbb{R}$ for regression or $Y \in \{0, 1\}$ (or other finite set) for classification.

© The Author(s), under exclusive license to Springer Nature Switzerland AG 2025
M. Nguyen, *Advanced Modeling and Data Challenges*,
https://doi.org/10.1007/978-3-032-01719-2_11

The yardstick for success is the function's accuracy in out-of-sample predictions, often measured by a loss function $L(\hat{y}, y)$. We typically choose $\hat{f}$ to minimize expected loss:

$$\text{(Predictive Problem)} \quad \hat{f} = \arg\min_{f \in \mathcal{F}} \mathbb{E}[L(f(X), Y)],$$

where $\mathcal{F}$ is a class of functions (models) and $\mathbb{E}[\cdot]$ is taken over the joint distribution of (X, Y).

11.1.2 Estimation or Causal Inference

By contrast, estimation or causal inference generally aims to uncover the underlying mechanism: *how does X (or a particular component $T \subseteq X$) cause changes in Y?* The canonical problem is to estimate parameters β in a model $m_\beta(x)$ such that:

$$Y = m_\beta(X) + \varepsilon,$$

or, in linear form,

$$Y = X\beta + \varepsilon.$$

A variety of statistical properties—**consistency, unbiasedness, efficiency, confidence intervals, hypothesis tests**—are relevant here. *Causal* interpretations usually require assumptions beyond typical i.i.d. sampling: unconfoundedness, exogeneity, or random assignment, so that β indeed captures how changes in X cause changes in Y.

Key Distinction

- **Prediction** does not require that the parameters used in $\hat{f}$ reflect any real-world mechanism. As long as out-of-sample predictive performance is good, the model is deemed successful—even if it's a "black box."
- **Causal inference** demands interpretability in terms of structural or exogenous relationships. The main objective is consistent estimation of the true (or theoretically defined) parameter β, which has an economic, biomedical, or policy interpretation.

11.2 Mathematical Setup

11.2.1 Probability Space and Data

We posit a probability space $(\Omega, \mathcal{F}, P)$ and random variables (X, Y) on it. We typically have an i.i.d. sample $\{(X_i, Y_i)\}_{i=1}^n$ from the true distribution $\mathcal{D}$. Let:

$$(X, Y) \sim \mathcal{D}, \quad (X_i, Y_i) \overset{\text{i.i.d.}}{\sim} \mathcal{D}.$$

In prediction, we train on $\{(X_i, Y_i)\}_{i=1}^n$ to obtain $\hat{f}$, and we evaluate on a test point $(\tilde{X}, \tilde{Y})$ drawn from $\mathcal{D}$. In causal inference, we scrutinize the data generating process carefully, ensuring that we can identify a causal effect. For example, we may require:

- Potential outcomes $\{Y_i(0), Y_i(1)\}$ for treatment effect settings.
- Unconfoundedness or randomization assumptions.

11.2.2 Loss Functions and Risk

A general framework for both tasks is the risk minimization approach. For a function f, define:

- The population (or expected) risk:

$$\mathcal{R}(f) = \mathbb{E}[L(f(X), Y)].$$

- The empirical risk (on a sample of size n):

$$\hat{\mathcal{R}}_n(f) = \frac{1}{n} \sum_{i=1}^{n} L(f(X_i), Y_i).$$

Prediction We often solve the empirical risk minimization (ERM) problem:

$$\hat{f} = \arg\min_{f \in \mathcal{F}} \hat{\mathcal{R}}_n(f),$$

possibly with regularization. The measure of success is $\mathcal{R}(\hat{f})$, i.e., how well $\hat{f}$ generalizes beyond the training sample.

Causal/Parameter Estimation We might define an M-estimator for β (Newey & McFadden, 1994). Consider a function $\psi(\beta; X, Y)$ such that the true parameter β_0 satisfies:

$$\mathbb{E}[\psi(\beta_0; X, Y)] = 0.$$

The empirical M-estimator solves

$$\hat{\beta} = \arg\min_{\beta} \left\| \frac{1}{n} \sum_{i=1}^{n} \psi(\beta; X_i, Y_i) \right\|,$$

or equivalently sets it to zero in a method-of-moments sense:

$$\frac{1}{n} \sum_{i=1}^{n} \psi(\hat{\beta}; X_i, Y_i) = 0.$$

Properties like consistency ($\hat{\beta} \xrightarrow{p} \beta_0$) or asymptotic normality ($\sqrt{n}(\hat{\beta} - \beta_0) \xrightarrow{d} N(0, \Sigma)$) are central. The emphasis is on *uncovering the true* β_0 rather than purely predictive accuracy.

11.3 Prediction in Detail

11.3.1 Empirical Risk Minimization and Generalization

In supervised learning, the goal is to find a function f from a class of candidate models $\mathcal{F}$ (e.g., linear models, neural networks, tree-based models) that accurately predicts an outcome Y given an input X. This is typically formulated as an **Empirical Risk Minimization problem**, where we seek to minimize the average loss over the training data:

$$\hat{f} = \arg\min_{f \in \mathcal{F}} \frac{1}{n} \sum_{i=1}^{n} L(f(X_i), Y_i).$$

where $L(\cdot, \cdot)$ is a **loss function** that quantifies the error between predictions and actual values. Common choices include:

- **Squared Error (Regression):** $L(\hat{y}, y) = (\hat{y} - y)^2$.
- **Absolute Error (Regression):** $L(\hat{y}, y) = |\hat{y} - y|$.
- **Logistic Loss (Classification):** $L(\hat{p}, y) = -[y \log \hat{p} + (1 - y) \log(1 - \hat{p})]$.

By minimizing empirical risk, we find a function $\hat{f}$ that best fits the observed data. However, **minimizing training error does not guarantee good generalization**—the ability of $\hat{f}$ to perform well on unseen data.

11.3.1.1 Overfitting and Regularization

If $\mathcal{F}$ is very large or expressive (e.g., deep neural networks with millions of parameters), $\hat{f}$ can become too **complex**, learning patterns that exist in the training set but do not generalize to new data. This is called **overfitting**.

To mitigate overfitting, we introduce **regularization**, modifying the optimization objective to penalize complex models:

$$\hat{f}_\lambda = \arg\min_{f \in \mathcal{F}} \left\{ \hat{\mathcal{R}}_n(f) + \lambda \Omega(f) \right\}.$$

where:

- $\hat{\mathcal{R}}_n(f) = \frac{1}{n} \sum_{i=1}^{n} L(f(X_i), Y_i)$ is the empirical risk.
- $\Omega(f)$ is a **complexity penalty** that discourages overly flexible models.
- λ controls the strength of regularization.

Common choices of $\Omega(f)$ include:

- **LASSO penalty:** $\|\beta\|_1$ (sparsity constraint in linear models).
- **Ridge penalty:** $\|\beta\|_2^2$ (shrinking coefficients to reduce variance).
- **Neural network weight decay:** $\sum w^2$ (prevents exploding weights).

Regularization encourages **simpler models**, which are more likely to generalize well.

11.3.1.2 Generalization and Statistical Learning Theory

A fundamental question in machine learning is: *How well does* $\hat{f}$ perform on unseen data? This is captured by the **expected risk**:

$$R(f) = \mathbb{E}[L(f(X), Y)].$$

Ideally, we want to minimize the gap between the **true risk** $R(\hat{f})$ and the best possible risk $R(f^*)$ within $\mathcal{F}$:

$$R(\hat{f}) - \min_{f \in \mathcal{F}} R(f).$$

This difference, called the **excess risk**, measures how well $\hat{f}$ generalizes beyond the training sample. **Statistical Learning Theory** provides theoretical tools to analyze this gap (Vapnik, 2013; Hastie et al., 2009). In particular, it establishes generalization bounds that depend on the **capacity** of the function class $\mathcal{F}$.

11.3.1.3 Complexity Measures

Two important ways to quantify the complexity of $\mathcal{F}$ are

- VC Dimension
- Rademacher Complexity

11.3.1.3.1 VC Dimension

The VC dimension measures the ability of a hypothesis class $\mathcal{F}$ to fit arbitrary labels. Formally, the VC dimension of $\mathcal{F}$, denoted as $\text{VC}(\mathcal{F})$, is the largest number of points that can be shattered by some function in $\mathcal{F}$.

- A set of points is shattered by $\mathcal{F}$ if, for every possible labeling of these points, there exists a function $f \in \mathcal{F}$ that perfectly classifies them.

Example 1: Linear Classifiers in 2D

- Consider a set of points in $\mathbb{R}^2$ (the plane).
- If $\mathcal{F}$ consists of linear decision boundaries, we can shatter at most three points in general position (because a single line can separate them in any way).
- However, four points cannot always be shattered (e.g., if arranged in an XOR pattern). Thus, the VC dimension of linear classifiers in $\mathbb{R}^2$ is 3.

Key Property

- A higher VC dimension means a more expressive model class (higher capacity).
- If $\text{VC}(\mathcal{F})$ is too large, the model can memorize the training set, leading to poor generalization.

11.3.1.3.2 Rademacher Complexity

VC dimension is a combinatorial measure, but Rademacher complexity is a more refined, data-dependent measure of function class flexibility.

Intuition Rademacher complexity quantifies how well functions in $\mathcal{F}$ can **correlate with random noise**. If a function class can fit random labels well, it is too flexible and likely to overfit.

Definition

Given n training samples, let $\sigma_1, \ldots, \sigma_n$ be **independent Rademacher variables** (i.e., random variables taking values ± 1 with equal probability). The **empirical**

Rademacher complexity of $\mathcal{F}$ is:

$$\hat{\mathcal{R}}_n(\mathcal{F}) = \mathbb{E}_\sigma \left[\sup_{f \in \mathcal{F}} \frac{1}{n} \sum_{i=1}^{n} \sigma_i f(X_i) \right].$$

Interpretation

- If $\hat{\mathcal{R}}_n(\mathcal{F})$ is large, then $\mathcal{F}$ can fit random noise well $\Rightarrow$ high risk of overfitting.
- If $\hat{\mathcal{R}}_n(\mathcal{F})$ is small, then $\mathcal{F}$ is more stable $\Rightarrow$ better generalization.

Example 2: Linear Models with Bounded Norm

- Suppose $\mathcal{F}$ consists of linear models $f(X) = w^\top X$, where $\|w\| \leq C$.
- The **Rademacher complexity** of this class scales as $\mathcal{O}(C/\sqrt{n})$.
- This suggests that controlling the norm of w (e.g., via [Ridge Regression]) improves generalization.

11.3.2 Bias-Variance Decomposition

For a regression problem with squared-error loss, a classic decomposition is:

$$\mathbb{E}_{\text{train}}[(\hat{f}(X) - Y)^2]$$
$$= \underbrace{(\mathbb{E}[\hat{f}(X)] - f^*(X))^2}_{\text{Bias}^2} + \underbrace{\mathbb{E}[(\hat{f}(X) - \mathbb{E}[\hat{f}(X)])^2]}_{\text{Variance}} + \underbrace{\sigma_\varepsilon^2}_{\text{Irreducible Error}}$$

where $f^*(X) = \mathbb{E}[Y \mid X]$. Minimizing the sum of bias^2 and variance is key.

In prediction, a small increase in bias is often acceptable if it yields a large reduction in variance—this can improve out-of-sample performance. However, for causal inference, any added bias is problematic if it distorts the interpretation of parameters.

11.3.3 Example: Linear Regression for Prediction

Consider a linear predictor:

$$\hat{y} = x^\top \hat{\beta}.$$

We choose $\hat{\beta}$ to minimize:

$$\sum_{i=1}^{n} (y_i - x_i^\top \beta)^2 \quad \text{or with a penalty:} \quad \sum_{i=1}^{n} (y_i - x_i^\top \beta)^2 + \lambda \|\beta\|_2^2.$$

Goal Achieve minimal prediction error on unseen data $(\tilde{x}, \tilde{y})$.

The estimated $\hat{\beta}$ might be biased if we use regularization (e.g., ridge). But from a purely predictive lens, that bias can be advantageous if it lowers variance substantially and thus lowers expected prediction error.

11.3.4 Applications in Economics

In economics (and related social sciences), **prediction** plays an increasingly prominent role (Mullainathan & Spiess, 2017; Athey & Imbens, 2019):

- **Measure Variables**: Predicting missing or proxy variables (e.g., predicting income from observable covariates, or predicting individual preferences from online behaviors).
- **Embed Prediction Tasks Within Parameter Estimation or Treatment Effects**: Sometimes, a first-stage prediction (e.g., imputing missing data or generating prognostic scores) is used as an input for subsequent causal analyses.
- **Control for Observed Confounders**: Machine learning methods—such as LASSO, random forests, or neural nets—can be used to control for high-dimensional X when doing partial-out adjustments or residualizing outcomes (Belloni et al., 2014; Chernozhukov et al., 2018).

11.4 Parameter Estimation and Causal Inference

11.4.1 Estimation in Parametric Models

In a simple parametric form:

$$Y = X\beta + \varepsilon, \quad \mathbb{E}[\varepsilon \mid X] = 0, \quad \mathrm{Var}(\varepsilon \mid X) = \sigma^2 I.$$

The [Ordinary Least Squares] estimator is:

$$\hat{\beta}_{\mathrm{OLS}} = \arg\min_{\beta} \|Y - X\beta\|_2^2 = (X^\top X)^{-1} X^\top Y.$$

Under classical assumptions (e.g., no perfect collinearity, homoskedastic errors), $\hat{\beta}_{\mathrm{OLS}}$ is BLUE—the Best Linear Unbiased Estimator.

In a more general form, **parameter estimation**, denoted $\hat{\beta}$, focuses on **estimating the relationship** between y and x, often with a view toward **causality**. In many econometric or statistical settings, we write:

$$y = x^\top \beta + \varepsilon,$$

or more generally $y = g(x; \beta) + \varepsilon$, where β encodes the structural or causal parameters we wish to recover.

The core aim is **consistency**—that is, for large n, we want $\hat{\beta}$ to converge to the true β that defines the underlying relationship. In other words:

$$\hat{\beta} \xrightarrow{p} \beta, \quad \text{as } n \to \infty.$$

Some texts phrase it informally as requiring that

$$\mathbb{E}[\hat{f}] = f,$$

meaning the estimator is (asymptotically) unbiased for the true function or parameters.

However, **consistency** alone may not suffice for scientific inference. One often also examines:

- **Asymptotic Normality**: $\sqrt{n}(\hat{\beta} - \beta) \xrightarrow{d} \mathcal{N}(0, \Sigma)$.
- **Confidence Intervals**: $\hat{\beta}_j \pm z_{\alpha/2}\,\mathrm{SE}(\hat{\beta}_j)$.
- **Hypothesis Tests**: $H_0 : \beta_j = 0 \quad \text{vs.} \quad H_1 : \beta_j \neq 0$.

11.4.2 Causal Inference Fundamentals

To interpret β in $Y = X\beta + \varepsilon$ as "causal," we typically require that changes in X (or at least in one component of X) lead to changes in Y that are not confounded by omitted variables or simultaneity. In a prototypical potential-outcomes framework (for a binary treatment D):

- $Y_i(1)$: outcome if unit i receives treatment $D = 1$.
- $Y_i(0)$: outcome if unit i receives no treatment $D = 0$.

The observed outcome Y_i is

$$Y_i = D_i Y_i(1) + (1 - D_i) Y_i(0).$$

The **Average Treatment Effect** (ATE) is:

$$\tau = \mathbb{E}[Y(1) - Y(0)].$$

Identification of τ requires an assumption like unconfoundedness:

$$\{Y(0), Y(1)\} \perp D \mid X,$$

i.e., after conditioning on X, the treatment assignment is as-if random. Estimation strategies then revolve around properly adjusting for X.

Such assumptions are not necessary for raw prediction of Y: a black-box function can yield $\hat{Y} \approx Y$ without ensuring that $\hat{Y}(1) - \hat{Y}(0)$ is an unbiased estimate of τ.

11.4.3 Role of Identification

Identification means that the parameter of interest (β or τ) is uniquely pinned down by the distribution of observables (under assumptions). If β is not identified (e.g., because of endogeneity or insufficient variation in X), no matter how large the sample, we cannot estimate β consistently.

In prediction, "identification" is not usually the main concern. The function $\hat{f}(x)$ could be a complicated ensemble method that just fits well, without guaranteeing any structural or causal interpretation of its parameters.

11.4.4 Challenges

1. **High-Dimensional Spaces**: With large p (number of predictors), covariance among variables (multicollinearity) can hamper classical estimation. This is the setting of the well-known **bias-variance tradeoff** (Hastie et al., 2009; Bishop, 2006).
2. **Endogeneity**: If x is correlated with the error term ε, ordinary least squares (OLS) is biased. Causal inference demands identifying exogenous variation in x, which requires additional assumptions or designs (e.g., randomization).
3. **Model Misspecification**: If the functional form $g(x; \beta)$ is incorrect, parameter estimates can systematically deviate from capturing the true underlying mechanism.

11.5 Causation versus Prediction

Understanding the relationship between **causation** and **prediction** is crucial in statistical modeling. Building on Kleinberg et al. (2015) and Mullainathan and Spiess (2017), consider a scenario where Y is an outcome variable dependent on X, and we want to **manipulate** X to maximize some payoff function $\pi(X, Y)$. Formally:

$$\pi(X, Y) = \mathbb{E}\big[U(X, Y)\big] \quad \text{or some other objective measure.}$$

The decision on X depends on **how changes in** X influence π. Taking a derivative:

$$\frac{d\pi(X,Y)}{dX} = \frac{\partial \pi}{\partial X}(Y) + \frac{\partial \pi}{\partial Y}\frac{\partial Y}{\partial X}.$$

We can interpret the terms:

- $\frac{\partial \pi}{\partial X}$: The *direct* dependence of the payoff on X, which can be predicted if we can forecast how π changes with X.
- $\frac{\partial Y}{\partial X}$: The *causal* effect of X on Y, essential for understanding how interventions on X shift Y.
- $\frac{\partial \pi}{\partial Y}$: The marginal effect of Y on the payoff.

Hence, Kleinberg et al. (2015) frames this distinction as one between **predicting** Y effectively (for instance, "If I observe X, can I guess Y?") versus **managing** or **causing** Y to change via **interventions** on X. Empirically:

- To **predict** Y, we model $\mathbb{E}[Y \mid X]$.
- To **infer causality**, we require identification strategies that isolate exogenous variation in X.

Empirical work in economics or social science often aims to estimate partial derivatives of structural or reduced-form equations:

- $\frac{\partial Y}{\partial X}$: The **causal derivative**; tells us how Y changes if we **intervene** on X.
- $\frac{\partial \pi}{\partial X}$: The **effect of** X on payoff, partially mediated by changes in Y.

Without proper identification (e.g., **randomization, instrumental variables, difference-in-differences, or other quasi-experimental designs**), we risk **conflating association** ($\hat{f}$ that predicts Y) with **causation** ($\hat{\beta}$ that truly captures how X shifts Y).

To illustrate these concepts, consider the following directed acyclic graph (DAG) (Fig. 11.1):

```r
library(ggdag)
#>
#> Attaching package: 'ggdag'
#> The following object is masked from 'package:stats':
#>
#>     filter
library(dagitty)
library(ggplot2)
```

```
# Define the DAG structure with custom coordinates
dag <- dagitty('
dag {
  X0 [pos="0,1"]
  X [pos="1,2"]
  Y [pos="1,1"]
  II [pos="1,0"]

  X0 -> Y
  X0 -> II
  X -> Y
  Y -> II
}
')

# Convert to ggdag format with manual layout
dag_plot <- ggdag(dag) +
    theme_void() +
    geom_text(aes(x = 0.5, y = 1.2, label = "Causation"), size
    ↪ = 4) +
    geom_text(aes(x = 0.3, y = 0.5, label = "Prediction"),
    ↪ size = 4)
```

```
dag_plot
```

11.6 Illustrative Equations and Mathematical Contrasts

Below, we showcase a few derivations that highlight how predictive modeling vs. causal inference differ in their mathematical structure and interpretation.

11.6.1 *Risk Minimization vs. Consistency*

Consider a real-valued outcome Y and predictors X. Let $\ell(y, \hat{y})$ be a loss function, and define the Bayes regressor f^* as:

$$f^* = \arg\min_f \mathbb{E}[\ell(Y, f(X))].$$

For squared error loss, the Bayes regressor is $f^*(x) = \mathbb{E}[Y \mid X = x]$.

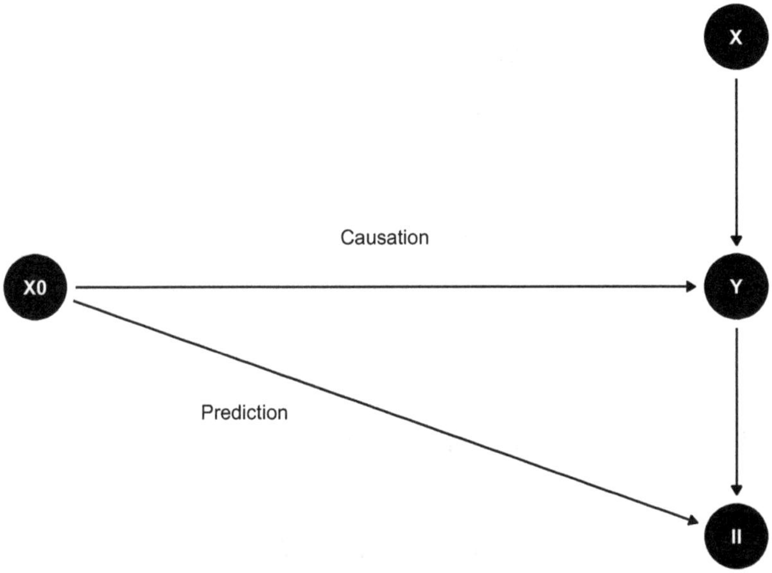

Fig. 11.1 Flow chart

A learning algorithm tries to approximate f^*. If we parametrize $f_\beta(x) = x^\top \beta$ and do empirical risk minimization with a large enough sample, β converges to the minimizer of:

$$\beta^* = \arg\min_\beta \mathbb{E}[(Y - X^\top \beta)^2].$$

Note that β^* is the solution to $\mathbb{E}[XX^\top]\beta = \mathbb{E}[XY]$. If $\mathrm{Cov}(X, X)$ is invertible, then

$$\beta^* = \mathrm{Cov}(X, X)^{-1}\mathrm{Cov}(X, Y).$$

This β^* is not necessarily the same as the "true" β_0 from a structural equation $Y = X\beta_0 + \varepsilon$ unless $\mathbb{E}[\varepsilon \mid X] = 0$.

From a predictive standpoint, β^* is the best linear predictor in the sense of mean squared error. From a causal standpoint, we want β_0 such that ε is mean-independent of X. If that fails, $\beta^* \neq \beta_0$.

11.6.2 Partial Derivatives vs. Predictions

A powerful way to see the difference is to compare:

- $\frac{\partial}{\partial x} f^*(x)$ – The partial derivative of the best predictor w.r.t. x. This is about how the model's *prediction* changes with x.
- $\frac{\partial}{\partial x} m_\beta(x)$ – The partial derivative of the structural function $m_\beta(\cdot)$. This is about how the *true outcome* Y changes with x, i.e., a causal effect if m_β is indeed structural.

Unless the model was identified and the assumptions hold (exogeneity, no omitted variables, etc.), the partial derivative from a purely predictive model does not represent the causal effect.

In short: "slopes" from a black-box predictive model are not guaranteed to reflect how interventions on X would shift Y.

11.6.3 Example: High-Dimensional Regularization

Suppose we have a large number of predictors p, possibly $p \gg n$. A common approach in both prediction and inference is LASSO:

$$\hat{\beta}_{\text{LASSO}} = \arg\min_{\beta} \left\{ \frac{1}{n} \sum_{i=1}^{n} (y_i - x_i^\top \beta)^2 + \lambda \|\beta\|_1 \right\}.$$

- **Prediction:** Choose λ to optimize out-of-sample MSE. Some bias is introduced in $\hat{\beta}$, but the final model might predict extremely well, especially if many true coefficients are near zero.
- **Causal Estimation:** We must worry about whether the LASSO is shrinking or zeroing out confounders. If a crucial confounder's coefficient is set to zero, the resulting estimate for a treatment variable's coefficient will be biased. Therefore, special procedures (like the double/debiased machine learning approach (Chernozhukov et al., 2018)) are introduced to correct for the selection bias or to do post-selection inference (Belloni et al., 2014).

The mathematics of "best subset" for prediction vs. valid coverage intervals for parameters diverges significantly.

11.6.4 Potential Outcomes Notation

Let $D \in \{0, 1\}$ be a treatment indicator, and define potential outcomes:

$$Y_i(0), Y_i(1).$$

The observed outcome is:

$$Y_i = D_i Y_i(1) + (1 - D_i) Y_i(0).$$

- **Prediction:** One might train a model $\hat{Y} = \hat{f}(X, D)$ to guess Y from (X, D). That model could be a black box with no guarantee that $\hat{Y}(1) - \hat{Y}(0)$ is an unbiased estimate of $Y_i(1) - Y_i(0)$.
- **Causal Inference:** We want to estimate $\mathbb{E}[Y(1) - Y(0)]$ or $\mathbb{E}[Y(1) - Y(0) \mid X = x]$. Identification typically requires $\{Y(0), Y(1)\} \perp D \mid X$, i.e., after conditioning on X, the treatment assignment is as-if random. Under such an assumption, the difference $\hat{f}(x, 1) - \hat{f}(x, 0)$ can be interpreted as a causal effect.

11.7 Extended Mathematical Points

We now delve deeper into some mathematical nuances that are especially relevant when distinguishing between predictive vs. causal modeling.

11.7.1 M-Estimation and Asymptotic Theory

M-Estimators unify many approaches: maximum likelihood, method of moments, generalized method of moments, and quasi-likelihood estimators. Let β_0 be the true parameter and define the population criterion function:

$$Q(\beta) = \mathbb{E}[m(\beta; X, Y)],$$

for some function m. The M-estimator $\hat{\beta}$ solves:

$$\hat{\beta} = \arg\max_{\beta \in \Theta} \frac{1}{n} \sum_{i=1}^{n} m(\beta; X_i, Y_i).$$

(Or arg min, depending on convention.)

Under regularity conditions (Newey & McFadden, 1994; White, 1980), we have:

- **Consistency:** $\hat{\beta} \xrightarrow{p} \beta_0$.
- **Asymptotic Normality:** $\sqrt{n}(\hat{\beta} - \beta_0) \xrightarrow{d} N(0, \Sigma)$,

where Σ is derived from derivatives of $m(\cdot; \cdot, \cdot)$ and the distribution of (X, Y).

For prediction, such classical asymptotic properties may be of less interest unless we want to build confidence intervals around predictions. For causal inference, the entire enterprise revolves around these properties to ensure valid inference about β_0.

11.7.2 The Danger of Omitted Variables

Consider a structural equation:

$$Y = \beta_1 X_1 + \beta_2 X_2 + \varepsilon, \quad \mathbb{E}[\varepsilon \mid X_1, X_2] = 0.$$

If we ignore X_2 and regress Y on X_1 only, the resulting $\hat{\beta}_1$ can be severely biased:

$$\hat{\beta}_1 = \arg\min_b \sum_{i=1}^{n} (y_i - b\,x_{i1})^2.$$

The expected value of $\hat{\beta}_1$ in large samples is:

$$\beta_1 + \beta_2 \frac{\mathrm{Cov}(X_1, X_2)}{\mathrm{Var}(X_1)}.$$

This extra term $\beta_2 \frac{\mathrm{Cov}(X_1, X_2)}{\mathrm{Var}(X_1)}$ is the **omitted variables bias**. For **prediction**, omitting X_2 might sometimes be acceptable if X_2 has little incremental predictive value or if we only care about accuracy in some domain. However, for **inference** on β_1, ignoring X_2 invalidates the causal interpretation.

11.7.3 Cross-Validation vs. Statistical Testing

- **Cross-Validation**: Predominantly used in **prediction** tasks. We split the data into training and validation sets, measure out-of-sample error, and select hyperparameters that minimize CV error.
- **Statistical Testing**: Predominantly used in **inference** tasks. We compute test statistics (e.g., t-test, Wald test), form confidence intervals, or test hypotheses about parameters ($H_0 : \beta_j = 0$).

They serve different objectives:

1. **CV** is about **predictive model selection**.
2. **Testing** is about **scientific or policy conclusions** on whether β_j differs from zero (i.e., "Does a particular variable have a causal effect?").

11.8 Putting It All Together: Comparing Objectives

As an overarching illustration, let $\hat{f}$ be any trained predictor (ML model, regression, etc.) and let $\hat{\beta}$ be a parameter estimator from a structural or causal model. Their respective tasks differ:

- **Form of Output**
  - $\hat{f}$ is a *function* from $\mathcal{X} \to \mathcal{Y}$.
  - $\hat{\beta}$ is a *vector of parameters* with theoretical meaning.
- **Criterion**
  - **Prediction**: Minimizes predictive loss $\mathbb{E}[L(Y, \hat{f}(X))]$.
  - **Causal Inference**: Seeks β such that $Y = m_\beta(X)$ is a correct *structural* representation. Minimizes bias in β, or satisfies orthogonality conditions in method-of-moments style, etc.
- **Validity**
  - **Prediction**: Usually validated by out-of-sample experiments or cross-validation.
  - **Estimation**: Validated by theoretical identification arguments, assumptions about exogeneity, randomization, or no omitted confounders.
- **Interpretation**
  - **Prediction**: "$\hat{f}(x)$ is our best guess of Y for new x."
  - **Causal Inference**: "β measures how Y changes if we intervene on X."

11.9 Conclusion

Prediction and **Estimation/Causal Inference** serve distinctly different roles in data analysis:

- **Prediction**: The emphasis is on **predictive accuracy**. The final model $\hat{f}$ may have uninterpretable parameters (e.g., deep neural networks) yet excel at forecasting Y. Bias in parameter estimates is not necessarily problematic if it reduces variance and improves out-of-sample performance.
- **Estimation/Causal Inference**: The emphasis is on obtaining **consistent** and **unbiased** estimates of parameters (β, or a treatment effect τ). We impose stronger assumptions about data collection and the relationship between X and ε. The success criterion is whether $\hat{\beta} \approx \beta_0$ in a formal sense, with valid confidence intervals and robust identification strategies.

Key Takeaway

- If your question is "How do I **predict** Y for new X as accurately as possible?", you prioritize **prediction**.
- If your question is "How does **changing** X (or assigning treatment D) affect Y in a **causal** sense?", you focus on **estimation** with a fully developed identification strategy.

Bibliography

Agarwal, A., Dahleh, M., Shah, D., & Shen, D. (2023). Causal matrix completion. In *The Thirty Sixth Annual Conference on Learning Theory* (pp. 3821–3826). PMLR.

Ahrens, H., & Pincus, R. (1981). On two measures of unbalancedness in a one-way model and their relation to efficiency. *Biometrical Journal, 23*(3), 227–235.

Aiken, L. S., & West, S. G. (2005). Interaction effects. In *Encyclopedia of statistics in behavioral science*.

Amemiya, T. (1971). The estimation of the variances in a variance-components model. *International Economic Review, 12*(1), 1–13.

Amemiya, T., & MaCurdy, T. E. (1986). Instrumental-variable estimation of an error-components model. *Econometrica: Journal of the Econometric Society* **54**(4), 869–880.

Assmann, S. F., Hosmer, D. W., Lemeshow, S., & Mundt, K. A. (1996). Confidence intervals for measures of interaction. *Epidemiology, 7*(3), 286–90.

Athey, S., & Imbens, G. W. (2019). Machine learning methods that economists should know about. *Annual Review of Economics, 11*, 685–725.

Azoulay, P., Fons-Rosen, C., & Zivin, J. S. G. (2019). Does science advance one funeral at a time? *American Economic Review, 109*(8), 2889–2920.

Babcock, P. (2010). Real costs of nominal grade inflation? New evidence from student course evaluations. *Economic Inquiry, 48*(4), 983–996.

Balestra, P., & Varadharajan-Krishnakumar, J. (1987). Full information estimations of a system of simultaneous equations with error component structure. *Econometric Theory, 3*(2), 223–246.

Baltagi, B. H. (1981). Simultaneous equations with error components. *Journal of Econometrics, 17*(2), 189–200.

Baltagi, B. H., & Li, Q. (1991). A joint test for serial correlation and random individual effects. *Statistics & Probability Letters, 11*(3), 277–280.

Baltagi, B. H., & Li, Q. (1995). Testing ar (1) against ma (1) disturbances in an error component model. *Journal of Econometrics, 68*(1), 133–151.

Baron, R. M., & Kenny, D. A. (1986). The moderator–mediator variable distinction in social psychological research: Conceptual, strategic, and statistical considerations. *Journal of Personality and Social Psychology, 51*(6), 1173.

Bartlett, M. S. (1947). The use of transformations. *Biometrics, 3*(1), 39–52.

Bauer, D. J., & Curran, P. J. (2005). Probing interactions in fixed and multilevel regression: Inferential and graphical techniques. *Multivariate Behavioral Research, 40*(3), 373–400.

Bauer, D. J., Preacher, K. J., & Gil, K. M. (2006). Conceptualizing and testing random indirect effects and moderated mediation in multilevel models: New procedures and recommendations. *Psychological Methods, 11*(2), 142.

© The Author(s), under exclusive license to Springer Nature Switzerland AG 2025

M. Nguyen, *Advanced Modeling and Data Challenges*,
https://doi.org/10.1007/978-3-032-01719-2

Beerli, A., Ruffner, J., Siegenthaler, M., & Peri, G. (2021). The abolition of immigration restrictions and the performance of firms and workers: Evidence from Switzerland. *American Economic Review, 111*(3), 976–1012.

Belloni, A., Chernozhukov, V., & Hansen, C. (2014). High-dimensional methods and inference on structural and treatment effects. *Journal of Economic Perspectives, 28*(2), 29–50.

Bendel, R. B., & Afifi, A. A. (1977). Comparison of stopping rules in forward "stepwise" regression. *Journal of the American Statistical Association, 72*(357), 46–53.

Benjamini, Y., & Hochberg, Y. (1995). Controlling the false discovery rate: A practical and powerful approach to multiple testing. *Journal of the Royal Statistical Society: Series B (Methodological), 57*(1), 289–300.

Benjamini, Y., & Yekutieli, D. (2001). The control of the false discovery rate in multiple testing under dependency. *Annals of Statistics, 29*(4), 1165–1188.

Bera, A. K., Sosa-Escudero, W., & Yoon, M. (2001). Tests for the error component model in the presence of local misspecification. *Journal of Econometrics, 101*(1), 1–23.

Berkouwer, S. B., & Dean, J. T. (2022). Credit, attention, and externalities in the adoption of energy efficient technologies by low-income households. *American Economic Review, 112*(10), 3291–3330.

Bickel, P. J., & Doksum, K. A. (1981). An analysis of transformations revisited. *Journal of the American Statistical Association, 76*(374), 296–311.

Bishop, C. M. (2006). *Pattern recognition and machine learning*. Springer.

Blau, D. M. (1999). The effect of income on child development. *Review of Economics and Statistics, 81*(2), 261–276.

Bollen, K. A., & Stine, R. (1990). Direct and indirect effects: Classical and bootstrap estimates of variability. *Sociological Methodology, 20* , 115–140.

Box, G. E., & Cox, D. R. (1981). *An analysis of transformations revisited, rebutted*. University of Wisconsin-Madison. Mathematics Research Center.

Box, G. E., & Pierce, D. A. (1970). Distribution of residual autocorrelations in autoregressive-integrated moving average time series models. *Journal of the American Statistical Association, 65*(332), 1509–1526.

Breusch, T. S. (1978). Testing for autocorrelation in dynamic linear models. *Australian Economic Papers, 17*(31), 334–355.

Breusch, T. S., Mizon, G. E., & Schmidt, P. (1989). Efficient estimation using panel data. *Econometrica: Journal of the Econometric Society, 57*(3), 695–700.

Breusch, T. S., & Pagan, A. R. (1979). A simple test for heteroscedasticity and random coefficient variation. *Econometrica: Journal of the Econometric Society, 47*(5), 1287–1294.

Breusch, T. S., & Pagan, A. R. (1980). The lagrange multiplier test and its applications to model specification in econometrics. *The Review of Economic Studies, 47*(1), 239–253.

Cabral, M., Cui, C., & Dworsky, M. (2022). The demand for insurance and rationale for a mandate: Evidence from workers' compensation insurance. *American Economic Review, 112*(5), 1621–1668.

Carranza, E., Garlick, R., Orkin, K., & Rankin, N. (2022). Job search and hiring with limited information about workseekers' skills. *American Economic Review, 112*(11), 3547–3583.

Chatterjee, S. (2015). Matrix estimation by universal singular value thresholding. *The Annals of Statistics, 43*(1), 177–214.

Chen, J., & Roth, J. (2024). Logs with zeros? Some problems and solutions. *The Quarterly Journal of Economics, 139*(2), 891–936.

Chernozhukov, V., Chetverikov, D., Demirer, M., Duflo, E., Hansen, C., Newey, W., & Robins, J. (2018). Double/debiased machine learning for treatment and structural parameters. *The Econometrics Journal, 21*(1), C1–C68.

Conley, T. G. (1999). Gmm estimation with cross sectional dependence. *Journal of Econometrics, 92*(1), 1–45.

Davidson, R., & MacKinnon, J. G. (1981). Several tests for model specification in the presence of alternative hypotheses. *Econometrica: Journal of the Econometric Society, 49*(3), 781–793.

Dawson, J. F., & Richter, A. W. (2006). Probing three-way interactions in moderated multiple regression: Development and application of a slope difference test. *Journal of Applied Psychology, 91*(4), 917.

Driscoll, J. C., & Kraay, A. C. (1998). Consistent covariance matrix estimation with spatially dependent panel data. *Review of Economics and Statistics, 80*(4), 549–560.

Ekwaru, J. P., & Veugelers, P. J. (2018). The overlooked importance of constants added in log transformation of independent variables with zero values: A proposed approach for determining an optimal constant. *Statistics in Biopharmaceutical Research, 10*(1), 26–29.

Esarey, J., & Sumner, J. L. (2018). Marginal effects in interaction models: Determining and controlling the false positive rate. *Comparative Political Studies, 51*(9), 1144–1176.

Faber, B., & Gaubert, C. (2019). Tourism and economic development: Evidence from Mexico's coastline. *American Economic Review, 109*(6), 2245–2293.

Fossum, J. L., & Montoya, A. K. (2023). When to use different inferential methods for power analysis and data analysis for between-subjects mediation. *Advances in Methods and Practices in Psychological Science, 6*(2), 25152459231156606.

Fritz, M. S., Taylor, A. B., & MacKinnon, D. P. (2012). Explanation of two anomalous results in statistical mediation analysis. *Multivariate Behavioral Research, 47*(1), 61–87.

Furnival, G. M., & Wilson, R. W. (2000). Regressions by leaps and bounds. *Technometrics, 42*(1), 69–79.

Glasser, M. (1964). Linear regression analysis with missing observations among the independent variables. *Journal of the American Statistical Association, 59*(307), 834–844.

Glejser, H. (1969). A new test for heteroskedasticity. *Journal of the American Statistical Association, 64*(325), 316–323.

Godfrey, L. G. (1978). Testing against general autoregressive and moving average error models when the regressors include lagged dependent variables. *Econometrica: Journal of the Econometric Society, 46*(6), 1293–1301.

Goldfeld, S. M., & Quandt, R. E. (1965). Some tests for homoscedasticity. *Journal of the American statistical Association, 60*(310), 539–547.

Gould, W., Pitblado, J., & Poi, B. (2010). Maximum likelihood estimation with stata: Statacorp lp. Technical report, Stata Press.

Gourieroux, C., Holly, A., & Monfort, A. (1982). Likelihood ratio test, Wald test, and Kuhn-Tucker test in linear models with inequality constraints on the regression parameters. *Econometrica: Journal of the Econometric Society, 50*(1), 63–80.

Haitovsky, Y. (1968). Missing data in regression analysis. *Journal of the Royal Statistical Society: Series B (Methodological), 30*(1), 67–82.

Harvey, A. C., & Collier, P. (1977). Testing for functional misspecification in regression analysis. *Journal of Econometrics, 6*(1), 103–119.

Hastie, T., Mazumder, R., Lee, J. D., & Zadeh, R. (2015). Matrix completion and low-rank SVD via fast alternating least squares. *The Journal of Machine Learning Research, 16*(1), 3367–3402.

Hastie, T., Tibshirani, R., & Friedman, J. H. (2009). *The elements of statistical learning: Data mining, inference, and prediction* (Vol. 2). Springer.

Hayes, A. F., & Preacher, K. J. (2014). Statistical mediation analysis with a multicategorical independent variable. *British Journal of Mathematical and Statistical Psychology, 67*(3), 451–470.

Hayes, A. F., & Scharkow, M. (2013). The relative trustworthiness of inferential tests of the indirect effect in statistical mediation analysis: Does method really matter? *Psychological Science, 24*(10), 1918–1927.

Heckman, J. J. (1976). The common structure of statistical models of truncation, sample selection and limited dependent variables and a simple estimator for such models. In *Annals of economic and social measurement* (Vol. 5(4), pp. 475–492). NBER.

Hjort, J., & Poulsen, J. (2019). The arrival of fast internet and employment in Africa. *American Economic Review, 109*(3), 1032–1079.

Honda, Y. (1985). Testing the error components model with non-normal disturbances. *The Review of Economic Studies, 52*(4), 681–690.

Hosmer, D. W., & Lemeshow, S. (1992). Confidence interval estimation of interaction. *Epidemiology, 3*(5), 452–456.

Hoyle, R. H. (1999). *Statistical strategies for small sample research*. Sage.

Imai, K., Keele, L., & Tingley, D. (2010a). A general approach to causal mediation analysis. *Psychological Methods, 15*(4), 309.

Imai, K., Keele, L., & Yamamoto, T. (2010b). Identification, inference and sensitivity analysis for causal mediation effects. *Statistical Science, 25*(1), 51–71.

James, G., Witten, D., Hastie, T., & Tibshirani, R. (2013). *An introduction to statistical learning* (Vol. 112). Springer.

Johnson, M. S. (2020). Regulation by shaming: Deterrence effects of publicizing violations of workplace safety and health laws. *American Economic Review, 110*(6), 1866–1904.

Johnson, N. L. (1949). Systems of frequency curves generated by methods of translation. *Biometrika, 36*(1/2), 149.

Johnson, P. O., & Neyman, J. (1936). Tests of certain linear hypotheses and their applications to some educational problems. *Statistical Research Memoirs, 1*, 57.

Jones, M. P. (1996). Indicator and stratification methods for missing explanatory variables in multiple linear regression. *Journal of the American Statistical Association, 91*(433), 222–230.

Kenny, D. A., & Judd, C. M. (2014). Power anomalies in testing mediation. *Psychological Science, 25*(2), 334–339.

King, M. L., & Wu, P. X. (1997). Locally optimal one-sided tests for multiparameter hypotheses. *Econometric Reviews, 16*(2), 131–156.

Kleinberg, J., Ludwig, J., Mullainathan, S., & Obermeyer, Z. (2015). Prediction policy problems. *American Economic Review, 105*(5), 491–495.

Knol, M. J., & VanderWeele, T. J. (2012). Recommendations for presenting analyses of effect modification and interaction. *International Journal of Epidemiology, 41*(2), 514–520.

Ledgerwood, A., & Shrout, P. E. (2011). The trade-off between accuracy and precision in latent variable models of mediation processes. *Journal of Personality and Social Psychology, 101*(6), 1174.

Little, R. J. (1992). Regression with missing x's: A review. *Journal of the American Statistical Association, 87*(420), 1227–1237.

Ljung, G. M., & Box, G. E. (1978). On a measure of lack of fit in time series models. *Biometrika, 65*(2), 297–303.

MacKinnon, D. P., Fairchild, A. J., & Fritz, M. S. (2007). Mediation analysis. *Annual Review of Psychology, 58*, 593–614.

MacKinnon, D. P., Warsi, G., & Dwyer, J. H. (1995). A simulation study of mediated effect measures. *Multivariate Behavioral Research, 30*(1), 41–62.

Mallows, C. L. (1995). More comments on cp. *Technometrics, 37*(4), 362–372.

Manly, B. F. (1976). Exponential data transformations. *Journal of the Royal Statistical Society Series D: The Statistician, 25*(1), 37–42.

Mirenda, L., Mocetti, S., & Rizzica, L. (2022). The economic effects of mafia: Firm level evidence. *American Economic Review, 112*(8), 2748–2773.

Mullainathan, S., & Spiess, J. (2017). Machine learning: An applied econometric approach. *Journal of Economic Perspectives, 31*(2), 87–106.

Narayanan, S., & Nair, H. S. (2013). Estimating causal installed-base effects: A bias-correction approach. *Journal of Marketing Research, 50*(1), 70–94.

Nerlove, M. (1971). Further evidence on the estimation of dynamic economic relations from a time series of cross sections. *Econometrica: Journal of the Econometric Society, 39*(2), 359–382.

Newey, W. K., & McFadden, D. (1994). Large sample estimation and hypothesis testing. *Handbook of Econometrics, 4*, 2111–2245.

Newey, W. K., & West, K. D. (1986). A simple, positive semi-definite, heteroskedasticity and autocorrelationconsistent covariance matrix.

Nickell, S. (1981). Biases in dynamic models with fixed effects. *Econometrica: Journal of the Econometric Society, 49*(6), 1417–1426.

Norris, S., Pecenco, M., & Weaver, J. (2021). The effects of parental and sibling incarceration: Evidence from ohio. *American Economic Review, 111*(9), 2926–2963.

Park, R. E. (1966). Estimation with heteroscedastic error terms. *Econometrica*, 34(4).

Preacher, K. J., & Hayes, A. F. (2004). SPSS and SAS procedures for estimating indirect effects in simple mediation models. *Behavior Research Methods, Instruments, & Computers, 36*, 717–731.

Ramsey, J. B. (1969). Tests for specification errors in classical linear least-squares regression analysis. *Journal of the Royal Statistical Society Series B: Statistical Methodology, 31*(2), 350–371.

Rubin, D. B. (1996). Multiple imputation after 18+ years. *Journal of the American statistical Association, 91*(434), 473–489.

Schuirmann, D. J. (1987). A comparison of the two one-sided tests procedure and the power approach for assessing the equivalence of average bioavailability. *Journal of Pharmacokinetics and Biopharmaceutics, 15*, 657–680.

Shrout, P. E., & Bolger, N. (2002). Mediation in experimental and nonexperimental studies: New procedures and recommendations. *Psychological Methods, 7*(4), 422.

Smith, E. R. (1982). Beliefs, attributions, and evaluations: Nonhierarchical models of mediation in social cognition. *Journal of Personality and Social Psychology, 43*(2), 248.

Sobel, M. E. (1982). Asymptotic confidence intervals for indirect effects in structural equation models. *Sociological Methodology, 13*, 290–312.

Spiller, S. A., Fitzsimons, G. J., Lynch Jr, J. G., & McClelland, G. H. (2013). Spotlights, floodlights, and the magic number zero: Simple effects tests in moderated regression. *Journal of Marketing Research, 50*(2), 277–288.

Sun, B., Liu, L., Miao, W., Wirth, K., Robins, J., & Tchetgen, E. J. T. (2018). Semiparametric estimation with data missing not at random using an instrumental variable. *Statistica Sinica, 28*(4), 1965.

Swamy, P., & Arora, S. S. (1972). The exact finite sample properties of the estimators of coefficients in the error components regression models. *Econometrica: Journal of the Econometric Society, 40*(2), 261–275.

Tchetgen Tchetgen, E. J., & Wirth, K. E. (2017). A general instrumental variable framework for regression analysis with outcome missing not at random. *Biometrics, 73*(4), 1123–1131.

Tibbe, T. D., & Montoya, A. K. (2022). Correcting the bias correction for the bootstrap confidence interval in mediation analysis. *Frontiers in Psychology, 13*, 810258.

Utts, J. M. (1982). The rainbow test for lack of fit in regression. *Communications in Statistics-Theory and Methods, 11*(24), 2801–2815.

Vach, W., & Vach, W. (1994). Missing values and subsampling. *Logistic regression with missing values in the covariates* (pp. 98–102).

VanderWeele, T. (2015). *Explanation in causal inference: Methods for mediation and interaction*. Oxford University Press.

Vapnik, V. (2013). *The nature of statistical learning theory*. Springer.

Von Hippel, P. T. (2009). 8. how to impute interactions, squares, and other transformed variables. *Sociological Methodology, 39*(1), 265–291.

Vuong, Q. H. (1989). Likelihood ratio tests for model selection and non-nested hypotheses. *Econometrica: Journal of the Econometric Society, 57*(2), 307–333.

Wallace, T. D., & Hussain, A. (1969). The use of error components models in combining cross section with time series data. *Econometrica: Journal of the Econometric Society, 37*(1), 55–72.

Westfall, J., & Yarkoni, T. (2016). Statistically controlling for confounding constructs is harder than you think. *PloS One, 11*(3), e0152719.

White, H. (1980). A heteroskedasticity-consistent covariance matrix estimator and a direct test for heteroskedasticity. *Econometrica: Journal of the Econometric Society, 48*(4), 817–838.

Zou, G. Y. (2008). On the estimation of additive interaction by use of the four-by-two table and beyond. *American Journal of Epidemiology, 168*(2), 212–224.

GPSR Compliance
The European Union's (EU) General Product Safety Regulation (GPSR) is a set of rules that requires consumer products to be safe and our obligations to ensure this.

If you have any concerns about our products, you can contact us on

ProductSafety@springernature.com

In case Publisher is established outside the EU, the EU authorized representative is:

Springer Nature Customer Service Center GmbH
Europaplatz 3
69115 Heidelberg, Germany

www.ingramcontent.com/pod-product-compliance
Lightning Source LLC
Chambersburg PA
CBHW061656240326
41796CB00003B/72
* 9 7 8 3 0 3 2 0 1 7 1 8 5 *